Epigenetics and Human Health

Volume 16

This books series concerns the quickly expanding field of epigenetics. It evokes novel research questions and how these are linked to human health. Each of its volumes covers a particular topic that is important to clinical practice and research, and to researchers in epigenetics who study the relevant medical issues of today. The individual books of the series complement each other and underline the importance of epigenetics in health-related disciplines such as toxicology, reproduction, psychiatry, and nutritional and environmental medicine.

Epigenetics and Human Health was founded in 2011. Prof. Dr. Robert Feil, PD Dr. Mario Noyer-Weidner and Prof. Dr. Jörn Walter were the founding editors and were responsible as series editors for volumes 1-10 until 2016. In 2023, the series was relaunched together with Prof. Rasime Kalkan and Dr. Luis M. Vaschetto, who are the current series editors.

Epigenetics and Human Health is indexed in SCOPUS, EMBASE, Reaxys, EMCare, EMBiology and SCImago.

Luis M. Vaschetto
Editor

Long Non-Coding RNAs, Human Health, and Diseases

Springer

Editor
Luis M. Vaschetto
Ronin Institute (RIIS 2.0)
Sacramento, CA, USA

National Coalition of Independent Scholars
South Bend, IN, USA

ISSN 2191-2262 ISSN 2191-2270 (electronic)
Epigenetics and Human Health
ISBN 978-3-032-19667-5 ISBN 978-3-032-19668-2 (eBook)
https://doi.org/10.1007/978-3-032-19668-2

This Springer imprint is published by the registered company Springer Nature Switzerland AG
The registered company address is: Gewerbestrasse 11, 6330 Cham, Switzerland

Preface

The traditional central dogma of molecular biology focused intensely on the canonical pathway of genetic information where it flows from deoxyribonucleic acid (DNA) to messenger RNA (mRNA), and ultimately to protein. The approximately 2% of the human genome responsible for encoding proteins was then historically regarded as uniquely essential for phenotypic characterization. Conversely, the vast majority of noncoding DNA, often transcribed as RNA without protein-coding capacity, was frequently dismissed as "junk," and its product as transcriptional noise. Over the last decades, however, this view has been challenged by the discovery and subsequent study of noncoding RNAs (ncRNAs).

In this new Volume of the internationally acclaimed Epigenetics and Human Health book series, we focused on a critical class of ncRNA referred to as long noncoding RNAs (lncRNAs). LncRNAs are defined as structurally and functionally diverse RNA transcripts exceeding 200 nucleotides in length, which are typically not translated into proteins. Recent research has become obvious that lncRNAs play essential regulatory functions and govern eukaryotic gene expression at different levels. Their functions hold major implications for understanding and addressing both human health and the pathogenesis of a wide spectrum of diseases.

LncRNA applications may range from biomarker discovery to therapeutic targets with an exceptionally high specificity and detectability in biological fluids (often encapsulated in exosomes). These RNA transcripts can be employed as sensitive and noninvasive diagnostic and prognostic biomarkers. Furthermore, lncRNAs have significant potential to modulate gene expression networks by targeting specific genes, which represents a major step toward highly targeted and personalized medicine.

Long Non-Coding RNAs, Human Health, and Diseases offers the reader a unique and comprehensive exploration into the future of genomics medicine, providing an updated panorama about the roles of lncRNAs in human health and disease. A special mention and deep gratitude are extended to Drs. Samuel Jeyasingh, Roberta Tarallo, Prashanth N Suravajhala, Dhirodatta Senapati, and Soumyadev Sarkar, whose insightful and critical comments have enhanced the quality and scientific rigor of this book.

Sacramento, CA, USA — Luis María Vaschetto

Contents

About the Editor and Contributors

About the Editor

Luis M. Vaschetto obtained his Ph.D. in Biological Sciences from the Universidad Nacional de Córdoba (Argentina). He was a postdoctoral researcher at the Consejo Nacional de Investigaciones Científicas y Técnicas (CONICET) and a Research Associate at the South Dakota State University (SDSU). Dr. Vaschetto has first-authored over ten academic articles in top-tier journals such as *WIREs RNA*, *Briefings in Functional Genomics*, *Journal of Cellular Biochemistry*, *Molecular Breeding*, *Current Gene Therapy*, *RNA Biology*, and *Current Genetics*. He has also edited and authored ten international books, including three for the Epigenetics and Human Health (EHH) book series and three for Methods in Molecular Biology (Springer Nature, USA). His research mainly focuses on understanding the epigenetic mechanisms involved in plant and animal development, as well as their effects on human health and diseases.

Contributors

Daniel Scott Algara Unité de Biologie Cellulaire des Lymphocytes and Direction of International Affairs, Institut Pasteur, Paris, France

Letícia Ferreira Alves Josep Carreras Leukaemia Research Institute (IJC), Badalona, Barcelona, Catalonia, Spain

Mateus Nóbrega Aoki Laboratory for Applied Science and Technology in Health, Carlos Chagas Institute, Oswaldo Cruz Foundation (Fiocruz), Curitiba, Brazil

Miriam Emily Avendaño National Institute of Technology of Mexico/IT. (TecNM) Oaxaca, Oaxaca, Mexico

Antara Banerjee Medical Biotechnology Lab, Faculty of Allied Health Sciences, Chettinad Academy of Research and Education (CARE), Chettinad Hospital and Research Institute, Chennai, India

Héctor Alejandro Cabrera-Fuentes National Institute of Technology of Mexico/IT. (TecNM) Oaxaca, Oaxaca, Mexico

UNAM-UABJO Faculty of Medicine Research Centre, Faculty of Medicine and Surgery, Benito Juárez Autonomous University of Oaxaca, Oaxaca, Mexico

R&D Group, Scientific Research and Innovation Imam Abdulrahman Bin Faisal University Dammam, Dammam, Saudi Arabia

National Institute of Technology of Mexico/IT. (TecNM), Tijuana, Baja California, Mexico

Socorro Pina Canseco UNAM-UABJO Faculty of Medicine Research Centre, Faculty of Medicine and Surgery, Benito Juárez Autonomous University of Oaxaca, Oaxaca, Mexico

Antonella Cicirelli Department of Precision and Regenerative Medicine and Ionian Area, University of Bari Aldo Moro, Piazza G. Cesare, Bari, Italy

Alessio Coppola Laboratory of Functional Genetics of Rare Diseases, Department of Microbiology and Genetics, University of Salamanca (USAL), Salamanca, Spain

GIR of Biomedicine of Rare Diseases, University of Salamanca, Salamanca, Spain

Institute of Biomedical Research of Salamanca (IBSAL), Salamanca, Spain

Holanda Isabel Cruz-Luis SECIHTI, Faculty of Medicine and Surgery, Benito Juárez Autonomous University of Oaxaca, Oaxaca, Mexico

Margarito Martínez Cruz National Institute of Technology of Mexico/IT. (TecNM) Oaxaca, Oaxaca, Mexico

Eli Cruz-Parada National Institute of Technology of Mexico/IT. (TecNM) Oaxaca, Oaxaca, Mexico

Rajesh Das Department of Bioinformatics, Pondicherry University, Kalapet, Puducherry, India

Bhaskar Datta Department of Biological Sciences & Engineering, Indian Institute of Technology Gandhinagar, Palaj, Gandhinagar, Gujarat, India

Department of Chemistry, Indian Institute of Technology Gandhinagar, Palaj, Gandhinagar, Gujarat, India

Bruno Bernardes de Jesus Department of Medical Sciences and Institute of Biomedicine – iBiMED, University of Aveiro, Aveiro, Portugal

Jaqueline Carvalho de Oliveira Department of Genetics, Federal University of Paraná (UFPR), Curitiba, PR, Brazil

Carlos Romero Díaz National Institute of Technology of Mexico/IT. (TecNM) Oaxaca, Oaxaca, Mexico

Loreto Gesualdo Department of Precision and Regenerative Medicine and Ionian Area, University of Bari Aldo Moro, Piazza G. Cesare, Bari, Italy

Arkadeep Ghosh Department of Biological Sciences & Engineering, Indian Institute of Technology Gandhinagar, Palaj, Gandhinagar, Gujarat, India

Sonia Guil Josep Carreras Leukaemia Research Institute (IJC), Badalona, Barcelona, Catalonia, Spain

Gloria Rubí Sánchez Hernández SECIHTI, Faculty of Medicine and Surgery, Benito Juárez Autonomous University of Oaxaca, Oaxaca, Mexico

María Teresa Hernández-Huerta SECIHTI, Faculty of Medicine and Surgery, Benito Juárez Autonomous University of Oaxaca, Oaxaca, Mexico

José Luis Hernández-Morales SECIHTI, Faculty of Medicine and Surgery, Benito Juárez Autonomous University of Oaxaca, Oaxaca, Mexico

María Isidoro-García Institute of Biomedical Research of Salamanca (IBSAL), Salamanca, Spain

Department of Clinical Biochemistry, University Hospital of Salamanca, Salamanca, Spain

Reference Unit for Rare Diseases (DiERCyL), Castilla León, Spain

Department of Medicine, University of Salamanca, Salamanca, Spain

Jesus Lacal Laboratory of Functional Genetics of Rare Diseases, Department of Microbiology and Genetics, University of Salamanca (USAL), Salamanca, Spain

GIR of Biomedicine of Rare Diseases, University of Salamanca, Salamanca, Spain

Institute of Biomedical Research of Salamanca (IBSAL), Salamanca, Spain

Giorgio Ladisa Department of Precision and Regenerative Medicine and Ionian Area, University of Bari Aldo Moro, Piazza G. Cesare, Bari, Italy

Carlos Lastre-Domínguez National Institute of Technology of Mexico/IT. (TecNM) Oaxaca, Oaxaca, Mexico

Anelis Maria Marin Laboratory for Applied Science and Technology in Health, Carlos Chagas Institute, Oswaldo Cruz Foundation (Fiocruz), Curitiba, Brazil

Shruti Mathur Department of Biological Sciences & Engineering, Indian Institute of Technology Gandhinagar, Palaj, Gandhinagar, Gujarat, India

Eduardo Pérez-Campos Mayoral UNAM-UABJO Faculty of Medicine Research Centre, Faculty of Medicine and Surgery, Benito Juárez Autonomous University of Oaxaca, Oaxaca, Mexico

Laura Pérez-Campos Mayoral UNAM-UABJO Faculty of Medicine Research Centre, Faculty of Medicine and Surgery, Benito Juárez Autonomous University of Oaxaca, Oaxaca, Mexico

Francesca Montenegro Department of Precision and Regenerative Medicine and Ionian Area, University of Bari Aldo Moro, Piazza G. Cesare, Bari, Italy

Parisa Naji Department of Cell and Molecular Biology, Semnan University, Semnan, Iran

Sibin Nambidi Medical Biotechnology Lab, Faculty of Allied Health Sciences, Chettinad Academy of Research and Education (CARE), Chettinad Hospital and Research Institute, Chennai, India

Alessio Nigro PersonGene – University of Bari Spinoff – Piazza G. Cesare, Bari, Italy

Sasi Nivruthi Department of Biotechnology, School of Bioengineering, SRM Institute of Science and Technology, Kattankulathur, Tamil Nadu, India

Alessandro Palma Division of Neuroscience, Department of Psychology, Sapienza University of Rome, Rome, Italy

Farshid Parvini Department of Cell and Molecular Biology, Semnan University, Semnan, Iran

Surajit Pathak Medical Biotechnology Lab, Faculty of Allied Health Sciences, Chettinad Academy of Research and Education (CARE), Chettinad Hospital and Research Institute, Chennai, India

Dahrii Paul Department of Bioinformatics, Pondicherry University, Kalapet, Puducherry, India

Eduardo Pérez-Campos National Institute of Technology of Mexico/IT. (TecNM) Oaxaca, Oaxaca, Mexico

"Dr. Eduardo Pérez Ortega" Clinical Pathology Laboratory, Oaxaca, Mexico

Angela Picerno Department of Precision and Regenerative Medicine and Ionian Area, University of Bari Aldo Moro, Piazza G. Cesare, Bari, Italy

Maria Augusta Poersch Department of Genetics, Federal University of Paraná (UFPR), Curitiba, PR, Brazil

Dilipkumar Preetha Department of Biotechnology, School of Bioengineering, SRM Institute of Science and Technology, Kattankulathur, Tamil Nadu, India

Pablo Prieto-Matos GIR of Biomedicine of Rare Diseases, University of Salamanca, Salamanca, Spain

Institute of Biomedical Research of Salamanca (IBSAL), Salamanca, Spain

Reference Unit for Rare Diseases (DiERCyL), Castilla León, Spain

Pediatrics Area, Department of Biomedical and Diagnostic Sciences, University of Salamanca (USAL), Salamanca, Spain

Edgar Ramos-Martínez UNAM-UABJO Faculty of Medicine Research Centre, Faculty of Medicine and Surgery, Benito Juárez Autonomous University of Oaxaca, Oaxaca, Mexico

Iván Ramos-Martínez Mycology Unit, Department of Microbiology and Parasitology, Faculty of Medicine, National Autonomous University of Mexico, Mexico City, Mexico

Fabio Sallustio Department of Precision and Regenerative Medicine and Ionian Area, University of Bari Aldo Moro, Piazza G. Cesare, Bari, Italy

Iyyappan Saranya Department of Biotechnology, School of Bioengineering, SRM Institute of Science and Technology, Kattankulathur, Tamil Nadu, India

Nagarajan Selvamurugan Department of Biotechnology, School of Bioengineering, SRM Institute of Science and Technology, Kattankulathur, Tamil Nadu, India

Md Mujibur Rahman Sheikh Department of Bioinformatics, Pondicherry University, Kalapet, Puducherry, India

Deepshikha Singh Department of Biological Sciences & Engineering, Indian Institute of Technology Gandhinagar, Palaj, Gandhinagar, Gujarat, India

Sony Snigdha Sinku Department of Biological Sciences & Engineering, Indian Institute of Technology Gandhinagar, Palaj, Gandhinagar, Gujarat, India

Vigneshwar Suriya Prakash Sinnarasan Department of Bioinformatics, Pondicherry University, Kalapet, Puducherry, India

Nivetha Suresh Department of Biotechnology, School of Bioengineering, SRM Institute of Science and Technology, Kattankulathur, Tamil Nadu, India

Carlota Tavares-Marcos Department of Medical Sciences and Institute of Biomedicine – iBiMED, University of Aveiro, Aveiro, Portugal

Amouda Venkatesan Department of Bioinformatics, Pondicherry University, Kalapet, Puducherry, India

Saghar Yousefnia Department of Cell and Molecular Biology, Semnan University, Semnan, Iran

Dalila Luciola Zanette Laboratory for Applied Science and Technology in Health, Carlos Chagas Institute, Oswaldo Cruz Foundation (Fiocruz), Curitiba, Brazil

Chapter 1
Long Noncoding RNAs and Human Genome

Alessandro Palma

Abstract Long noncoding RNAs (lncRNAs) represent a vast and dynamic component of the human genome, playing essential roles in gene regulation, chromatin organization, and cellular homeostasis. lncRNAs exhibit remarkable tissue- and cell-type specificity, influencing developmental processes and disease mechanisms. Despite their growing recognition, the genomic architecture, functional diversity, and evolutionary significance of lncRNAs remain incompletely understood. This chapter explores the distribution, classification, and structural features of lncRNA genes across the human genome, highlighting their regulatory potential and implications for genome organization. By integrating current knowledge with recent discoveries, this chapter aims to shed light on the intricate relationship between lncRNAs and the human genome, paving the way for future research in this rapidly evolving field.

Keywords Long noncoding RNAs · Noncoding RNAs · Human genome · Genes · Single nucleotide polymorphisms · Genomics

Abbreviations

asRNA	Antisense RNA
circRNA	Circular intronic RNA
GC	Guanine and cytosine
GWAS	Genome-wide association study
lncRNA	Long noncoding RNA
ncRNA	Noncoding RNA

A. Palma (✉)
Division of Neuroscience, Department of Psychology, Sapienza University of Rome, Rome, Italy
e-mail: alessandro.palma2@uniroma1.it

L. M. Vaschetto (ed.), *Long Non-Coding RNAs, Human Health, and Diseases*, Epigenetics and Human Health 16, https://doi.org/10.1007/978-3-032-19668-2_1

1.1 Introduction

Long noncoding RNAs (lncRNAs) represent a diverse and essential class of transcripts that have emerged as key regulators in numerous biological processes, despite their lack of protein-coding potential. These RNA molecules, which are typically over 500 nucleotides in length (Mattick et al. 2023), have been implicated in gene regulation, chromatin remodeling, and cellular signaling pathways. Unlike protein-coding genes, lncRNAs exhibit a complex genomic organization, showing variations in size, structure, and functional mechanisms. The classification and characterization of lncRNAs are still evolving, with various systems based on their genomic context, length, and interactions with other molecules. This chapter delves into the fundamental genomic features of human lncRNAs, highlighting their diversity in genomic distribution, gene and transcript length, splicing mechanisms, polyadenylation, guanine and cytosine (GC) content, and the role of genetic mutations in human diseases. We also explore their conservation across species, which offers insight into their functional importance and potential as therapeutic targets.

1.2 Genomic Classification of lncRNAs

Long noncoding RNAs can be classified based on different genomic features. One of the available classifications describes lncRNAs based on their context and location, which divides lncRNAs into intergenic lncRNAs and intronic lncRNAs. The first are lncRNAs transcribed intergenically from both strands, whereas the second are lncRNAs transcribed entirely from introns of protein-coding genes (Ma et al. 2013). Other types of classification take into consideration further genomic or empirical attributes of lncRNAs. From the perspective of their length, they could be divided into long noncoding RNAs (Wang and Chang 2011), long-intergenic noncoding RNAs (Guttman et al. 2009), very long intergenic noncoding RNAs (St Laurent et al. 2013), macroRNAs (Hackermüller et al. 2014), and promoter-associated long RNAs (Kapranov et al. 2007). However, one of the most diffused and used classifications of lncRNAs is based on their association with annotated protein-coding genes (Laurent et al. 2015). This classification categorizes lncRNAs into 10 distinct classes (Fig. 1.1):

1. Intronic noncoding RNA (ncRNA): lncRNAs located in intronic regions of protein-coding genes.
2. Circular intronic RNAs (circRNAs): lncRNAs localized in intronic regions that circle after their transcription through back-splicing mechanisms.
3. Sense ncRNA: lncRNAs transcribed from the sense strand of a genomic region.
4. Natural antisense ncRNA (asRNAs): long noncoding RNAs located on the opposite strand of a protein-coding gene.
5. Mirror antisense RNAs: spliced antisense transcripts with exon structures closely mirroring exon-intron structures of the sense transcripts.

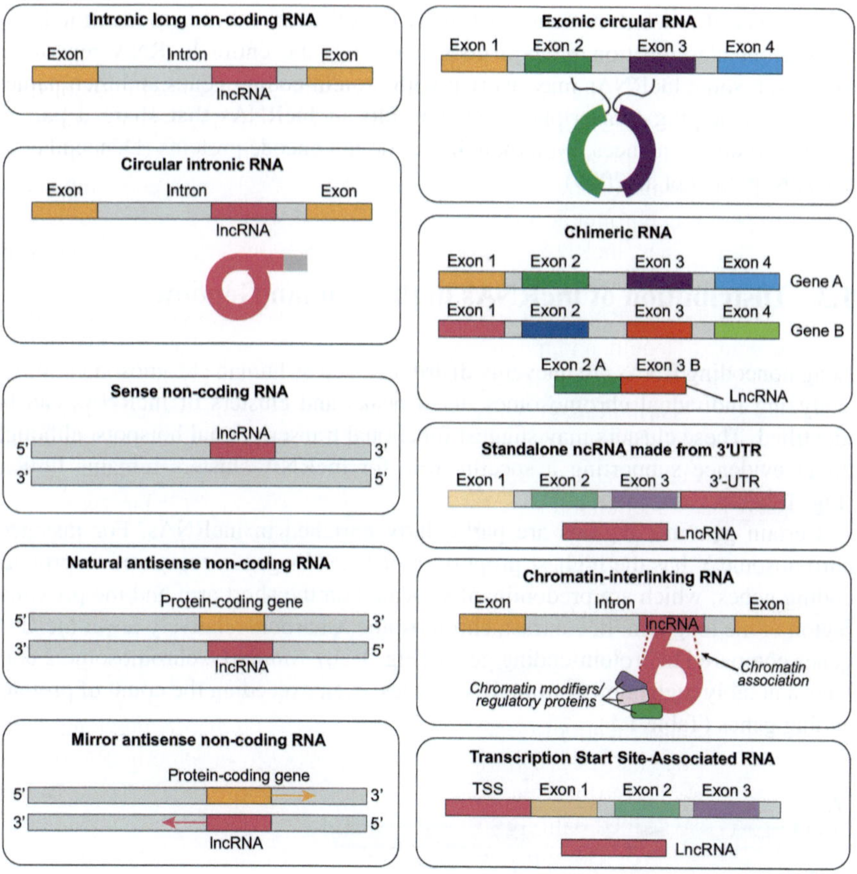

Fig. 1.1 lncRNA classification. Schematics of one of the most used classifications of lncRNAs

6. Exonic circular RNAs: lncRNAs localized in exonic regions that circle after their transcription through back-splicing mechanisms.
7. Chimeric RNAs: transcripts whose sequences are chimeras formed from sequences found in separate genes; this is often due to trans-splicing, read-through transcription, or chromosomal rearrangements.
8. Standalone ncRNAs made from 3′-UTRs (untranslated region-associated RNAs): originate from the cleavage, readthrough, or independent transcription of the 3′-UTR of protein-coding genes.
9. Chromatin-interlinking RNA: often act as scaffolds, bringing together chromatin-modifying complexes and regulatory proteins to influence gene expression.
10. Transcription start site-associated RNAs: noncoding RNAs located near transcription start sites of actively transcribed genes.

However, it is worth noticing that many lncRNAs are still being identified, and the current classification method does not cover the entire lncRNA repertoire. Moreover, some lncRNAs may overlap with protein-coding genes and then named "sense overlapping transcripts," which results in lncRNAs that share a part of protein-coding sequences, even though they do not encode proteins (Denoeud et al. 2007; Kapranov et al. 2005).

1.3 Distribution of lncRNAs in the Human Genome

Long noncoding RNAs are unevenly distributed across human chromosomes. When analyzing individual chromosomes, local peaks and clusters of lncRNAs can be identified. These clusters may suggest functional transcriptional hotspots, although direct evidence supporting a specific role for lncRNA clusters remains limited (Fig. 1.2a).

Certain genomic regions are particularly enriched in lncRNAs. For instance, chromosome Y has the highest proportion of lncRNA genes compared to protein-coding genes, which are predominantly located on the short arm and the proximal region of the long arm. In contrast, chromosome X harbors relatively fewer lncRNA genes compared to protein-coding genes (Fig. 1.2b). Similarly, chromosome 2 contains a notably high number of lncRNA genes, even exceeding the count of protein-coding genes (Table 1.1).

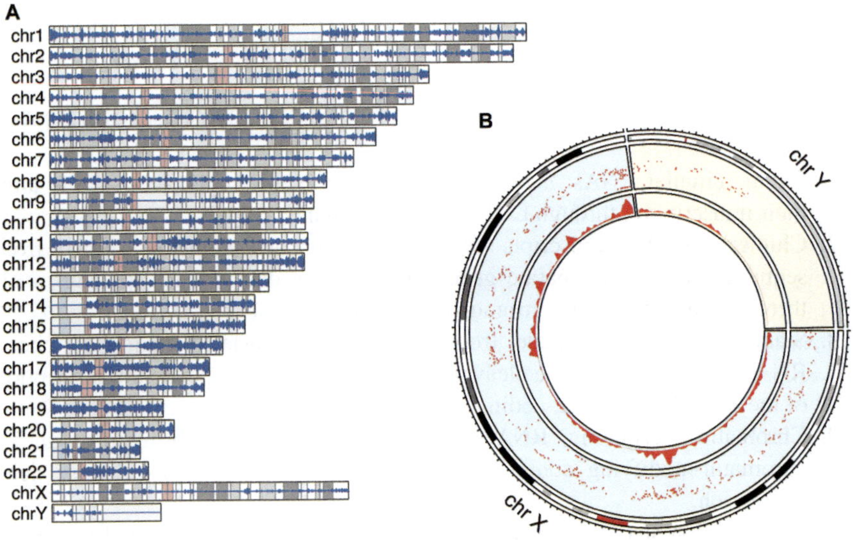

Fig. 1.2 lncRNA distribution across the human genome. (**a**) Peak distribution of lncRNAs in human chromosomes. (**b**) Rainfall plot showing the distribution of lncRNAs (red dots and red distribution) in human X and Y chromosomes. (Data from GENCODE v.47)

Table 1.1 lncRNA and protein-coding gene numbers. The table shows the lncRNA and protein-coding gene numbers for each human chromosome (GENCODE v.47)

Chromosome	Number of lncRNA genes	Number of protein-coding genes
Chr 1	1649	2066
Chr 2	1451	1248
Chr 3	1033	1082
Chr 4	928	758
Chr 5	1118	887
Chr 6	1011	1050
Chr 7	904	930
Chr 8	930	704
Chr 9	685	776
Chr 10	840	732
Chr 11	1010	1317
Chr 12	1150	1039
Chr 13	564	323
Chr 14	685	617
Chr 15	842	602
Chr 16	1102	856
Chr 17	1093	1186
Chr 18	607	267
Chr 19	845	1477
Chr 20	507	546
Chr 21	384	220
Chr 22	464	447
Chr X	396	859
Chr Y	112	63

However, because human chromosomes vary in length, it is essential to consider lncRNA density relative to chromosome size. A striking example is chromosomes 16 and 17, which exhibit a relatively high number of lncRNA genes despite their smaller sizes (Palma et al. 2025).

In the human genome, both protein-coding and lncRNA genes are distributed across both DNA strands. However, there is no significant difference in the number of lncRNA genes located on the plus versus minus strands, and their gene sizes appear comparable. This suggests that lncRNA gene presence and evolution are not influenced by strand orientation.

1.4 Genomic Features of LncRNAs

1.4.1 lncRNA Gene and Transcript Length

lncRNAs vary significantly in gene and transcript length. Originally, the noncoding transcripts longer than 200 nucleotides (nt) are classified as lncRNAs; a recent consensus statement highlighted the need to shift this threshold to 500 nt in order to exclude a range of small RNAs (e.g., RNA polymerase III transcripts, small nuclear RNAs, intron-derived small nucleolar RNAs) that are usually 50–500 nt in length (Mattick et al. 2023). However, the old classification type is still widely used, as documented by the annotation of databases such as GENCODE, which annotates as lncRNAs some genes with a length shorter than 500 nt.

In general, lncRNA genes are shorter than protein-coding genes, the latter being double in size on average (Palma et al. 2025). The mean length of lncRNA genes ranges from a few hundred to several thousand nucleotides. According to the current latest version of GENCODE (version 47 [GRCh38.p14]), the human genome contains 35,934 genes coding for lncRNAs, which produce 191,106 different transcript isoforms.

The longest lncRNA gene is *NRXN1-DT*, located on the plus strand of chromosome 2, with a width spanning 1,375,317 nt, coding for a 3939-nt-long mature divergent transcript. Conversely, the longest transcript isoform belonging to the lncRNA class measures 347,561 nt and is coded by the *XACT* gene, located on the minus strand of chromosome X. This transcript is made of two distinct exons.

Just like protein-coding genes, lncRNA genes tend to be longer than transcripts because they are composed of both introns and exons. The average length of lncRNA genes is 29,056.67 nt, with a median of 6721.5 nt. Table 1.2 provides a list of the longest human lncRNA genes and transcripts.

Table 1.2 List of the top five long noncoding genes and transcripts according to version 47 of GENCODE

Type	Name	Ensembl ID	Length (nt)[a]	Chromosome
Gene	NRXN1-DT	ENSG00000231918.2	1,375,317	Chr2
Gene	ENSG00000225689	ENSG00000225689.3	1,299,386	ChrX
Gene	LINC02055	ENSG00000254101.9	1,232,815	Chr8
Gene	ENSG00000271860	ENSG00000271860.10	1,179,835	Chr6
Gene	PTCHD1-AS	ENSG00000233067.5	1,170,958	ChrX
Transcript	XACT	ENST00000674361.1	347,561	ChrX
Transcript	HELLPAR	ENST00000626826.1	205,012	Chr12
Transcript	KCNQ1OT1	ENST00000597346.1	91,667	Chr11
Transcript	PPM1F-AS1	ENST00000458178.2	37,852	Chr22
Transcript	TSIX	ENST00000604411.1	37,027	ChrX

Chr chromosome, *nt* nucleotides

[a]Gene length refers to the entire length of the primary gene sequence, while transcript length refers to the mature (spliced) form of transcripts

1.4.2 Intron, Exons, and Splicing Isoforms of lncRNAs

Exons are traditionally associated with protein-coding sequences; however, in lncRNAs, exons remain untranslated and play crucial roles in their functionality (Aspden et al. 2023). While some lncRNAs exist as single-exon transcripts, others contain multiple introns, structurally resembling protein-coding genes. The presence of introns can influence lncRNA maturation, nuclear export, and functional properties. Additionally, some lncRNAs undergo regulated splicing events that generate distinct isoforms, adding another layer of complexity to their regulatory potential.

lncRNAs exhibit unique splicing characteristics that distinguish them from protein-coding genes. Variations in canonical splicing signals, including the branch point, 5′ and 3′ splice sites, and interactions with specific splicing factors, can modulate exon inclusion or exclusion (Abou Alezz et al. 2020). Consequently, both nascent and mature pre-lncRNAs often undergo inefficient splicing compared to pre-mRNAs of protein-coding genes, contributing to incomplete splicing, reduced transcript stability, and lower overall abundance. This aligns with the general observation that lncRNAs tend to be less stable and less efficiently processed than mRNAs.

Compared to protein-coding genes, lncRNAs typically have fewer exons and exhibit lower splicing efficiency, as observed in transcriptome annotations such as GENCODE. The lncRNA databases and related bioinformatic tools are detailed in Chap. 2 by Datta's research group. lncRNA genes also have a lower median number of exons per transcript and significantly fewer transcript isoforms per gene (Palma et al. 2025). However, their median exon and intron lengths tend to be longer than those of protein-coding genes (Khan et al. 2021).

Beyond the canonical GT-AG splice junction, lncRNAs show an enrichment in the GC-AG splice junction, which accounts for approximately 3% of all lncRNA splice junctions. This frequency is notably higher than in protein-coding genes, where GC-AG splicing represents only 0.8% of all junctions (Abou Alezz et al. 2020). This preference is particularly pronounced at the first exon-intron junction, potentially influencing alternative splicing and intron retention mechanisms.

A significant fraction of lncRNAs exists as single-exon transcripts, while others exhibit complex splicing patterns that may contribute to their tissue-specific expression. Notably, some lncRNAs retain introns within their mature transcripts, which can confer functional advantages such as enhanced protein, RNA, or DNA-binding capacity (Dey and Mattick 2021).

Additionally, some lncRNA sequences contain small open reading frames that can encode micropeptides (Chen et al. 2020; Prensner et al. 2021), while others undergo back-splicing events, giving rise to circular RNAs (circRNAs). These circRNAs can function as molecular sponges for microRNAs, thereby influencing post-transcriptional gene regulation (Qin et al. 2020; Di Timoteo et al. 2020).

1.4.3 Polyadenylation Sites and Signals of Human lncRNAs

Polyadenylation is a fundamental post-transcriptional modification that occurs at the 3′ end of nascent transcripts, playing a crucial role in mRNA stability, nuclear export, and translation efficiency. While many transcripts are stabilized through the addition of a polyadenylated (poly[A]) tail, some are maintained via alternative mechanisms such as RNase P cleavage (Brown et al. 2012; Wilusz et al. 2012), or capping by small nucleolar ribonucleoprotein complexes at both ends (Yin et al. 2012). The presence or absence of a poly(A) tail significantly impacts lncRNA stability, processing, and subcellular localization.

Although a substantial fraction of lncRNAs are polyadenylated, many lack canonical poly(A) tails, distinguishing them from typical mRNAs. Studies estimate that approximately 28% of annotated human lncRNAs contain either a poly(A) site (the precise genomic location where cleavage and poly[A] tail addition occur) or a poly(A) signal (the upstream sequence that recruits the cleavage and polyadenylation complex) (Palma et al. 2025), reflecting the broader trend that many lncRNAs remain nonpolyadenylated (Schlackow et al. 2017). This distinction has important experimental implications, particularly in RNA sequencing approaches. Standard poly(A)-enriched sequencing methods may fail to detect nonpolyadenylated lncRNAs, leading to biased transcriptome profiles.

Interestingly, no significant difference exists in the distance between poly(A) sites and signals between lncRNAs and protein-coding genes, which averages 22 nt (Palma et al. 2025). Overall, the biological significance of lncRNA polyadenylation remains an open question, requiring further molecular investigations to understand its functional consequences. One possible hypothesis is that polyadenylation of a subset of lncRNAs acts as a regulatory switch, modulating their stability and localization in a manner that fine-tunes gene expression programs in a context-dependent way, similar to but more dynamic than protein-coding transcripts.

1.4.4 Genomic GC Content of Human lncRNAs

Genomic regions with high guanine and cytosine content (GC-rich regions) are known to play a crucial role in forming stable structures, influencing genome functionality (Zhou et al. 2011). These GC-rich regions often contain a high density of CpG dinucleotides, which serve as promoters for many human genes (Illingworth and Bird 2009). Moreover, GC skew is commonly associated with unmethylated CpG island promoters (Ginno et al. 2012), and their transcription can lead to the formation of R-loops, which stabilize nascent RNA at its DNA locus, resulting in GC-rich RNA hybrids (Hartono et al. 2015). By anchoring RNA transcripts to their sites of origin, such structures can facilitate local recruitment of chromatin modifiers, alter DNA methylation patterns, and modulate transcriptional accessibility,

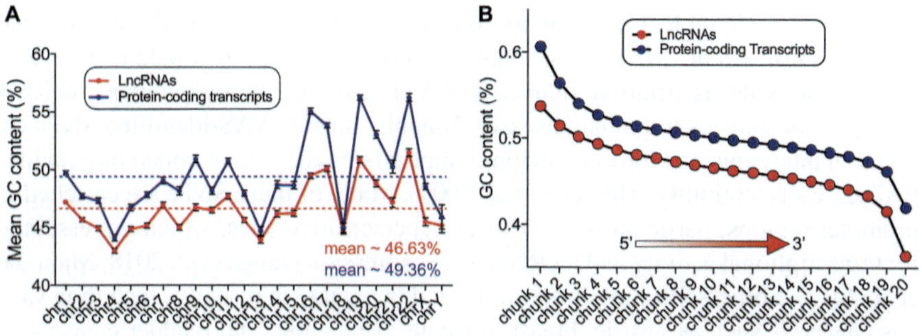

Fig. 1.3 GC content of lncRNA and protein-coding transcripts. (**a**) Average GC content for lncRNAs (red) and protein-coding transcripts (blue) for each human chromosome according to GENCODE v.47. (**b**) GC content on lncRNAs (red) and protein-coding transcripts (blue) sequences divided by two chunks, according to GENCODE v.47. (Adapted from Palma et al. (2025) under a Creative Commons Attribution (CC BY) license)

thereby allowing GC-rich lncRNAs to act as epigenetic regulators of gene expression.

Previous studies have shown that long noncoding transcripts generally have lower GC content compared to protein-coding transcripts (PCTs), with exons tending to have higher GC content than introns (Haerty and Ponting 2015). However, this distinction should be approached with caution, as analytical biases might arise, especially as some lncRNAs retain introns in their mature, functional form. The average GC content of the human genome is roughly 41% (Cohen et al. 2005; Lander et al. 2001; Piovesan et al. 2019), with protein-coding transcripts showing an average GC content of approximately 49.5%. One recent report showed a mean GC content of 46.6% for lncRNAs (Palma et al. 2025), further confirming that lncRNAs are characterized by a lower GC content compared to protein-coding genes (Fig. 1.3a). Such characteristics may contribute to differences in their transcriptional regulation and degradation pathways. Indeed, GC-rich lncRNAs may form stable secondary structures, influencing their ability to act as scaffolds for ribonucleoprotein complexes, secondary structure, stability, and interaction with RNA-binding proteins.

One typical feature of lncRNAs shared with protein-coding transcripts is their elevated GC content in the 5′ regions (Palma et al. 2025) (Fig. 1.3b). This feature may be linked to crucial molecular events, such as transcription regulation and sequence stabilization, given the abundance of regulatory sites in these areas.

1.4.5 Genomic Mutations of lncRNAs in Humans

lncRNAs, like protein-coding genes, are subject to various genetic mutations, including single nucleotide polymorphisms (SNPs), insertions, deletions, and structural variations. These mutations can impact lncRNA expression, stability, and interactions with molecular partners, potentially contributing to disease mechanisms (Esposito et al. 2023; Haraksingh and Snyder 2013; Karczewski et al. 2020).

Genome-wide association studies (GWAS) have increasingly linked lncRNA gene polymorphisms to human diseases. Notably, most GWAS-identified SNPs are found in intergenic and intronic regions, emphasizing the role of noncoding regions in disease susceptibility (Hindorff et al. 2009). Many of these SNPs are enriched in promoter regions, particularly in DNaseI hypersensitive sites, which are essential for transcriptional activity and lncRNA gene expression (Janga et al. 2018; Maurano et al. 2012). Some SNPs have been implicated in a range of diseases, including various cancers—such as gastric, breast, prostate, pancreatic, and bladder cancer—as well as neurological and psychiatric disorders (Sattari et al. 2020; Yang et al. 2023).

While a vast majority of SNPs in lncRNA loci do not appear to influence lncRNA functionality or lead to clinical phenotypes, a small fraction has been classified as clinically relevant, according to SNP annotation data from Ensembl. These disease-associated SNPs may affect lncRNA expression by altering regulatory DNA elements, such as transcription factor binding sites (Kulkarni et al. 2019). Additionally, structural variations within lncRNA loci—such as disruptions in secondary structure or alternative splicing—could further contribute to disease pathogenesis. Beyond SNPs, larger genomic alterations, including amplifications, deletions, and translocations, have also been implicated in disease by affecting lncRNA copy number and integrity (Aznaourova et al. 2020).

Importantly, several lncRNAs carrying disease-associated variants have been proposed as clinical biomarkers, aiding in patient stratification and survival prognosis, particularly in cancer (Arriaga-Canon et al. 2018; Zhang et al. 2017). However, despite growing evidence of their involvement in disease, human genetics studies have traditionally focused on mutations in protein-coding genes, which are more directly linked to monogenic or multifactorial diseases. Nevertheless, an increasing number of clinically relevant SNPs affecting lncRNA genes are being identified (Palma et al. 2025).

It is also important to note that some of these genetic variations may affect lncRNA genes that are antisense to or overlapping with protein-coding genes, or they may occur in regulatory regions critical for protein-coding gene expression. This suggests that certain genetic diseases or clinically relevant phenotypes involving lncRNAs remain undiscovered. Expanding genetic studies and diagnostic panels to include noncoding genes will be essential to fully understand their role in human disease.

1.5 Evolutionary Insights into lncRNA Conservation

lncRNA genes are less conserved compared to protein-coding genes, which is probably due to a lower level of constraint between structure and function (Kutter et al. 2012; Pang et al. 2006). However, when looking at specific genomic features of lncRNAs, a varying degree of conservation can be observed. For instance, promoter sequences of lncRNAs tend to be conserved across species at levels comparable to those of protein-coding genes, as well as exon structures and splice junctions (Derrien et al. 2012; Nitsche et al. 2015; Ponjavic et al. 2007). Moreover, some lncRNAs show conserver sequence domains that determine their functions, such as the most conserved regions of *XIST*—the gene responsible for chromosome dosage compensation in placental mammals—which correspond to low-copy repetitive elements (Brockdorff 2002; Brown et al. 1992). However, the lack of sequence conservation can be misleading, as secondary structure conservation of lncRNAs could play a major role in determining their function (Johnsson et al. 2014).

It is also worth mentioning the existence of a subclass of lncRNAs, named antisense lncRNA (asRNA), that share complementarity to sense-expressed transcripts and are expressed at their same genomic locus. This genomic proximity of sense-antisense pairs may suggest they evolved together, playing a role in tethering the asRNA transcript to the DNA during transcription (Johnsson et al. 2014).

1.6 Conclusion

Long noncoding RNAs have emerged as pivotal regulatory molecules with vast functional roles across a variety of cellular processes. While much progress has been made in understanding their genomic features and classifications, the full scope of their involvement in human health and disease remains an area of active research. The complexity of their transcriptional regulation, coupled with their genetic diversity and functional flexibility, underscores the need for continued exploration. Notably, the growing recognition of disease-associated lncRNAs provides exciting prospects for their use as biomarkers and therapeutic targets, particularly in cancer and other complex diseases. As genomic techniques evolve, future studies will likely uncover even more layers of lncRNA functionality, offering new avenues for precision medicine and therapeutic interventions.

Acknowledgments Corrected and edited with the assistance of artificial intelligence (AI) tools OpenAI, ChatGPT (GPT-5), 2025.

Every section was carefully reviewed, revised, and approved by the authors to ensure that the text is appropriate concerning content and scientific correctness while maintaining the integrity of the original content.

Funding Information This chapter received no funding.

Compliance with Ethical Standards

Conflict of Interest The author declares that he has no conflict of interest.

Ethical Approval This chapter does not contain any studies with human participants performed by any of the authors.

References

Abou Alezz M, Celli L, Belotti G, Lisa A, Bione S (2020) GC-AG introns features in long non-coding and protein-coding genes suggest their role in gene expression regulation. Front Genet 11:1–15

Arriaga-Canon C, De La Rosa-Velázquez IA, González-Barrios R, Montiel-Manríquez R, Oliva-Rico D, Jiménez-Trejo F, Cortés-González C, Herrera LA (2018) The use of long non-coding RNAs as prognostic biomarkers and therapeutic targets in prostate cancer. Oncotarget 9:20872–20890

Aspden JL, Wallace EWJ, Whiffin N (2023) Not all exons are protein coding: addressing a common misconception. Cell Genomics 3:100296

Aznaourova M, Schmerer N, Schmeck B, Schulte LN (2020) Disease-causing mutations and rearrangements in long non-coding RNA gene loci. Front Genet 11:527484

Brockdorff N (2002) X-chromosome inactivation: closing in on proteins that bind Xist RNA. Trends Genet 18:352–358

Brown CJ, Hendrich BD, Rupert JL, Lafrenière RG, Xing Y, Lawrence J, Willard HF (1992) The human XIST gene: analysis of a 17 kb inactive X-specific RNA that contains conserved repeats and is highly localized within the nucleus. Cell 71:527–542

Brown JA, Valenstein ML, Yario TA, Tycowski KT, Steitz JA (2012) Formation of triple-helical structures by the 3′-end sequences of MALAT1 and MENβ noncoding RNAs. Proc Natl Acad Sci USA 109:19202–19207

Chen J, Brunner A, Cogan JZ, Nuñez JK, Fields P, Adamson B, Itzhak DN, Li JY, Mann M, Leonetti D et al (2020) Pervasive functional translation of noncanonical human open reading frames. Science 367:1140–1146

Cohen N, Dagan T, Stone L, Graur D (2005) GC composition of the human genome: in search of isochores. Mol Biol Evol 22:1260–1272

Denoeud F, Kapranov P, Ucla C, Frankish A, Castelo R, Drenkow J, Lagarde J, Alioto T, Manzano C, Chrast J et al (2007) Prominent use of distal 5′ transcription start sites and discovery of a large number of additional exons in ENCODE regions. Genome Res 17:746–759

Derrien T, Johnson R, Bussotti G, Tanzer A, Djebali S, Tilgner H, Guernec G, Martin D, Merkel A, Knowles DG et al (2012) The GENCODE v7 catalog of human long noncoding RNAs: analysis of their gene structure, evolution, and expression. Genome Res 22:1775–1789

Dey P, Mattick JS (2021) High frequency of intron retention and clustered H3K4me3-marked nucleosomes in short first introns of human long non-coding RNAs. Epigenetics Chromatin 14:1–11

Di Timoteo G, Dattilo D, Centrón-Broco A, Colantoni A, Guarnacci M, Rossi F, Incarnato D, Oliviero S, Fatica A, Morlando M et al (2020) Modulation of circRNA metabolism by m6A modification. Cell Rep 31:107641

Esposito R, Lanzós A, Uroda T, Ramnarayanan S, Büchi I, Polidori T, Guillen-Ramirez H, Mihaljevic A, Merlin BM, Mela L et al (2023) Tumour mutations in long noncoding RNAs enhance cell fitness. Nat Commun 14:3342

Ginno P, Lott P, Christensen H, Korf I, Chédin F (2012) R-loop formation is a distinctive characteristic of unmethylated human CpG Island promoters. Mol Cell 45:814–825

Guttman M, Amit I, Garber M, French C, Lin MF, Feldser D, Huarte M, Zuk O, Carey BW, Cassady JP et al (2009) Chromatin signature reveals over a thousand highly conserved large non-coding RNAs in mammals. Nature 458:223–227

Hackermüller J, Reiche K, Otto C, Hösler N, Blumert C, Brocke-Heidrich K, Böhlig L, Nitsche A, Kasack K, Ahnert P et al (2014) Cell cycle, oncogenic and tumor suppressor pathways regulate numerous long and macro non-protein-coding RNAs. Genome Biol 15:R48

Haerty W, Ponting CP (2015) Unexpected selection to retain high GC content and splicing enhancers within exons of multiexonic IncRNA loci. RNA 21:320–332

Haraksingh RR, Snyder MP (2013) Impacts of variation in the human genome on gene regulation. J Mol Biol 425:3970–3977

Hartono SR, Korf IF, Chédin F (2015) GC skew is a conserved property of unmethylated CpG Island promoters across vertebrates. Nucleic Acids Res 43:9729–9741
Hindorff LA, Sethupathy P, Junkins HA, Ramos EM, Mehta JP, Collins FS, Manolio TA (2009) Potential etiologic and functional implications of genome-wide association loci for human diseases and traits. Proc Natl Acad Sci USA 106:9362–9367
Illingworth RS, Bird AP (2009) CpG islands—"A rough guide". FEBS Lett 583:1713–1720
Janga H, Aznaourova M, Boldt F, Damm K, Grünweller A, Schulte LN (2018) Cas9-mediated excision of proximal DNaseI/H3K4me3 signatures confers robust silencing of microRNA and long non-coding RNA genes. PLoS One 13:1–17
Johnsson P, Lipovich L, Grandér D, Morris KV (2014) Evolutionary conservation of long noncoding RNAs; sequence, structure, function. Biochim Biophys Acta 1840:1063–1071
Kapranov P, Drenkow J, Cheng J, Long J, Helt G, Dike S, Gingeras TR (2005) Examples of the complex architecture of the human transcriptome revealed by RACE and high-density tiling arrays. Genome Res 15:987–997
Kapranov P, Cheng J, Dike S, Nix DA, Duttagupta R, Willingham AT, Stadler PF, Hertel J, Hackermüller J, Hofacker IL et al (2007) RNA maps reveal new RNA classes and a possible function for pervasive transcription. Science 316:1484–1488
Karczewski KJ, Francioli LC, Tiao G, Cummings BB, Alföldi J, Wang Q, Collins RL, Laricchia KM, Ganna A, Birnbaum DP et al (2020) The mutational constraint spectrum quantified from variation in 141,456 humans. Nature 581:434–443
Khan MR, Wellinger RJ, Laurent B (2021) Exploring the alternative splicing of long noncoding RNAs. Trends Genet 37:695–698
Kulkarni S, Lied A, Kulkarni V, Rucevic M, Martin MP, Walker-Sperling V, Anderson SK, Ewy R, Singh S, Nguyen H et al (2019) CCR5AS lncRNA variation differentially regulates CCR5, influencing HIV disease outcome. Nat Immunol 20:824–834
Kutter C, Watt S, Stefflova K, Wilson MD, Goncalves A, Ponting CP, Odom DT, Marques AC (2012) Rapid turnover of long noncoding RNAs and the evolution of gene expression. PLoS Genet 8:e1002841
Lander ES, Linton LM, Birren B, Nusbaum C, Zody MC, Baldwin J, Devon K, Dewar K, Doyle M, Fitzhugh W et al (2001) Initial sequencing and analysis of the human genome: international human genome sequencing consortium (nature (2001) 409 (860-921)). Nature 412:565–566
Laurent GS, Wahlestedt C, Kapranov P (2015) The landscape of long non-coding RNA classification. Trends Genet 31:239–251
Ma L, Bajic VB, Zhang Z (2013) On the classification of long non-coding RNAs. RNA Biol 10:924–933
Mattick JS, Amaral PP, Carninci P, Carpenter S, Chang HY, Chen LL, Chen R, Dean C, Dinger ME, Fitzgerald KA et al (2023) Long non-coding RNAs: definitions, functions, challenges and recommendations. Nat Rev Mol Cell Biol 24:430–447
Maurano MT, Humbert R, Rynes E, Thurman RE, Wang H, Reynolds AP, Sandstrom R, Qu H, Shafer A, Neri F et al (2012) Systematic localization of common disease-associated variation in regulatory DNA. Science 337:1190–1195
Nitsche A, Rose D, Fasold M, Reiche K, Stadler PF (2015) Comparison of splice sites reveals that long noncoding RNAs are evolutionarily well conserved. RNA 21:801–812
Palma A, Buonaiuto G, Ballarino M, Laneve P (2025) Genome biology of long non-coding RNAs in humans: a virtual karyotype. Comput Struct Biotechnol J 27:575–584
Pang KC, Frith MC, Mattick JS (2006) Rapid evolution of noncoding RNAs: lack of conservation does not mean lack of function. Trends Genet 22:1–5
Piovesan A, Pelleri MC, Antonaros F, Strippoli P, Caracausi M, Vitale L (2019) On the length, weight and GC content of the human genome. BMC Res Notes 12:1–7
Ponjavic J, Ponting CP, Lunter G (2007) Functionality or transcriptional noise? Evidence for selection within long noncoding RNAs. Genome Res 17:556–565

Prensner JR, Enache OM, Luria V, Krug K, Clauser KR, Dempster JM, Karger A, Wang L, Stumbraite K, Wang VM et al (2021) Noncanonical open reading frames encode functional proteins essential for cancer cell survival. Nat Biotechnol 39:697–704

Qin T, Li J, Zhang KQ (2020) Structure, regulation, and function of linear and circular long non-coding RNAs. Front Genet 11:1–15

Sattari A, Nicknafs F, Noroozi R (2020) The role of single nucleotide polymorphisms within long non-coding RNAs in susceptibility to human disorders. Ecol Genet Genomics 17:100071

Schlackow M, Nojima T, Gomes T, Dhir A, Carmo-Fonseca M, Proudfoot NJ (2017) Distinctive patterns of transcription and RNA processing for human lincRNAs. Mol Cell 65:25–38

St Laurent G, Shtokalo D, Dong B, Tackett MR, Fan X, Lazorthes S, Nicolas E, Sang N, Triche TJ, McCaffrey TA et al (2013) VlincRNAs controlled by retroviral elements are a hallmark of pluripotency and cancer. Genome Biol 14:R73

Wang KC, Chang HY (2011) Molecular mechanisms of long noncoding RNAs. Mol Cell 43:904–914

Wilusz JE, JnBaptiste CK, Lu LY, Kuhn CD, Joshua-Tor L, Sharp PA (2012) A triple helix stabilizes the 3′ ends of long noncoding RNAs that lack poly(a) tails. Genes Dev 26:2392–2407

Yang Y, Wang D, Miao YR, Wu X, Luo H, Cao W, Yang W, Yang J, Guo AY, Gong J (2023) lncRNASNP v3: an updated database for functional variants in long non-coding RNAs. Nucleic Acids Res 51:D192–D198

Yin QF, Yang L, Zhang Y, Xiang JF, Wu YW, Carmichael GG, Chen LL (2012) Long noncoding RNAs with snoRNA ends. Mol Cell 48:219–230

Zhang X, Hamblin MH, Yin KJ (2017) The long noncoding RNA Malat1: Its physiological and pathophysiological functions. RNA Biol 14:1705–1714

Zhou VW, Goren A, Bernstein BE (2011) Charting histone modifications and the functional organization of mammalian genomes. Nat Rev Genet 12:7–18

Chapter 2
Bioinformatic Tools and Long Non-coding RNA Databases

Deepshikha Singh, Arkadeep Ghosh, Shruti Mathur, Sony Snigdha Sinku, and Bhaskar Datta

Abstract Long non-coding RNAs (lncRNAs), once thought to be mere transcriptional noise, have emerged as powerful regulators of gene expression. The study of lncRNAs is complicated by their low expression, structural complexity, and limited sequence conservation, creating difficulties in identification and understanding of function. This chapter describes the critical role of computational and bioinformatics tools in decoding the mysteries of lncRNAs. We walk through the wide range of available tools from those that predict whether an RNA is truly non-coding to advanced algorithms that model their intricate secondary and tertiary structures and correlations with disease progression. The chapter highlights the expanding landscape of lncRNA databases, covering both human and non-human species, and their support of research into lncRNA functions, interactions, and disease associations. Despite tremendous progress, challenges remain in predicting conformationally dynamic RNA structures, addressing inconsistent annotations, and validating computational results through wet laboratory-based experimentation. The chapter presents efforts to address such challenges, including possibilities in the future on the integration of experimental techniques, artificial intelligence, and high-throughput screening methods.

D. Singh · A. Ghosh · S. Mathur · S. S. Sinku
Department of Biological Sciences and Engineering, Indian Institute of Technology Gandhinagar, Gandhinagar, Gujarat, India
e-mail: Deepshikha.s@iitgn.ac.in; 24250022@iitgn.ac.in; 24210098@iitgn.ac.in; 24320008@iitgn.ac.in

B. Datta (✉)
Department of Biological Sciences and Engineering, Indian Institute of Technology Gandhinagar, Gandhinagar, Gujarat, India

Department of Chemistry, Indian Institute of Technology Gandhinagar, Gandhinagar, Gujarat, India
e-mail: bdatta@iitgn.ac.in

L. M. Vaschetto (ed.), *Long Non-Coding RNAs, Human Health, and Diseases*, Epigenetics and Human Health 16, https://doi.org/10.1007/978-3-032-19668-2_2

Keywords Non-coding RNA · lncRNA · Secondary structure · Tertiary structure · G-quadruplex · Computational tools · Machine learning · Deep learning · LncRNA-protein interactions · Bioinformatics databases · Disease biomarkers

Abbreviations

2D-BLSTM	Two-Dimensional Bidirectional Long Short-Term Memory
AFM	Atomic Force Microscopy
AGO-CLIP-Seq	Argonaute Cross-Linking Immunoprecipitation Sequencing
ANN	Artificial Neural Network
BIGTranscriptome	Broad Institute Genome Transcriptome
BiLSTM	Bidirectional Long Short-Term Memory
BLASTX	Basic Local Alignment Search Tool (nucleotide to protein)
BLAT	BLAST-Like Alignment Tool
CD Spectroscopy	Circular Dichroism Spectroscopy
ChAR-seq	Chromatin-Associated RNA sequencing
CHESS	Comprehensive Human Expressed Sequences
ChIP	Chromatin Immunoprecipitation
circRNA	Circular RNA
ClinVar	Clinical Variant Database
CLIP-seq	Crosslinking Immunoprecipitation Sequencing
CLLM	Conditional Log-Linear Model
CNCI	Coding-Non-Coding Index
CNN	Convolutional Neural Network
COME	Coding potential calculator using multiple features
COSMIC	Catalogue of Somatic Mutations in Cancer
CPAT	Coding Potential Assessment Tool
CPC	Coding Potential Calculator
CPPred	Coding Potential Prediction tool
CRISPR	Clustered Regularly Interspaced Short Palindromic Repeats
Cryo-EM	Cryogenic Electron Microscopy
dbSNP	Database of Single Nucleotide Polymorphisms
DL	Deep Learning
DNN	Deep Neural Network
DO	Disease Ontology
DP	Dynamic Programming
EMSA	Electrophoretic Mobility Shift Assay
Ensembl	Genome Browser for Vertebrate Genomes
FANTOM CAT	FANTOM CAGE Associated Transcriptome
FANTOM	Functional Annotation of Mammalian Genome
FCS	Fluorescence Correlation Spectroscopy
G4	G-Quadruplex
G4NN	G4 Neural Network
G4RIP-qPCR	G4 RNA Immunoprecipitation followed by qPCR

G4RIP-seq	G4 RNA Immunoprecipitation Sequencing
GCN	Graph Convolutional Network
GloCNN	Global Convolutional Neural Network
GO	Gene Ontology
GRID-seq	Global RNA Interactions with DNA by deep sequencing
GU	Guanine-Uracil (wobble base pairing)
iM	Intercalated Motif
iMARGI	In situ Mapping of RNA–Genome Interactions
lncRNA	Long non-coding RNA
LocCNN	Local Convolutional Neural Network
LSTM	Long Short-Term Memory
MEA	Maximum Expected Accuracy
MeSH	Medical Subject Headings
MFE	Minimum Free Energy
microCLIP	CLIP-Seq-Based MicroRNA Capture
miRNA	MicroRNA
MiTranscriptome	Michigan Transcriptome Project
ML	Machine Learning
mRNA	Messenger RNA
MSA	Multiple Sequence Alignment
NCBI	National Center for Biotechnology Information
ncRNA	Non-coding RNA
NGS	Next Generation Sequencing
NK cells	Natural Killer Cells
NMR	Nuclear Magnetic Resonance (Spectroscopy)
NNTM	Nearest Neighbor Thermodynamic Model
NPInter	Noncoding RNA Protein Interaction Database
ORF	Open reading frame
PARS	Parallel Analysis of RNA Structure
PDB	Protein Data Bank
piRNA	PIWI-interacting RNA
PLEK	Predictor of Long non-coding RNAs and mRNAs based on an improved k-mer scheme
PQS	Putative Quadruplex-forming Sequence
qRT-PCR	Quantitative Reverse Transcription Polymerase Chain Reaction
RBP	RNA-Binding Protein
RefSeq	Reference Sequence Database (NCBI)
ResNet	Residual Neural Network
RF	Random Forest
Rfam	RNA Families Database
rG4	RNA G-Quadruplex
RMDB	RNA Mapping Database
RNA-seq	RNA Sequencing
RNN	Recurrent Neural Network
RNP	Ribonucleoprotein

RT-stop Assay	Reverse Transcription Stop Assay
SANS	Small-Angle Neutron Scattering
SAXS	Small-Angle X-ray Scattering
SCFG	Stochastic Context-Free Grammar
SF2/ASF	Splicing Factor 2/Alternative Splicing Factor
SHAPE	Selective 2″-Hydroxyl Acylation analyzed by Primer Extension
SHAPE-seq	Selective 2″-Hydroxyl Acylation analyzed by Primer Extension sequencing
siRNA	Small interfering RNA
SKF	Similarity Kernel Fusion
snoRNA	Small nucleolar RNA
snRNA	Small non-coding RNA
SSE	Secondary Structure Element
SSVM	Structured Support Vector Machine
SVM	Support Vector Machine
TCGA	The Cancer Genome Atlas
TERC	Telomerase RNA Component
TERT	Telomerase Reverse Transcriptase
TF	Transcription Factor
ThT Assay	Thioflavin T Fluorescence Assay
tRF	Transfer RNA-derived fragment
XGBoost	Extreme Gradient Boosting

2.1 Introduction

Most of the human genome comprises non-coding sections (Karam et al. 2024). A growing body of knowledge indicates that non-coding RNAs (ncRNAs) can exert transcriptional and post-transcriptional regulation of gene expression (Rintala-Maki and Sutherland 2009; Ørom et al. 2010). NcRNAs are classified into two groups based on their length. Small non-coding RNAs (snRNAs) are less than 200 nucleotides (nt) in length and include transfer RNA-derived fragments (tRFs), small interfering RNA (siRNA), piwi-interacting RNA (piRNA), small nucleolar RNA (snoRNA), and microRNA (miRNA) (Diamantopoulos et al. 2018). NcRNAs longer than 200 nt are termed long non-coding RNAs (lncRNAs). LncRNAs were previously suspected as transcriptional noise but are increasingly implicated in various physiological processes and correlated with disease progression (Uszczynska-Ratajczak et al. 2018). LncRNAs are transcribed by RNA polymerase II and undergo post-transcriptional modifications, including 5′ capping, splicing, and 3′ polyadenylation. However, the absence of an open reading frame (ORF) prevents their translation (Maroni et al. 2023). Compared to messenger RNA (mRNAs), lncRNAs are expressed at lower levels and exhibit highly tissue-specific expression patterns (Beermann et al. 2016). LncRNAs are distinguished by their heterogeneity, with sequence conservation ranging from very high to none across species. LncRNAs display distinctive structure-function relationships as part of their interaction with proteins, miRNA, and mRNA (Ramírez-Colmenero et al. 2020).

Dysregulated expressions of certain lncRNAs have been correlated with different facets of cancer, including tumor proliferation, invasion, metastasis, and resistance to therapy (Rupaimoole et al. 2015; Li et al. 2021a; Singh et al. 2022; Irfan et al. 2022; Zangouei et al. 2023; Ahmad et al. 2023). A substantial body of literature has highlighted specific lncRNA-protein interactions that modulate oncogenic and tumor-suppressive pathways. Prominent examples include the lncRNA MALAT1 that interacts with the splicing factor SF2/ASF to promote metastasis in lung cancer. LncRNA HOTAIR interacts with the PRC2 complex to alter chromatin states, impacting gene silencing and promoting breast cancer progression (Gupta et al. 2010; Xu et al. 2024). The lncRNA PVT1 interacts with and stabilizes MYC, resulting in enhanced MYC-driven oncogenic pathways (Jin et al. 2019).

Misclassification of lncRNAs can lead to inaccurate functional annotations and hinder research on regulatory RNAs. Notably, lncRNAs and mRNAs share several similarities, such as transcription by RNA polymerase II, the presence of 5′ cap and 3′ poly(A) tail, and compatibility with splicing. LncRNAs and mRNAs can both fold into intricate secondary structures (Djebali et al. 2012; Quinn and Chang 2016). Accordingly, the accurate identification of lncRNAs relies on the effective integration of sequence, structural, and functional data via the use of advanced computational tools coupled with experimental validation. Bioinformatics has significantly advanced the study of ncRNAs, aiding novel RNA discovery, prediction of structures, evolutionary studies, functional analysis, and regulatory network mapping. In this chapter, we provide an updated comprehensive account of in silico resources that facilitate the study of lncRNAs. We discuss the challenges underlying the development and use of in silico tools and datasets, thereby suggesting areas for future research.

2.2 Distinctive Features of lncRNAs

LncRNAs can regulate cellular processes by acting as decoys, scaffolds, guides, signaling molecules, and molecular sponges for miRNAs. As decoys, they block signaling pathways by binding to essential proteins or RNAs. For example, lncRNA PANDA binds to NF-YA, affecting p53-mediated apoptosis (Wan et al. 2011). As scaffolds, they facilitate the assembly of protein complexes. The lncRNA telomerase RNA component (TERC) serves as a scaffold for ribonucleoprotein (RNP) assembly in telomerase by providing a template for DNA polymerization by telomerase reverse transcriptase (TERT). Mutations in TERT disrupt this function and are linked to the progression of diseases such as cancer and dyskeratosis congenita (Mattick et al. 2023). LncRNA p21 acts as a guide by recruiting the protein hnRNP-K to specific sites, resulting in repression of gene transcription (Huarte et al. 2010; Rinn and Chang 2012). LncRNAs can act as signaling molecules by interacting with RNAs, proteins, and lipids to integrate developmental cues. For example, lncRNA LINC00673 enhances Wnt/β-catenin signaling by promoting the interaction between DDX3 and CK1ε (Zhao et al. 2021b). The lncRNA TRPM2-AS functions as a molecular sponge for miRNAs by binding to miR-612, promoting gastric cancer progression (Xiao et al. 2020).

LncRNAs can be categorized into five types based on their genomic organization and location relative to nearby protein-coding genes. Sense lncRNAs are transcribed in the same direction as adjacent protein-coding genes, while antisense lncRNAs are transcribed in the opposite direction, overlapping at least one protein-coding exon. Bidirectional or divergent lncRNAs originate from the promoter of a protein-coding gene but are transcribed in a divergent manner. Intronic lncRNAs arise from the intronic regions of protein-coding genes and terminate without overlapping introns. Lastly, intergenic lncRNAs are transcribed from regions located between two protein-coding genes, independent of their sequences (Rinn and Chang 2012; Beermann et al. 2016). Notwithstanding their origins as represented by the above five categories, lncRNAs can leverage their secondary and tertiary structures towards their functional roles, such as scaffolds, guides, decoys, signaling molecules, and molecular sponges in different cellular processes. LncRNAs influence nearly every stage of gene expression regulation. Specific binding motifs within lncRNAs facilitate interactions with proteins, RNA, or DNA, while evolutionarily conserved regions often indicate critical roles in gene regulation or RNA processing (McCown et al. 2019). The structure-function relationships underlying certain ncRNAs such as tRNAs are well understood. While lncRNAs are a very broad class, the structural domains that govern their functionality and define their interactome remain poorly understood. These lacunae have spurred the development of informatics and computational tools that can help in predicting the secondary structure of lncRNAs and their interactions with other biomolecules.

2.3 Computational Tools and Methodologies for the Identification of lncRNAs

The completion of human genome sequencing in the early 2000s thrust ncRNAs into the limelight and has continued to drive intense research over the past two decades. The accurate identification of a specific lncRNA from among an ocean of ncRNAs is a cumbersome process. In this section, we discuss methods that have proven useful for qualifying ncRNA molecules as lncRNAs.

2.3.1 Computational Tools for Distinguishing lncRNAs from Coding RNAs

The experimental identification of lncRNAs from transcriptome data is not straightforward. Computational methods attempt to screen ncRNAs by factoring singular characteristics of lncRNAs and mRNAs. Notable lncRNA identification algorithms include the logistic regression model, support vector machines (SVM), random forests (RF), and deep learning (DL) (Li et al. 2020).

2.3.1.1 Logistic Regression Model Tools

Tools like the coding potential assessment tool (CPAT) and lncScore use logistic regression models to identify lncRNAs (Li et al. 2020). CPAT focuses on features such as ORF length, ORF coverage, the Fickett Score, and hexamer frequencies (Wang et al. 2013). The Fickett Score is a measurement that evaluates how nucleotides (A, T, G, C) are positioned differently in coding versus non-coding regions, providing a statistical basis for distinguishing between them. Hexamer frequencies refer to how often specific six-nucleotide sequences appear in the RNA, which differ between coding and non-coding sequences. While another tool lncScore employs ORF features similar to CPAT, it also includes an exon feature and the maximum coding subsequence to improve the accuracy of its logistic regression model (Zhao et al. 2016). The maximum coding subsequence is the longest RNA segment translatable into a protein, typically shorter in lncRNAs than in mRNAs.

2.3.1.2 Support Vector Machine (SVM) Based Tools

The SVM is a stable, flexible, and widely used algorithm. SVM separates datasets by identifying an optimal hyperplane, a decision boundary maximizing the margin between classes in the feature space and employing kernel functions to implicitly map data into higher dimensions. These strategies enable effective classification of non-linearly separable patterns. SVM forms the basis of tools such as CPC, PLEK, CNCI, COME, LncRScan-SVM, and CPPred (Li et al. 2020; Ballarino et al. 2023). PLEK is a publicly available computational tool that uses a k-mer scheme and SVM to identify lncRNAs without genomic sequences or annotations (Li et al. 2014). A k-mer scheme involves analyzing how often shorter sequences of a specific length (k) appear in the RNA, creating a frequency pattern that differs between coding and ncRNAs. Among the other tools, CNCI uses SVM, adjoining nucleotide triplets (three-base patterns having coding-like signature across all reading frames), sequence intrinsic composition (compositional signatures like codon preference), and hexamer frequency for distinguishing non-coding and coding RNAs (Sun et al. 2013). COME uses a decompose-compose method with the help of sequence-derived features, experimental features, and secondary structure conservation to identify lncRNAs. The decompose-compose method involves breaking down RNA sequences into various component features (decompose) and then reconstructing predictions based on those features (compose). LncRScan-SVM calculates coding or non-coding potential to identify lncRNA using an SVM algorithm in combination with gene structure, transcript sequence, potential codon sequence, and conservation (Sun et al. 2015; Hu et al. 2017b; Ballarino et al. 2023).

2.3.1.3 Random Forest Algorithm Tools

RF operates via multiple decision trees, which help in avoiding bad samples or noise, ensuring accuracy (Li et al. 2020). LncRNA-ID, LncRNApred, FEELnc, and PLncPRO use RF algorithms to find lncRNA from sequence data. LncRNA-ID uses ORF features, ribosome interaction features (e.g., ribosome release score), and protein conservation-like features (Achawanantakun et al. 2015). In contrast, FEELnc uses an RF method to extract aspects of ORF coverage, codon use, nucleotide frequency, and multi-k-mer frequencies (Wucher et al. 2017). Multi-k-mer frequencies refer to the analysis of patterns pertaining to multiple different lengths of nucleotide sub-sequences (not just one specific length) appearing in the RNA. PLncPRO uses sequence-based features, protein homology searching, and 3-mer frequency-like features in its RF algorithm for the plant transcriptome (Singh et al. 2017). 3-mer frequency-like features specifically examine patterns of three-nucleotide sequences in the RNA, which can reveal coding potential.

2.3.1.4 Deep Learning Approaches

Recent advances in machine learning (ML), particularly DL, have significantly enhanced the identification and functional annotation of lncRNAs. DL models based on artificial neural networks (ANNs) process raw input data through multiple hidden layers to capture intricate patterns and produce high-confidence outputs. Tools such as LncRNA-MFDL utilize deep neural networks (DNNs) by integrating features like ORF characteristics, k-mer composition, and RNA secondary structure (Fan and Zhang 2015). Similarly, DeepLNC transforms RNA sequences into quantitative representations using Shannon entropy to distinguish between coding and non-coding transcripts (Tripathi et al. 2016). Other tools, such as LNCAdeep, leverage ORF length, GC content, and homology features, while NCResNet incorporates diverse attributes, including protein and physicochemical properties alongside RNA structural information. The lncRNAnet model combines recurrent neural networks (RNNs) for sequence modeling with convolutional neural networks (CNNs) for stop codon detection to enhance lncRNA prediction accuracy (Yang et al. 2018). Moving further, integration methods like Linc2RNA blend traditional ANN-based models with advanced frameworks such as lncRNAnet2 (LRN2) to improve cross-species identification and isoform-level annotation of lncRNAs (Ramakrishnaiah et al. 2023).

2.3.2 Compilation and Annotation of lncRNA Datasets

Advancements in transcriptomic data analysis and sequencing have enabled accurate identification of non-coding sequences from protein-coding genes. While numerous lncRNA datasets have been compiled using informatics and computational tools, inconsistencies or errors in source data could ultimately affect the accuracy of such datasets. The integration of multiple databases has sought to streamline the annotation of lncRNAs and minimize genomic redundancy. For example, LncRNAKB is a comprehensive repository that was developed through the rigorous integration of annotations from widely used lncRNA databases, including CHESS (v2.1), LNCipedia (v5.2), NONCODE (v5.0), FANTOM (5.0.v3), MiTranscriptome (v2), and BIGTranscriptome (v1). Such integration follows a cumulative stepwise intersection approach, ensuring the retention of high-confidence lncRNA annotations while eliminating ambiguous and redundant records (Seifuddin et al. 2020).

LncRNAWiki 2.0 is another example of a repository that integrates multiple inputs on lncRNAs. The original LncRNAWiki was limited by its MediaWiki-based system, which hindered structured data management and customized functions. LncRNAWiki 2.0 addresses these limitations with an enhanced database system using MySQL/Java and a new curation model. This enables structured content organization and curators to submit and edit annotations with more annotation items. Community curation is a key feature, and expert review and user error reporting ensure annotation quality. LncRNAWiki 2.0 reports 2512 lncRNAs and 106,242 associations related to diseases, functions, drugs, interacting partners, molecular signatures, experimental samples, and CRISPR designs. It also offers online tools to help users identify lncRNAs with potentially important functions and provides user-friendly web interfaces to facilitate data curation, retrieval, and visualization (Liu et al. 2022).

LNCipedia 5 is a comprehensive database that is projected as a manual for human lncRNAs. The repository encompasses sequences and annotations from several sources, including Ensembl, RefSeq, and FANTOM CAT (O'Leary et al. 2016; Hon et al. 2017a; Zerbino et al. 2018). Stringent filtering procedures eliminate dubious lncRNAs that overlap protein-coding genes (Yates et al. 2017). Nearly 2400 PubMed articles have been manually curated, describing approximately 1500 functionally characterized lncRNAs that are relevant across disease conditions (Volders et al. 2019).

NONCODEV6 is an extensive database with more than 600,000 entries that annotates lncRNAs encompassing 23 animal and plant species. NONCODEV6 combines transcriptome sequencing data to examine interactions between lncRNAs and oncogenic pathways. The recent version includes the annotation of plant lncRNAs, sequence conservation across species, tissue expression profiles, and anticipated roles in five common plants. The database also incorporates functional prediction methods and comparative analysis to assist researchers in investigating the evolutionary importance of lncRNAs. Its uniformity and thorough labeling make data more reliable (Zhao et al. 2021a). The integrated and curated databases such as LncRNAKB, LncRNAWiki 2.0, LNCipedia 5, and NONCODEV6 address the challenges of redundancy and inconsistent annotations in lncRNA research.

Figure 2.1 (Panel I) provides a summary overview of computational tools and databases that are reported for lncRNA identification and annotation. Panel II

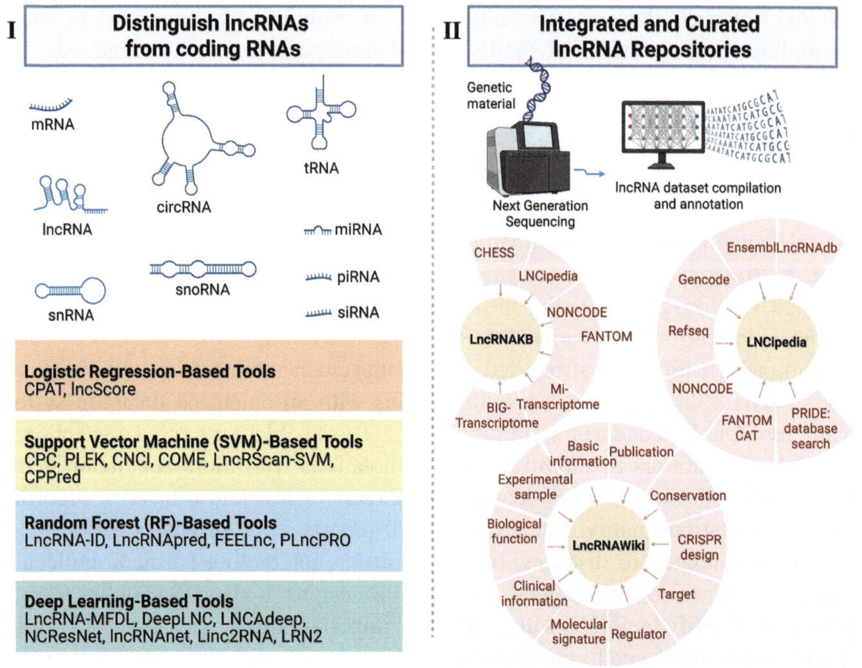

Fig. 2.1 A comprehensive overview of the computational tools and databases for lncRNA identification and annotation (Panel I) and integrated and curated lncRNA repositories for studying the structure, function, and role of lncRNAs in various biological processes and diseases (Panel II)

(Fig. 2.1) summarizes lncRNA repositories that have been reported for studying the structure and function of lncRNAs.

2.3.3 RNA-Seq Data Analysis Pipelines

Technologies such as next generation sequencing (NGS), droplet-based RNA sequencing, and microarray methods have permitted the facile and rapid discovery of RNA molecules (Lee and Kikyo 2012; Macosko et al. 2015). Tools like FastQC, Trimmomatic, and Kraken are used on the raw RNA data obtained by sequencing to remove poor-quality sequences, contaminants, and unwanted adapter sequences (Bolger et al. 2014; Chen et al. 2018; Wingett and Andrews 2018; Wood et al. 2019, Betti et al. 2022; Chauviere et al. 2023; Distefano et al. 2024). A reference file containing known genes from the species being studied is used for comparing the cleaned RNA reads. Tools such as TopHat2, STAR, HISAT2, and Bowtie2 have been developed to facilitate this comparison (Langmead and Salzberg 2012; Dobin et al. 2013; Kim et al. 2013, 2019; Iwakiri et al. 2016). Once aligned, software such as Cufflinks, StringTie, and Trinity help in the identification of RNA transcripts

(Trapnell et al. 2010; Grabherr et al. 2011; Pertea et al. 2015). The accurate identification of lncRNAs is challenging based on the similar features shared by protein-coding RNAs. The understanding of lncRNAs is continually evolving, with recent discoveries revealing novel characteristics such as the presence of short open reading frames (sORFs) capable of producing micropeptides. Such discoveries pose additional challenges, necessitating the refinement of gene classification and functional annotation strategies. To address these, it is essential to integrate high-throughput data, explore innovative models like fuzzy classification for bifunctional RNAs, and reassess coding potential from an evolutionary perspective. Approaches such as BLASTX analysis of transcriptome sequences can help identify protein-coding segments, thereby facilitating the isolation of true ncRNAs (Li et al. 2020). Additionally, tools like PhyloCSF apply evolutionary conservation models to evaluate codon usage and coding potential, whereas PhastWeb aids in assessing phylogenetic conservation post-identification (Lin et al. 2011; Ramani et al. 2019). Comprehensive platforms such as LncFinder further streamline this process by integrating multiple ML algorithms, offering a robust framework for lncRNA discovery and analysis (Han et al. 2019).

2.4 lncRNA Structure Prediction

The interaction of lncRNAs with proteins, DNA, RNA, and other ncRNAs relies on the ability of lncRNAs to fold into specific secondary or tertiary structures. The spatial configuration of lncRNAs is both unique and dynamic and can assist in the formation of ribonucleoprotein hubs that exert modulation of gene expression. Secondary structures of lncRNAs constitute a fundamental structural entity that has a direct bearing on the functional roles of the lncRNA. The secondary structural attributes of lncRNAs are often conserved across species and persist during the evolution of species (Ramírez-Colmenero et al. 2020). These structural units can help in cellular localization by determining subcellular trafficking and protecting the lncRNA from nuclease degradation. LncRNA secondary structures fulfill significant roles in chromosome inactivation, genomic imprinting, chromatin dynamics, protein modifications, and stability (Bugnon et al. 2022). Information about the secondary structures of lncRNAs with atom-level resolutions can be experimentally obtained via X-ray crystallography and nuclear magnetic resonance (NMR) spectroscopy. Chemical probe-based methods like SHAPE-seq and PARS help determine base-pairing details of lncRNAs. However, such experimental techniques are expensive and time-consuming and have a low thought. In the following section, we discuss informatics and computational tools that have been used for predicting the secondary structure of lncRNAs. Figure 2.2 summarizes RNA structure prediction tools and databases that are also discussed in the following sections.

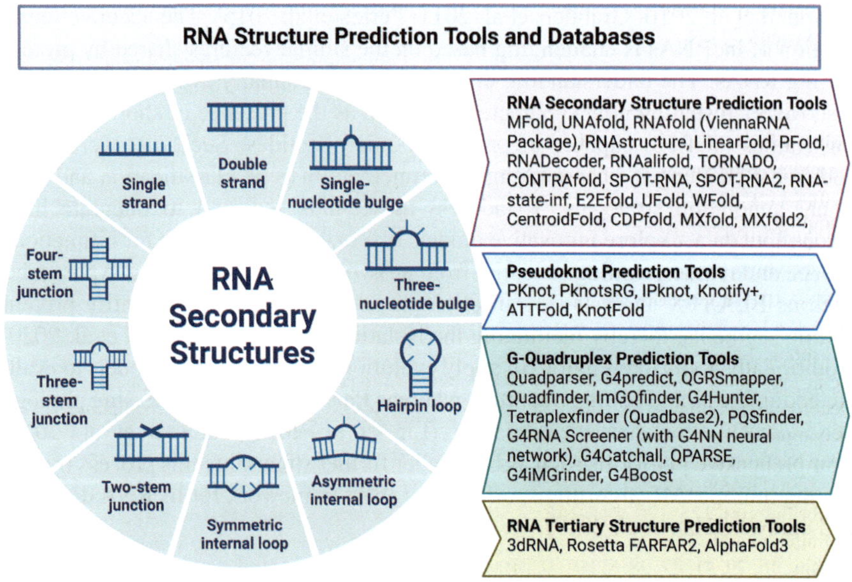

Fig. 2.2 RNA structure prediction tools and databases. Computational tools for predicting secondary, tertiary, pseudoknot, and G4s

2.4.1 Computational Tools for Predicting Secondary Structure of lncRNAs

Early tools for RNA secondary structure prediction, such as MFold, relied on minimum free energy (MFE) calculations, which account for all possible RNA secondary structures and their abundances based on the Boltzmann distribution. The predicted secondary structures sought to minimize the total free energy of the RNA in thermodynamic equilibrium. Over time, more advanced tools began combining Dynamic Programming (DP) with MFE methods. These include UNAfold (a modified version of MFold), RNAfold from the ViennaRNA package, which uses Zuker's and McCaskill's algorithms, and RNAstructure, which predicts MFE structures and base pair probabilities. Tools like LinearFold have further improved prediction speed by using linear-time and linear-space algorithms, making them suitable for long RNA sequences like lncRNAs (Zuker et al. 1999; Markham and Zuker 2008; Bellaousov et al. 2013; Gruber et al. 2015; Huang et al. 2019a; Pinkney et al. 2020; Ali et al. 2023; Yang et al. 2025).

Another approach uses evolutionary conservation through stochastic context-free grammars (SCFGs), which identify base pair co-variations across species (Yang et al. 2025). Examples include PFold, which uses the KH99 algorithm, and RNADecoder (Knudsen and Hein 2003; Pedersen 2004). While SCFG-based tools often outperform thermodynamics-based models due to their comparative nature, they depend heavily on high-quality multiple sequence alignments (MSA).

RNAalifold is not SCFG-based and uses MSA to predict MFE structures by allowing compensatory mutations that maintain base pairing (Bernhart et al. 2008; Hamada et al. 2011). TORNADO also employs SCFGs with DP and maximum likelihood training to optimize prediction (Rivas et al. 2012).

More recently, ML and DL models have outperformed traditional methods by using large RNA structure datasets. CONTRAfold applies a conditional log-linear model (CLLM) to bridge thermodynamic and probabilistic approaches (Do et al. 2006). Fully DL-based tools like SPOT-RNA and its updated version SPOT-RNA2 use advanced neural network architectures, such as ResNets, 2D-BLSTM, and CNNs, to predict complex RNA base-pair interactions, including non-canonical and pseudoknots (Singh et al. 2019, 2021). RNA-state-inf uses BiLSTM with a nearest-neighbor thermodynamic model (NNTM) for accurate predictions (Willmott et al. 2020). E2Efold combines a transformer-based Deep Score Network and a multi-layer post-processing network, whereas UFold uses U-Net-based CNNs to transform RNA sequences into image-like formats for faster prediction (Ronneberger et al. 2015; Chen et al. 2020; Fu et al. 2022). WFold integrates U-Net with transformer models to capture both local and long-range interactions, improving predictions for long RNAs (Yuan et al. 2024).

Hybrid tools such as CentroidFold use the γ-centroid estimator based on maximum expected accuracy (MEA) and integrate multiple models (e.g., CONTRAfold, McCaskill, RNAalifold, PFold) for accurate structure prediction (Hamada et al. 2009; Sato et al. 2009; Yang et al. 2025). Similarly, CDPfold combines CNNs with DP, whereas MXfold uses structured support vector machines (SSVMs) for hybrid thermodynamic-ML modeling (Akiyama et al. 2018; Zhang et al. 2019). Its successor, MXfold2, replaces SSVMs with DNN, combining CNN and BiLSTM layers for improved performance (Sato et al. 2021).

Notwithstanding these advances, the prediction of pseudoknot structures, in which one stem-loop intercalates within another, remains a major challenge. These structures are crucial for lncRNA function (Yuan et al. 2024). Attempts have been made to improve prediction of pseudoknots based on their relevance to lncRNA function. Tools such as PKnot deploy thermodynamic calculations but are limited by high time complexity (Rivas and Eddy 1999). PknotsRG reduced this time complexity while maintaining thermodynamic accuracy (Reeder and Giegerich 2004). IPknot further improved by combining ML and thermodynamics for faster and more accurate pseudoknot prediction (Sato et al. 2011). Other tools such as Knotify+, ATTFold, and KnotFold have developed more efficient algorithms using SCFGs, MFE, attention mechanisms, and minimum-cost flow models (Wang et al. 2020; Makris et al. 2023; Gong et al. 2024).

Newer DL tools such as SPOT-RNA2, E2Efold, UFold, and WFold can now predict pseudoknots without excessive time costs. Traditional tools such as RNAfold, PFold, MFold, RNAalifold, CONTRAfold, and CentroidFold exhibit a time complexity of $O(S^3)$ due to dynamic programming algorithms that exhaustively evaluate all possible base-pairing combinations. This makes them inefficient for long sequences, such as lncRNAs. Even $O(S^2)$ tools struggle with long RNAs, whereas modern approaches such as E2Efold, SPOT-RNA2, MXfold2, WFold, and

UFold offer significantly greater efficiency . LinearFold and Knotify+, with their O(S) (linear) time complexity, are among the fastest. Databases such as RMDB, Rfam, and Rfam 15.0 offer valuable sequence and structural data that support the training and validation of these advanced models (Griffiths-Jones 2004; Cordero et al. 2012; Ontiveros-Palacios et al. 2025).

2.4.2 *Tertiary Structure Prediction of lncRNAs with the Help of Computational Tools*

The paradigms of secondary and tertiary structures are not as well understood for RNA as compared to proteins. The primary structure of RNA is its linear sequence of nucleotides. The secondary structure arises from local base pairing interactions, primarily Watson-Crick and occasionally Hoogsteen base pairs, leading to structural motifs, such as hairpins, stem-loops, bulges, internal loops, pseudoknots, and RNA G4s (rG4s) (Tassinari et al. 2021). While RNA secondary structures are predictable when considered in isolation, the interplay between potentially competing structures that would ultimately result in a single three-dimensional spatial arrangement is a complex phenomenon.

RNA tertiary structures are also stabilized by interactions, such as base triples, A-minor interactions, tetraloop–receptor motifs, kissing loops, ribose zippers, and coaxial stacking. These interactions bring distant parts of the RNA chain into proximity, forming compact, functionally active structures (Chillón and Marcia 2020). A systematic understanding of lncRNA tertiary structures requires significant amounts of high-resolution structural data in conjunction with suitable structure-prediction tools. A greater quotient of experimental evidence is required to assess the prevalence of specific tertiary motifs in lncRNAs.

There is a dearth of tools for RNA tertiary structure prediction compared with the numerous approaches that have been developed for RNA secondary structure prediction. RNA tertiary structure prediction is based on methodologies such as template/fragment-based prediction or ab initio and deep-learning-based methods (Bernard et al. 2024). Further, RNA tertiary structure prediction tools have proven inefficient for predicting tertiary structures of ncRNA, including lncRNAs. LncRNAs display robust structure conservation that is not necessarily contingent on sequence conservation. Dynamic changes in the conformation of lncRNAs based on their interactions with other biomolecules result in weakly defined static conformations, thereby posing challenges for tertiary structure prediction (Jones and Sattler 2019; Fallon and Jones 2024). The longer length of lncRNAs with at least 200 or more nucleotides also creates difficulties in RNA tertiary structure prediction. LncRNA tertiary structure prediction relies on structural information that has been obtained via techniques, such as Cryo-EM, NMR spectroscopy, X-ray crystallography, small-angle X-ray/neutron scattering (SAXS/SANS), atomic force microscopy (AFM), fluorescence correlation spectroscopy (FCS), chemical probing using

SHAPE (Selective 2′-Hydroxyl Acylation by Primer Extension), and PARS (Parallel Analysis of RNA Structure) (Sanbonmatsu 2022). Computational tools such as 3dRNA have used a template-based approach to construct three-dimensional RNA structures by assembling fragments from known structural elements in the Protein Data Bank (PDB). The process involves breaking down the target RNA into smaller secondary structure elements (SSEs), such as helices and loops, followed by identifying suitable matches of the SSEs with those stored in a database (Wang et al. 2019; Zhang et al. 2022). Another tool, Rosetta FARFAR2, uses a fragment assembly method guided by energy minimization to predict RNA structures. It breaks down the RNA sequence into segments of three residues each and samples different conformations. The tool then iteratively refines these conformations using a scoring function that accounts for base pairing, p-stacking, and electrostatic interactions (Watkins et al. 2020).

Recent tools like AlphaFold3, which is an extension of protein-centric AlphaFold2, integrate DL with MSAs to predict RNA structures directly from primary sequences (Abramson et al. 2024). AlphaFold3 has emerged as more successful compared to its peers in RNA tertiary structure prediction (Mao et al. 2024). DL methods for RNA secondary structure prediction, while powerful, are constrained by overfitting and generalization. Overfitting occurs when a model learns the training data too well, capturing noise and specific patterns that are not applicable to new data. This can lead to high accuracy on structurally similar datasets but poor performance on dissimilar ones (Rivas et al. 2012).

A major concern in RNA tertiary structure prediction is the rich parameterization of DNNs, which can easily lead to overfitting, thereby hindering robust structural prediction. While techniques such as thermodynamic regularization and ensemble methods can mitigate overfitting, the risk remains substantial (Zakov et al. 2011). Another limitation arises from the difficulty in obtaining sufficient training data with empirically determined secondary structures. Notwithstanding advances in sequencing techniques, the experimental elucidation of RNA structures remains challenging, making it difficult to train models that are generalizable to a wide variety of sequences. Methods that rely solely on memorizing sequences and their secondary structures within a large number of neural network parameters are impractical for predicting new structural motifs in hitherto unexplored sequences (Rivas et al. 2012).

Furthermore, while some DNN-based methods can predict pseudoknotted structures, methods like MXfold2 are limited to pseudoknot-free secondary structure prediction. Additionally, if neither ML-based methods nor thermodynamic-based methods can accurately predict the secondary structure, MXfold2 may also struggle in making accurate predictions (Sato et al. 2021). In the future, the incorporation of experimental data within the ML model and improvements in modification-based modeling may enhance the ability of computational tools to accurately predict the tertiary structure of lncRNAs.

2.4.3 Computational Tools for G-Quadruplex Prediction

G-quadruplexes (G4s) are distinctive secondary structural motifs that are increasingly being linked to disease progression, including cancer. Identifying G4s in lncRNAs follows similar principles as in DNA, often beginning with the detection of consensus motifs such as: $\mathbf{G_3 + N_{1-7}G_3 + N_{1-7}G_3 + N_{1-7}G_3 +}$ (Puig Lombardi and Londoño-Vallejo 2020). One of the earliest computational tools developed for this purpose was Quadparser (2005), designed to scan DNA sequences for the G4 motif. Quadparser was subsequently adapted into G4predict, expanding its ability to detect both intra- and intermolecular G4s, thereby proving useful for predicting putative quadruplex-forming sequences (PQSs) in lncRNAs (Huppert and Balasubramanian 2005).

QGRS mapper was the first tool capable of identifying G4 motifs in both DNA and RNA. It introduced the G-score, a measure of a sequence's likelihood to form G4 structures (Kikin et al. 2006). While widely used for assessing G4 formation in DNA, the G-score's correlation with experimentally verified quadruplex formation in RNA remains limited (Lyu et al. 2021). Nonetheless, QGRS mapper data have been included in datasets like CanLncG4 (Sharma et al. 2025). Several other tools have emerged since QGRS Mapper, offering refinements in prediction strategies. Quadfinder allowed customization of motif parameters such as G-tract length and loop size. The cG/cC scoring system evaluated guanine (cG) and cytosine (cC) content surrounding a motif, with a higher cG/cC ratio suggesting stronger G4-forming potential. However, this method lacked scalability for genome-wide applications (Scaria et al. 2006; Beaudoin et al. 2014; Puig Lombardi and Londoño-Vallejo 2020). ImGQfinder was developed to recognize both canonical and non-canonical G4 structures but was less flexible in adapting to varying sequence types (Varizhuk et al. 2014, 2017). G4Hunter improved prediction accuracy by incorporating G-richness and G-skewness into a single propensity score (127). G4Hunter was later adapted into a web tool to allow input of user-defined sequences (Bedrat et al. 2016; Brázda et al. 2019). Tetraplexfinder, included in the Quadbase2 server, was valuable for large-scale G4 prediction in lncRNAs due to its ability to handle extensive sequence data (Dhapola and Chowdhury 2016). PQSfinder, an R-based algorithm, searched for G-run quartets and assigned scores based on structural and positional features, with an updated scoring scheme to better suit lncRNA sequences (Hon et al. 2017b).

Among ML methods, G4RNA Screener incorporated a neural network, G4NN, alongside G4Hunter and cG/cC scores to assess how closely a given RNA sequence resembled known G4s (Garant et al. 2017). Tools like G4Catchall further improved prediction by recognizing atypical G4 features, such as bulged or mismatched tracts and long loops, that are often present in lncRNA structures (Doluca 2019). QPARSE took a graph-based approach to detect and score non-canonical G-islands, useful for identifying multimeric or long-looped PQSs (Berselli et al. 2020). G4iMGrinder allowed prediction of both G4s and intercalated motifs (iMs), though lncRNAs generally favor G4 formation over iMs (Belmonte-Reche and Morales 2020; Bhatt et al. 2024). G4Boost leveraged the power of XGBoost, an ML model, to predict G4 topology, folding energy, and probability from input sequences. This architecture made it suitable for evaluating both mRNAs and lncRNAs. Similarly, rG4detector used a CNN to identify potential rG4s from transcriptomic data by incorporating sequence and secondary structure features (Turner et al. 2022; Cagirici et al. 2022). Qufind was recently introduced as a user-friendly web server capable of predicting both canonical and non-canonical G4s. Qufind benefits from integration with major databases such as ENSEMBL and NCBI, enhancing its predictive potential (Kaur et al. 2023). G4attention has combined CNNs, BiLSTM models, and attention mechanisms to enhance the accuracy of RNA G4 prediction, outperforming several conventional tools (Mukherjee et al. 2024). Despite the range of computational and informatics tools now available for predicting G4 structures in lncRNAs, experimental validation often lags behind. Techniques such as circular dichroism (CD) spectroscopy, Thioflavin T (ThT) fluorescence assays, NMR spectroscopy, and molecular and cellular methods like RT-stop assay, Electrophoretic Mobility Shift Assay (EMSA), G4 RNA Immunoprecipitation (G4RIP)-qPCR, and G4RIP-sequencing are essential for confirming G4 formation and function in lncRNAs. The successful convergence of computational and informatics-based predictions with robust experimental validation is key to uncovering the biological relevance of lncRNA G4s and their prospective roles in disease progression.

2.5 lncRNA Interaction Prediction

LncRNAs exert their biological functions through complex interactions with various biomolecules. Elucidating these molecular interactions using computational approaches presents challenges that are comparable to those encountered in structure determination. LncRNAs display a remarkable range of binding affinities, rendering them capable of associating with biomolecular partners across the cellular milieu. Figure 2.3 (Panel I) summarizes databases and computational tools that have been reported for the study of lncRNA-RNA and lncRNA-protein interactions.

The computational prediction of lncRNA interactions has evolved from early thermodynamic models (e.g., RNAduplex 1994) and sequence-based tools (e.g., RNAhybrid 2004) to sophisticated ML frameworks integrating heterogeneous networks (LncPNet 2022), DL architectures (TEC-LncMir 2024), and ensemble methods combining experimental data with computational approaches (LncPTPred 2025). Recent advancements leverage the capabilities of DL frameworks, including CNNs, RNNs, and capsule networks. Neural networks can predict those interactions with the help of previously known established interactions automatically and can adapt to the structural and evolutionary contexts with the help of the integration of a large amount of data provided to them. The effectiveness of this method is conveyed by its name as it mimics the function of the human brain and stores, analyzes, and predicts from a large amount of data provided to it. The field continues to progress with recent innovations in graph neural networks and template-based approaches. Databases like NPInter v5.0, RAIN, and DIANA-LncBase v3.0 now aggregate multi-omics data,

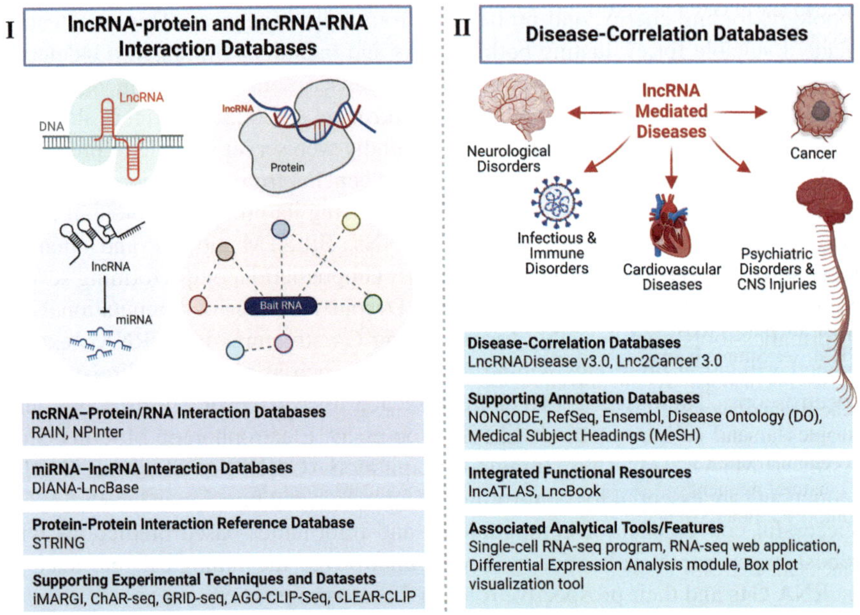

Fig. 2.3 Key databases and tools for studying lncRNA-protein and lncRNA-RNA interactions (Panel I), and resources for disease correlation, functional annotation, and experimental validation of lncRNA functions (Panel II)

CLIP-seq validations, and tissue-specific interaction networks, enabling systems-level analyses. Table 2.1 lists various computational and informatics tools for predicting and analyzing lncRNA-RNA and lncRNA-protein interactions.

Nonetheless, several gaps remain to be addressed: (1) lack of training data for non-model organisms and rare lncRNAs, (2) poor handling of RNA modifications, isoform-specific interactions, and dynamic binding contexts, (3) excessive dependence on static sequence features instead of 3D structural or single-cell spatial data, and (4) lack of integration with functional validation pipelines. Future developments necessitate unified systems integrating interaction prediction with CRISPR screening validation, explainable artificial intelligence frameworks to interpret prediction reasoning, and uniform benchmarking datasets. Species-adaptive models and drug-target interaction modules are required since plant-specific techniques and clinical translation are still lacking.

2.6 Disease-Correlation Databases

Databases that enable storage and retrieval of information pertaining to lncRNAs have gained value based on the prospective diagnostic and therapeutic value of these biomolecules. Figure 2.3 (Panel II) lists various databases and computational tools

Table 2.1 Tools used for assessing lncRNA-RNA and lncRNA-protein interactions

Tools used for lncRNA-RNA interactions					
Tool	*Year*	*Algorithm*	*Function*	*Output format*	*Ref.*
RNAduplex	1994	MFE calculation	Predict RNA-RNA duplex formation	RNA duplex structures with MFE	Hofacker et al. (1994)
RNAhybrid	2004	Sequence alignment + thermodynamics	miRNA target prediction	miRNA-mRNA interaction sites	Rehmsmeier et al. (2004)
miRanda	2004	GU complementarity scoring	miRNA target identification	miRNA binding sites	John et al. (2004)
RNAup	2005	Accessibility + MFE calculation	RNA-RNA interaction prediction	Target sites with interaction energy	Mückstein et al. (2006)
IntaRNA	2008	DP + hybrid energy	RNA-RNA interaction (bacterial focus)	Interaction sites with energy scores	Smith et al. (2010)
RNAplex	2008	Simplified energy model	Fast RNA-RNA interaction	Potential binding regions	Tafer and Hofacker (2008)
RactIP	2010	Integer programming	RNA-RNA interaction prediction	Base-pair probabilities	Kato et al. (2010)
PETcofold	2011	Phylogenetic alignment + co-folding	Conserved RNA interactions	Co-evolved interaction sites	Seemann et al. (2011)
RIsearch	2012	Turner energy model	Rapid RNA-RNA hybridization scan	Hybridization potential	Wenzel et al. (2012)
LncTar	2015	Sliding window + MFE	lncRNA-RNA interaction	Binding regions with normalized energy	Li et al. (2015c)
GCLMI	2019	GCN + autoencoder	lncRNA-miRNA interaction (plant based)	Interaction network topology	Huang et al. (2019b)

(continued)

Table 2.1 (continued)

Tools used for lncRNA-RNA interactions					
Tool	*Year*	*Algorithm*	*Function*	*Output format*	*Ref.*
LncRRIsearch	2020	RIblast algorithm	lncRNA-lncRNA/mRNA interaction	Interaction pairs	Fukunaga et al. (2019)
LncMiRnet	2020	CNN + fused features	lncRNA-miRNA prediction	Binding probability scores	Yang et al. (2020)
LncRTPred	2023	ML + BLAT filtering	lncRNA-mRNA interaction	Cumulative model scores	Das et al. (2025)
TEC-LncMir	2024	Transformer + CNN	lncRNA-miRNA interaction	Deep-learning interaction scores	Yang et al. (2024)
Tools for lncRNA-protein interactions					
Tool	*Year*	*Algorithm*	*Function*	*Output format*	*Ref*
RNAbindR	2007	Electrostatic/H-bond analysis	RNA-protein binding sites	Binding residue predictions	Terribilini et al. (2007)
PiRaNhA	2010	SVM + amino acid composition	RNA-protein interaction	Classification scores	Murakami et al. (2010)
catRAPID	2011	Physicochemical profiles + Fourier transforms	Predicts interaction propensities via sequence/structure compatibility	Interaction propensity scores (0–1)	Bellucci et al. (2011)
RPISeq	2011	SVM/RF sequences	RNA-protein interaction	Probability scores	Muppirala et al. (2011)
LncPro	2013	Fisher's linear discriminant analysis	Predicts interactions using sequence-derived features	Discriminant score; binary classification (yes/no)	Lu et al. (2013)
LPIHN	2015	Heterogeneous network + random walk with restart (RWR)	Prioritizes candidate proteins for query lncRNAs	Interaction probability scores (0–1)	Li et al. (2015a)

RPI-pred	2015	SVM with sequence/structure features	Predicts RNA-protein interactions (including lncRNAs)	Binary classification	Suresh et al. (2015)
LPBNI	2016	Bipartite network resource allocation	Scores lncRNA-protein pairs via iterative resource distribution	Ranked interaction probabilities	Ge et al. (2016)
IPMiner	2016	Stacked autoencoders + RF ensemble	Identifies ncRNA-protein interactions (supports lncRNAs)	Probability scores (0–1)	Pan et al. (2016)
RNAcommender	2016	Domain similarity	RNA-protein matching	Compatibility scores	Corrado et al. (2016)
PLPIHS	2017	HeteSim metric + SVM	Predicts interactions using path-based relevance in networks	Binary classification	Xiao et al. (2017)
OmiXcore	2017	Non-linear ML	lincRNA-protein prioritization	Transcriptome-level predictions	Armaos et al. (2017)
LPI-ETSLP	2017	Semi-supervised learning	lncRNA-protein interaction	Eigenvalue predictions	Hu et al. (2017a)
3dRPC	2018	3D scoring function along with RPDOCK	RNA-protein complexes	Structural interaction models	Huang et al. (2018)
HLPI-ensemble	2018	Ensemble of SVM, RF, and XGBoost	Integrates sequence features for lncRNA-protein prediction	Probability scores (0–1)	Hu et al. (2018)

(continued)

Table 2.1 (continued)

Tools used for lncRNA-RNA interactions					
Tool	*Year*	*Algorithm*	*Function*	*Output format*	*Ref.*
IRWNRLPI	2018	Random walk + matrix factorization	Predicts interactions using network regularization	Interaction likelihood scores	Zhao et al. (2018)
SFPEL-LPI	2019	Sequence-based feature projection	lncRNA-protein prediction	Ensemble learning scores	Zhang et al. (2018)
RPITER	2019	CNN + stacked auto-encoder (SAE) + conjoint triad feature (CTF)	ncRNA-protein interaction	Conjoint triad predictions	Peng et al. (2019)
LPI-pred	2019	RNA2vec/Pro2vec with Word2vec & RF	lncRNA-protein interaction	Word embedding scores	Yi et al. (2020a)
LPI-BLS	2019	Broad learning system	lncRNA-protein prediction	Stacked classifier outputs	Fan and Zhang (2019)
BGFE	2019	Improved seq. Features	ncRNA-protein interaction	Deep feature scores	Zhan et al. (2019)
PLRPIM	2019	Hybrid deep learning (LSTM + RF)	Plant-specific lncRNA-protein interaction prediction	Probability scores (0–1)	Wekesa et al. (2019)
LPI-KTASLP	2019	Kernel target alignment	lncRNA-protein interaction	Semi-supervised scores	Shen et al. (2019)
DM-RPIs	2019	Deep stacking auto-encoder networks (DSANs) + ensemble	ncRNA-protein interaction	Stacked predictions	Cheng et al. (2019)
LPI-CNNCP	2020	CNN + copy-padding	lncRNA-protein interaction	CNN feature maps	Zhang et al. (2020a)
DRPLPI	2020	Deep forest along with sequence structural features	lncRNA-protein prediction	Structural feature scores	Wekesa et al. (2020b)

LPI-SKF	2020	SKF (similarity kernel fusion) and LapRLS (Laplacian regularized least squares)	lncRNA-protein interaction	Similarity fusion scores	Zhou et al. (2020)
RPI-SE	2020	Gradient boosting decision tree (GBDT, implemented by XGBoost), SVM and extremely randomized trees (ExtraTree)	ncRNA-protein interaction	Ensemble scores	Yi et al. (2020b)
LPI-RNMF	2020	Regularized nonnegative matrix factorization method	lncRNA-protein prediction	Matrix factorization	Xu et al. (2021)
LPGNMF	2020	Graph-regularized nonnegative matrix factorization	lncRNA-protein interaction	Network-regularized scores	Zhang et al. (2020b)
LPI-DL	2020	Sequence features and compact long short-term memory (LSTM)	Predicts plant lncRNA-protein interactions	Sequence-based predictions	Wekesa et al. (2020a)
GPLPI	2020	Graph attention autoencoder + CatBoost	Predicts plant lncRNA-protein interactions with evolutionary features	Binary classification	Wekesa et al. (2020c)
LGFC-CNN	2021	GloCNN and LocCNN models, hand-designed features, and structural features	lncRNA-protein interaction	Multi-source predictions	Huang et al. (2021)
EDLMFC	2021	CNN with a BLSTM	lncRNA-protein prediction	Temporal feature scores	Wang et al. (2021)
Capsule-LPI	2021	Capsule networks combining sequence features, motif information, physicochemical properties, and secondary structure features	lncRNA-protein interaction	Motif-structure scores	Li et al. (2021b)
LPIDF	2021	Deep-forest-based model	lncRNA-protein prediction	Feature importance scores	Tian et al. (2021)
NPI-DGCN	2022	Dual GCN architecture	lncRNA-protein interaction	Topology-sequence fusion	Zhao et al. (2022)

(continued)

Table 2.1 (continued)

Tools used for lncRNA-RNA interactions					
Tool	*Year*	*Algorithm*	*Function*	*Output format*	*Ref.*
LncPNet	2022	Heterogeneous network embedding + SVM	Integrates lncRNA/protein similarities for interaction prediction	Probability scores (0–1)	Zhuo et al. (2022)
LPLSG	2023	Hybrid of resource allocation and improved collaborative filtering methods	lncRNA-protein prediction	Local network scores	Wang et al. (2023)
LPI-MFF	2024	RF + CNN	lncRNA-protein interaction	Multi-feature fusion	Liang et al. (2024)
LncPTPred	2025	CLIP-seq data + LightGBM + XGBoost ensemble	Predicts interaction loci (20-nt windows) with CLIP-seq validation	Genomic coordinates + probability	Das et al. (2025)
Databases containing the data regarding the interaction of lncRNA with other biomolecules					
Database	*Year*	*Key features*	*Data content*	*Functionality*	*Ref*
NPInter v5.0	2023	iMARGI, ChAR-seq, GRID-seq integration	Over 2.5 million ncRNA interactions	Chromatin RNA mapping, SARS-CoV-2 module	Li et al. (2017), Bell et al. (2018), Wu et al. (2019), Van Nostrand et al. (2020), Schmidt et al. (2020), Zheng et al. (2023)
RAIN	2017	STRING integration	More than 10,500 experimental interactions	ncRNA-protein network visualization	Pafilis et al. (2013), Junge et al. (2017), Szklarczyk et al. (2023)
DIANA-LncBase v3.0	2019	Manual literature curation and more than 300 high-throughput datasets, including AGO-CLIP-Seq, microarrays, RNA sequencing and microCLIP framework	Over 500 K miRNA-lncRNA	Variant annotation, which comes from dbSNP, ClinVar, COSMIC + expression profiling	Sherry et al. (2001), Moore et al. (2015), Landrum et al. (2018), Paraskevopoulou et al. (2018), Tate et al. (2019), Fernandes et al. (2019), Karagkouni et al. (2019)

that are reported for disease correlation, functional annotation, and experimental validation of lncRNA functions. LncRNADisease v3.0 is a specialized database concentrating on experimentally validated links between lncRNAs, circular RNAs (circRNAs), and diseases (Lin et al. 2024). The database is an effective tool for studying the correlation of lncRNAs in human health and disorders. The most recent update doubles the size of its previous version with 25,440 entries encompassing 6066 lncRNAs, 10,732 circRNAs, and 566 illnesses. The database categorizes these interactions into general and causal relationships, providing deeper insight into how lncRNAs contribute to disease progression, notably in cancer and cardiovascular disease (Jia et al. 2020). One of the most significant enhancements to its latest version is the addition of confidence scores for each lncRNA-disease pair, which are experimentally validated using qRT-PCR, ChIP, and RNA-seq methods. It also includes standard annotations from NONCODE, RefSeq, and Ensembl, with disease classification based on Disease Ontology (DO) and Medical Subject Headings (MeSH) (Schriml et al. 2022). These enhancements improve data accuracy and dependability for researchers studying lncRNA functions in human illnesses.

A similarly thorough and methodically maintained repository Lnc2Cancer 3.0 is a comprehensive database that offers experimentally confirmed correlations between the lncRNAs and different cancers. Its dataset covers 2659 lncRNAs, 743 circRNAs, and 216 cancer subtypes with 9254 lncRNA-cancer connections and 1049 circRNA-cancer interactions. It analytically integrates high-throughput sequencing data (RNA-seq and single-cell RNA-seq) to directly check the regulatory roles of ncRNAs in cancer development (Gawronski and Kim 2017). This repository emphasizes functional annotations and regulatory mechanisms. A deeper understanding of lncRNA and circRNA-mediated regulation is provided by its cataloging of interactions involving miRNAs, transcription factors (TFs), genetic variations, methylation, and enhancer elements. It also characterizes biological functions while highlighting the clinical relevance of lncRNAs and circRNAs, identifying their roles in metastasis, recurrence, drug resistance, circulation in blood, and prognosis. The database introduces refined tools for easy data exploration and analysis. With 49 single-cell datasets spanning 20 cancer types, a single-cell RNA-seq program allows for cluster analysis, heatmaps, and differential expression analysis (Gao et al. 2021). In parallel, the RNA-seq web application uses information from The Cancer Genome Atlas (TCGA) to offer features for TF motif prediction, co-expression networks, expression analysis, and survival analysis. It is designed with nine robust functionalities to bridge the gap between low- and high-throughput cancer-related lncRNA experiments. The general function provides comprehensive information on each lncRNA, including its subcellular localization sourced from lncATLAS (Mas-Ponte et al. 2017), detailed functional annotations and GO information from LncBook (Ma et al. 2019), as well as mean expression levels in both cancerous and normal tissues. Additionally, it also generates box plots for detailed information underlining the specific lncRNAs across various cancers and facilitates a better understanding of their differential expression patterns. The tool's Differential Expression Analysis feature enables researchers to perform reliable and flexible expression analyses.

2.7 Comparative Analysis of lncRNA Databases

2.7.1 Comparative Analysis of Human lncRNA Databases

Prominent lncRNA databases, such as DIANA-LncBase v3.0, LncRNADisease v3.0, Lnc2Cancer 3.0, NPInter v5.0, and NONCODE v6, address unique facets of lncRNA biology from regulatory mechanisms to disease associations and structural interactions (Karagkouni et al. 2019; Gao et al. 2021; Zhao et al. 2021a; Zheng et al. 2023; Lin et al. 2024). A comparative analysis covering these and other lncRNA datasets reveals key differences in data collection strategies, curation depth, functional annotations, and intended applications. With a focus on miRNA-lncRNA interactions, LncBase v3.0 provides more than 500,000 empirically verified entries from 52 tissues and 192 cell types. A high-accuracy map of Argonaute (AGO)-bound miRNA targets is produced by using CLIP-Seq-based microCLIP analysis, which is essential for comprehending post-transcriptional regulation. In contrast, LncRNADisease v3.0 emphasizes the clinical and pathological roles of lncRNAs, documenting over 25,000 associations backed by qRT-PCR, ChIP, and RNA-Seq validations, while distinguishing between causal and general disease relationships. RNA-Seq and single-cell RNA-Seq techniques for dynamic expression profiling are integrated into Lnc2Cancer 3.0, a cancer-focused program that provides a curated perspective of lncRNA and circRNA dysregulation in 216 cancer subtypes. It includes expression comparisons across 33 cancer types, covering 9664 tumors and 711 normal controls from TCGA, and supports differential expression, survival analysis, and functional enrichment tools. By recording more than 2.5 million ncRNA interactions—including RNA-DNA, RNA-RNA, and RNA-protein interactions—and enriching them with SARS-CoV-2 data and RNA-binding protein (RBP) annotations, NPInter v5.0 broadens the field and offers vital information on ncRNA-mediated regulation.

One of the most extensive and easily accessible lncRNA compendiums is NONCODE v6. It provides high-confidence annotation for nearly 548,640 lncRNAs from 17 different species. NONCODE v6 allows comparative genomics and functional inference through tissue specificity, expression data, and conservation data. Structural databases that focus on rG4 motifs, which are essential for RNA folding, stability, and translational regulation such as DSSR-G4DB and G4Atlas offer a structural perspective (Hanson and Lu 2017; Yu et al. 2023). Therefore, each database provides unique strengths tailored to different research needs—from regulatory interactions (LncBase, NPInter) to disease mechanisms (LncRNADisease, Lnc2Cancer) and genome-wide annotation (NONCODE). The development of datasets to address structure-function associations, integrative approaches, and experimental verification reflects their growing significance in functional genomics, biomarker discovery, and drug development. The choice of a suitable database ultimately depends on alignment of the study aim with the themes of lncRNA structure prediction, expression analysis, interaction mapping, or disease association studies. The interplay between lncRNAs and the human genome is detailed in Chap. 1 by Palma.

2.7.2 *Comparative Analysis of Non-human lncRNA Databases*

A growing interest in comprehending the functions of lncRNAs in different species of plants and animals has resulted in the development of non-human lncRNA databases. These databases differ from one another in terms of data coverage, annotation types, update frequency, and user interface design. Such variations impact the overall utility of the datasets and constrain academic research in the domain of non-human lncRNAs. Figure 2.4 summarizes non-human lncRNA databases. NONCODE, one of the most comprehensive resources, includes over 548,640 lncRNA transcripts from 17 species, covering humans, animals (e.g., mouse, rat, cow, pig), and plants (e.g., Arabidopsis, rice, maize) (Zhao et al. 2021a). It provides rich annotations such as genomic location, expression profiles, conservation, functional annotation, and potential interactions with proteins and other RNAs. The database is regularly updated, with the latest version (v5.0) significantly expanding its content. While the interface is data-dense, it is well-organized and supports complex queries, making it powerful for in-depth analysis.

RNACentral serves as a meta-database that aggregates ncRNA data from over 40 expert sources, including Ensembl, RefSeq, and miRBase, across a broad range of species including human, model organisms, livestock, and many plant species (Sweeney et al. 2021). It offers functional classification, sequence data, identifiers, and cross-references to other databases. Its modern and responsive interface supports advanced search, visualization tools, and sequence similarity search, and it is updated on a rolling basis, ensuring up-to-date information.

For plant research, CANTATAdb 2.0 is a valuable resource, covering over 239,631 lncRNAs across 39 plant species. GO annotations, co-expression networks,

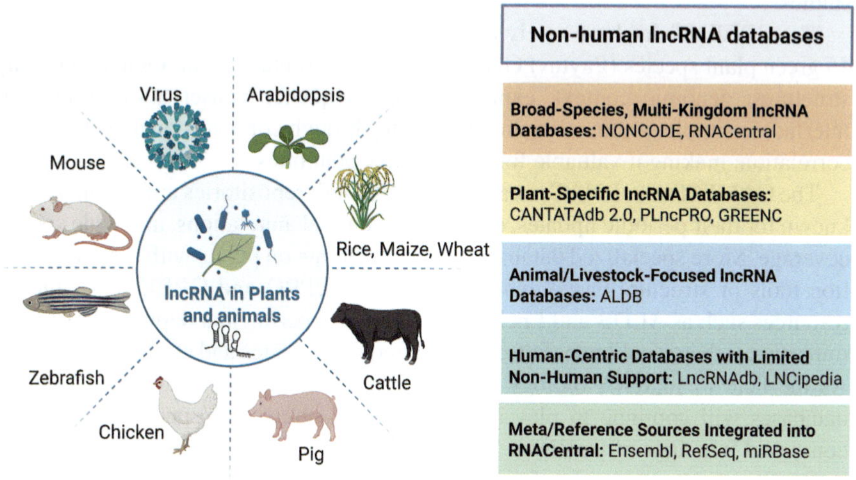

Fig. 2.4 Overview of non-human lncRNA databases covering diverse species including plants, animals, and viruses

tissue-specific expression patterns, and projected lncRNA-miRNA interactions are included in the database (Szcześniak et al. 2019). The graphic output is useful for examining expression trends, and the interface is user-friendly and searchable by species or gene ID. The growing plant genomics community is reflected in its active maintenance.

A pipeline based on ML is provided by PLncPRO to predict and annotate lncRNAs in agricultural crop species such as wheat and rice (Singh et al. 2017). PLncPRO focuses on ab initio predictions with coding potential scores, transcript architecture, expression evidence, and anticipated targets. While it does not have the highest number of species or transcripts, its methodological focus on crop species and integration of sequence features and expression data makes it unique. Using RF classifiers and coding potential features, it helps identify high-confidence lncRNAs from the available transcriptomic data. However, it is not frequently updated, and the interface is more technical and suited for computational users.

ALDB (Animal LncRNA Database) was amongst the first of its kind to provide lncRNA annotations in livestock species such as pigs, cattle, and chickens (Li et al. 2015b). It includes experimentally supported transcripts, with features like sequence conservation, gene synteny, and expression profiles across developmental stages and tissues. Despite its historical significance, its appeal to modern users is limited by the dated interface and the dataset's lack of significant updates since its debut.

Despite being predominantly human-focused, LncRNAdb and LNCipedia provide some assistance for lncRNAs that are not human (Quek et al. 2015). LncRNAdb provides experimentally validated functional lncRNAs across several species and includes literature links, structural data, and biological roles. While the data is rich in biological context, updates are sporadic, and coverage for non-human species remains limited. LNCipedia, though very user-friendly and frequently updated, has minimal content for non-human organisms, reducing its utility outside human studies.

The GREENC database solely focuses on plant species, with lncRNA data from 45 green plant species (Paytuví Gallart et al. 2016). It includes predicted secondary structures, genomic context, expression, and sequence conservation. While the interface is simple, it offers functional insights through conservation and expression correlation, making it valuable for comparative genomics.

The CANTATAdb, RNACentral, and NONCODE repositories are distinctive and known for their periodic updates, excellent functional annotations, and wide species coverage. More specialized databases that concentrate on plants with special prediction tools or structure-based annotations are PLncPRO and GREENC. Although resources such as ALDB and LncRNAdb provide experimental depth and historical importance, they would benefit from more frequent updates and expansion of scope. As the field of lncRNA biology grows beyond human studies, these non-human databases will continue to play a pivotal role in unraveling species-specific and conserved functions of lncRNAs.

2.7.3 *Case Studies on the Use of lncRNA Datasets and Informatics Tools*

A description of advances in innovations in the development of bioinformatic tools and datasets for studying lncRNAs would be incomplete without adequate illustration of their use. The computational tools and informatics resources pertinent to lncRNAs covered in the previous sections provide valuable information for researchers. These insights range from the identification of putative lncRNA sequences, prediction of secondary structures, description of expression profiles across tissues and disease states, and information regarding potential interaction partners. Such information is helpful for designing targeted experiments and for validating and interpreting experimental outcomes. Acha-Sagredo et al. (2020) employed an Agilent Custom microarray and statistical tools such as the *Limma* package for normalization and differential expression analysis to identify 949 differentially expressed lncRNAs in non-small cell lung carcinoma (NSCLC). The findings were validated using qPCR and cross-validated with TCGA datasets. Their integrative approach, which also included DNA methylation analysis, provided insights into the epigenetic regulation of key lncRNAs like FEZF1-AS1 and FENDRR. This methodology could serve as a model for similar studies in other cancer types. The newly identified lncRNAs warrant functional studies to understand their precise contribution in the initiation and progression of cancer. The identified lncRNAs represent potent biomarkers that can be used in diagnosis, as well as possible therapeutic targets (Acha-Sagredo et al. 2020).

Zheng et al. (2020) studied HUMT (LINC00857), an lncRNA upregulated in lymph node-metastatic triple-negative breast cancer (TNBC). A genome-wide microarray revealed its differential expression in metastatic versus parental cell lines. A range of bioinformatics tools was employed to analyze the molecular features and clinical significance of HUMT. The tools used are detailed in Table 2.2. These methods addressed dysregulation by identifying HUMT's upregulation in metastatic TNBC cells and tumors and correlating its expression with DNA methylation, concluding that HUMT stimulates the growth and metastasis of tumors in vivo. The suite of tools also helped in predicting the subcellular localization of HUMT, identifying its interacting protein partners, assessing the clinical relevance of HUMT and FOXK1, and linking HUMT expression with immune cell infiltration. The integration of multiple bioinformatics tools and databases, combined with experimental validation, provides strong evidence for HUMT's role in promoting lymphangiogenesis and metastasis through the YBX1/FOXK1 axis, suppressing NK cell recruitment, and highlighting its potential as a target for metastatic TNBC (Zheng et al. 2020). Beyond identification and functional annotation, bioinformatics tools have been used for prediction of secondary structures of specific RNAs. Mou, Liew, and Kwok deployed several bioinformatics tools to identify and characterize G4 structures in MALAT1 lncRNA. QGRS Mapper was used to detect G-rich motifs with high G-scores, identifying strong rG4 candidates in the 3′ region. G4RNA Screener, incorporating cGcC, G4Hunter, and G4NN algorithms, validated

Table 2.2 Bioinformatics tools used for analyzing the molecular properties and clinical relevance of HUMT

Bioinformatics approach	Description	Purpose	Data source/ tool(s)
Microarray analysis	Genome-wide expression profiling of lymph node-invasive cells (231LNM3 and 549LNM3) as compared to parental cells	Identify differentially expressed transcripts, including lncRNAs like HUMT, that are associated with metastasis	Microarray data
Coding potential assessment	Evaluating the protein-coding capacity of HUMT based on its sequence characteristics	Confirm that HUMT is a ncRNA and doesn't produce a protein	NCBI ORFfinder, PhyloCSF, CPAT
TCGA data analysis	Data from TCGA was analyzed for breast cancer and other cancers	Establish correlation between HUMT expression and DNA methylation levels to ascertain the levels of HUMT expression in tumor tissues relative to normal tissues	TCGA
CCLE data analysis	Analysis of the cancer cell line Encyclopedia (CCLE) data	Analyze the relationship between HUMT expression and methylation state of HUMT in cancer cell lines	CCLE
lncATLAS	Prediction of the subcellular localization of HUMT	Determine whether HUMT is primarily located in the nucleus or cytoplasm	lncATLAS
GEO dataset analysis	Analysis of GEO dataset GSE133620 using PAR-CLIP	Predict potential HUMT-interacting proteins	GSE133620
ENCODE database analysis	Analysis of ChIP-seq data from the ENCODE database	Identify binding sites of YBX1 protein on DNA, leading to the identification of FOXK1 as a target gene	ENCODE
Bc-GenExMiner v4.3	Breast cancer gene-expression miner analysis	Determine the prognostic significance of FOXK1 expression in basal-like breast cancer	Bc-GenExMiner v4.3
KM plotter	Kaplan-Meier plotter analysis	Evaluate the prognostic value of HUMT in different cancers by assessing its correlation with overall survival (OS) and relapse-free survival (RFS)	KM plotter ()
CIBERSORT	Estimation of immune cell proportions	Estimate the percentage of various immune cell types in TNBC samples from TCGA and GSE58812, and correlate HUMT expression with immune cell infiltration (e.g., NK cells)	CIBERSORT
Statistical analysis in R 3.5.3	R studio for data analysis of various data types to create statistical significance	Statistically validate HUMT/ YBX1/FOXK1 and other targets	R studio

rG4-forming potential based on guanine content, sequence skewness, and neural network predictions. Additionally, rG4-seq data from HeLa cells identified rG4 motifs in MALAT1, providing experimental support to computational predictions (Mou et al. 2022). Krishnamoorthy et al. investigated the role of MALAT1 in regulating breast cancer, with a special focus on its interaction with miR-150-3p. Bioinformatics tools were employed to predict and validate these molecular interactions and support experimental findings. The StarBase v3.0 platform was used to predict the binding relationship between MALAT1 and miR-150-3p. This web-based tool integrates data from CLIP-Seq, degradome sequencing, and RNA-RNA interaction studies, enabling researchers to identify potential binding sites and regulatory interactions between lncRNAs and miRNAs, such as that displayed by MALAT1 with miR-150-3p. The informatics-derived predictions were experimentally validated, confirming the tumor-promoting role of MALAT1 and revealing its regulatory control over miR-150-3p. The insights into the MALAT1/miR-150-3p axis underscore the potential of developing targeted therapies, such as antisense oligonucleotides or miRNA mimics, to suppress tumor progression and overcome drug resistance in breast cancer (Vignesh and Thangamalar 2024).

2.8 Conclusion

The convergence of computational and informatics tools with wet-laboratory experiments has provided a modern paradigm for research into lncRNAs. While traditional bioinformatics tools have provided foundational knowledge into lncRNA classification and sequence annotation, they often fall short in capturing the entire gamut of structural and functional complexity of these molecules. Recent advances in DL are bridging this gap by enabling the inference of RNA secondary and tertiary structures with greater speed and precision. These models excel at identifying nuanced features such as base-pairing probabilities, structural entropy, and accessibility profiles, which are the key parameters that influence lncRNA interactions and their regulatory potential.

Nevertheless, significant issues remain to be addressed. Figure 2.5 illustrates some of the challenges and opportunities in studying lncRNAs. Predicting the behavior of long and conformationally dynamic lncRNAs is challenging, especially for resolving pseudoknots and other non-canonical motifs. The discovery of sORFs within lncRNAs further complicates transcript classification and blurs the boundary between coding and non-coding regions. sORFs can pose challenges when alternative splicing yields isoforms with both coding and non-coding potential. Methodological shortcomings, including the lack of consensus on benchmarking practices, inconsistent annotations across reference databases, and limited access to high-confidence, experimentally validated lncRNA datasets, further compound the problems at hand.

To address these and other limitations, future research efforts must focus on embedding experimental constraints derived from techniques such as SHAPE-Seq,

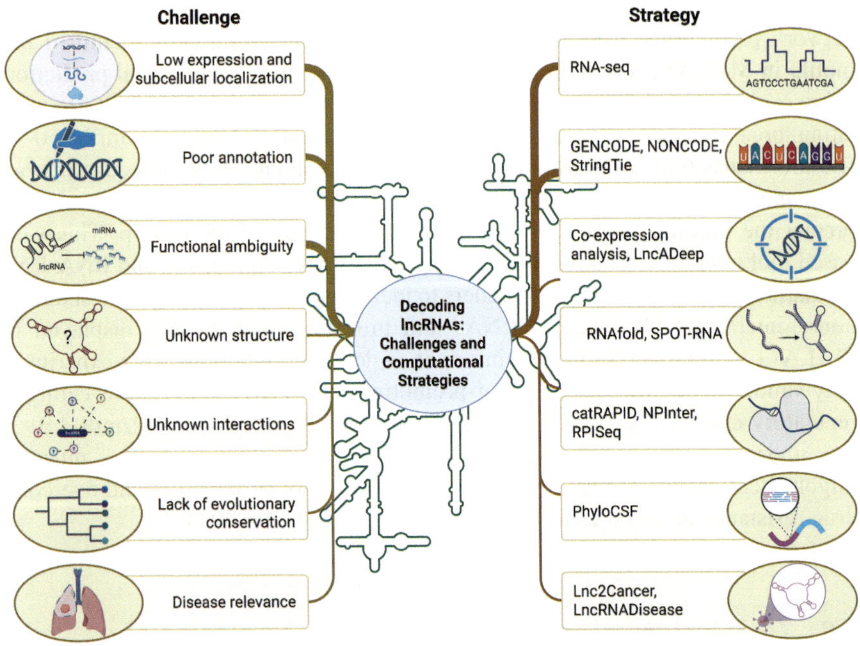

Fig. 2.5 Summary of the major hurdles in studying lncRNAs, which play diverse roles in gene regulation and disease. Major challenges include their low expression, poor annotation, structural ambiguity, and limited evolutionary conservation. The accompanying image outlines these obstacles and highlights computational tools developed to address them—such as RNA-seq for expression profiling, GENCODE and NONCODE for annotation, LncADeep for function prediction, RNAfold for structure modeling, and databases like Lnc2Cancer for disease associations. These integrative strategies are crucial for advancing our understanding of lncRNA biology

PARS, and NMR spectroscopy into computational models to improve structural conformity. Building comprehensive, high-quality databases that integrate multi-layered omics data such as RNA-protein interaction networks, cell-type-specific expression patterns, and RNA modification landscapes are vital for training more versatile predictive models. Extending the diversity of training datasets to include lncRNAs from a broader phylogenetic spectrum would also aid in identifying conserved functional elements and elucidating evolutionary trends.

The adoption of explainable artificial intelligence represents another promising frontier. By making model outputs interpretable, explainable artificial intelligence can bridge the gap between black-box predictions and mechanistic understanding. This could foster greater trust and utility in biomedical research. In parallel, CRISPR-based perturbation screens could be instrumental in functionally validating computationally predicted lncRNA roles, thereby closing the loop between in silico analysis and experimental biology. As cloud-based platforms mature and analytical channels become more standardized, the lncRNA research community will benefit from greater reproducibility, scalability, and inclusivity.

Finally, the maturation of specialized tools, such as those predicting G4s and other dynamic structural elements, will enable deeper exploration of lncRNA-mediated regulation at the subcellular level. Comparative genomics, powered by a richer catalogue of non-human lncRNAs, will reveal evolutionarily conserved functions that may serve as biomarkers and therapeutic targets. Together, these advancements would set the stage for AI-driven discovery pipelines capable of identifying lncRNA-based signatures of disease, accelerating the development of personalized diagnostics, and paving the way for novel RNA-targeted therapies.

Acknowledgments The authors acknowledge the help extended by the Library of IIT Gandhinagar. All the illustrations are created with BioRender.com.

Funding This study was funded by Gujarat State Biotechnology Mission (GSBTM) vide GSBTM/JD(R&D)/626/22-23/00006262.

Compliance with Ethical Standards

Conflict of Interest The authors declare that they have no competing interests.

Ethical Approval This chapter is a review of previously published accounts, as such, no animal or human studies were performed.

References

Abramson J, Adler J, Dunger J et al (2024) Accurate structure prediction of biomolecular interactions with AlphaFold 3. Nature 630:493–500. https://doi.org/10.1038/s41586-024-07487-w

Acha-Sagredo A, Uko B, Pantazi P et al (2020) Long non-coding RNA dysregulation is a frequent event in non-small cell lung carcinoma pathogenesis. Br J Cancer 122:1050–1058. https://doi.org/10.1038/s41416-020-0742-9

Achawanantakun R, Chen J, Sun Y, Zhang Y (2015) LncRNA-ID: long non-coding RNA identification using balanced random forests. Bioinformatics 31:3897–3905. https://doi.org/10.1093/bioinformatics/btv480

Ahmad M, Weiswald L-B, Poulain L et al (2023) Involvement of lncRNAs in cancer cells migration, invasion and metastasis: cytoskeleton and ECM crosstalk. J Exp Clin Cancer Res 42:173. https://doi.org/10.1186/s13046-023-02741-x

Akiyama M, Sato K, Sakakibara Y (2018) A max-margin training of RNA secondary structure prediction integrated with the thermodynamic model. J Bioinforma Comput Biol 16:1840025. https://doi.org/10.1142/S0219720018400255

Ali SE, Mittal A, Mathews DH (2023) RNA secondary structure analysis using RNAstructure. Curr Protoc 3:e846. https://doi.org/10.1002/cpz1.846

Armaos A, Cirillo D, Gaetano Tartaglia G (2017) *omiXcore*: a web server for prediction of protein interactions with large RNA. Bioinformatics 33:3104–3106. https://doi.org/10.1093/bioinformatics/btx361

Ballarino M, Pepe G, Helmer-Citterich M, Palma A (2023) Exploring the landscape of tools and resources for the analysis of long non-coding RNAs. Comput Struct Biotechnol J 21:4706–4716. https://doi.org/10.1016/j.csbj.2023.09.041

Beaudoin J-D, Jodoin R, Perreault J-P (2014) New scoring system to identify RNA G-quadruplex folding. Nucleic Acids Res 42:1209–1223. https://doi.org/10.1093/nar/gkt904

Bedrat A, Lacroix L, Mergny J-L (2016) Re-evaluation of G-quadruplex propensity with G4Hunter. Nucleic Acids Res 44:1746–1759. https://doi.org/10.1093/nar/gkw006

Beermann J, Piccoli M-T, Viereck J, Thum T (2016) Non-coding RNAs in development and disease: background, mechanisms, and therapeutic approaches. Physiol Rev 96:1297–1325. https://doi.org/10.1152/physrev.00041.2015

Bell JC, Jukam D, Teran NA et al (2018) Chromatin-associated RNA sequencing (ChAR-seq) maps genome-wide RNA-to-DNA contacts. elife 7:e27024. https://doi.org/10.7554/eLife.27024

Bellaousov S, Reuter JS, Seetin MG, Mathews DH (2013) RNAstructure: web servers for RNA secondary structure prediction and analysis. Nucleic Acids Res 41:W471–W474. https://doi.org/10.1093/nar/gkt290

Bellucci M, Agostini F, Masin M, Tartaglia GG (2011) Predicting protein associations with long noncoding RNAs. Nat Methods 8:444–445. https://doi.org/10.1038/nmeth.1611

Belmonte-Reche E, Morales JC (2020) G4-iM Grinder: when size and frequency matter. G-Quadruplex, i-Motif and higher order structure search and analysis tool. NAR Genomics Bioinf 2:Iqz005. https://doi.org/10.1093/nargab/lqz005

Bernard C, Postic G, Ghannay S, Tahi F (2024) State-of-the-RNArt: benchmarking current methods for RNA 3D structure prediction. NAR Genomics Bioinf 6:Iqae048. https://doi.org/10.1093/nargab/lqae048

Bernhart SH, Hofacker IL, Will S et al (2008) RNAalifold: improved consensus structure prediction for RNA alignments. BMC Bioinformatics 9:474. https://doi.org/10.1186/1471-2105-9-474

Berselli M, Lavezzo E, Toppo S (2020) QPARSE: searching for long-looped or multimeric G-quadruplexes potentially distinctive and druggable. Bioinformatics 36:393–399. https://doi.org/10.1093/bioinformatics/btz569

Betti SB, Tahmoorespur M, Javadmanesh A (2022) Identification of lncRNAs expression and their regulatory networks associated with development and growth of skeletal muscle in sheep using RNA-Seq. Agric Nat Resour 56:bbac339. https://doi.org/10.34044/j.anres.2022.56.2.15

Bhatt U, Cucchiarini A, Luo Y et al (2024) Preferential formation of Z-RNA over intercalated motifs in long noncoding RNA. Genome Res 34:217–230. https://doi.org/10.1101/gr.278236.123

Bolger AM, Lohse M, Usadel B (2014) Trimmomatic: a flexible trimmer for Illumina sequence data. Bioinformatics 30:2114–2120. https://doi.org/10.1093/bioinformatics/btu170

Brázda V, Kolomazník J, Lýsek J et al (2019) G4Hunter web application: a web server for G-quadruplex prediction. Bioinformatics 35:3493–3495. https://doi.org/10.1093/bioinformatics/btz087

Bugnon LA, Edera AA, Prochetto S et al (2022) Secondary structure prediction of long non-coding RNA: review and experimental comparison of existing approaches. Brief Bioinform 23:bbac205. https://doi.org/10.1093/bib/bbac205

Cagirici HB, Budak H, Sen TZ (2022) G4Boost: a machine learning-based tool for quadruplex identification and stability prediction. BMC Bioinform 23:240. https://doi.org/10.1186/s12859-022-04782-z

Chauviere L, Hoffbeck L, Shoaib M et al (2023) Firalink: a bioinformatics pipeline for long non-coding RNA data analysis. Noncoding RNA Res 8:602–604. https://doi.org/10.1016/j.ncrna.2023.09.002

Chen S, Zhou Y, Chen Y, Gu J (2018) Fastp: an ultra-fast all-in-one FASTQ preprocessor. Bioinformatics 34:i884–i890. https://doi.org/10.1093/bioinformatics/bty560

Chen X, Li Y, Umarov R et al (2020) RNA secondary structure prediction by learning unrolled algorithms. arXiv. https://doi.org/10.48550/arXiv.2002.05810

Cheng S, Zhang L, Tan J et al (2019) DM-RPIs: predicting nc RNA-protein interactions using stacked ensembling strategy. Comput Biol Chem 83:107088. https://doi.org/10.1016/j.compbiolchem.2019.107088

Chillón I, Marcia M (2020) The molecular structure of long non-coding RNAs: emerging patterns and functional implications. Crit Rev Biochem Mol Biol 55:662–690. https://doi.org/10.1080/10409238.2020.1828259

Cordero P, Lucks JB, Das R (2012) An RNA mapping DataBase for curating RNA structure mapping experiments. Bioinformatics 28:3006–3008. https://doi.org/10.1093/bioinformatics/bts554

Corrado G, Tebaldi T, Costa F et al (2016) RNAcommender: genome-wide recommendation of RNA–protein interactions. Bioinformatics 32:3627–3634. https://doi.org/10.1093/bioinformatics/btw517

Das G, Das T, Ghosh Z (2025) LncPTPred: predicting lncRNA-protein interaction based on cross-linking and immunoprecipitation (CLIP-Seq) data. Brief Bioinform 26:bbaf432

Dhapola P, Chowdhury S (2016) QuadBase2: web server for multiplexed guanine quadruplex mining and visualization. Nucleic Acids Res 44:W277–W283. https://doi.org/10.1093/nar/gkw425

Diamantopoulos MA, Tsiakanikas P, Scorilas A (2018) Non-coding RNAs: the riddle of the transcriptome and their perspectives in cancer. Ann Transl Med 6:241. https://doi.org/10.21037/atm.2018.06.10

Distefano R, Ilieva M, Rennie S, Uchida S (2024) Recommendations for bioinformatic tools in lncRNA research. Curr Bioinform 19:14–20. https://doi.org/10.2174/1574893618666230707103956

Djebali S, Davis CA, Merkel A et al (2012) Landscape of transcription in human cells. Nature 489:101–108. https://doi.org/10.1038/nature11233

Do CB, Woods DA, Batzoglou S (2006) CONTRAfold: RNA secondary structure prediction without physics-based models. Bioinformatics 22:e90–e98. https://doi.org/10.1093/bioinformatics/btl246

Dobin A, Davis CA, Schlesinger F et al (2013) STAR: ultrafast universal RNA-seq aligner. Bioinformatics 29:15–21. https://doi.org/10.1093/bioinformatics/bts635

Doluca O (2019) G4Catchall: a G-quadruplex prediction approach considering atypical features. J Theor Biol 463:92–98. https://doi.org/10.1016/j.jtbi.2018.12.007

Fallon L, Jones AN (2024) Alternative conformations of lncRNAs identified through structural deconvolution of SHAPE- and DMS-MaP datasets. bioRxiv

Fan X-N, Zhang S-W (2015) lncRNA-MFDL: identification of human long non-coding RNAs by fusing multiple features and using deep learning. Mol BioSyst 11:892–897. https://doi.org/10.1039/C4MB00650J

Fan X-N, Zhang S-W (2019) LPI-BLS: predicting lncRNA–protein interactions with a broad learning system-based stacked ensemble classifier. Neurocomputing 370:88–93. https://doi.org/10.1016/j.neucom.2019.08.084

Fernandes JCR, Acuña SM, Aoki JI et al (2019) Long non-coding RNAs in the regulation of gene expression: physiology and disease. Noncoding RNA 5:17. https://doi.org/10.3390/ncrna5010017

Fu L, Cao Y, Wu J et al (2022) UFold: fast and accurate RNA secondary structure prediction with deep learning. Nucleic Acids Res 50:e14–e14. https://doi.org/10.1093/nar/gkab1074

Fukunaga T, Iwakiri J, Ono Y, Hamada M (2019) LncRRIsearch: a web server for lncRNA-RNA interaction prediction integrated with tissue-specific expression and subcellular localization data. Front Genet 10:462. https://doi.org/10.3389/fgene.2019.00462

Gao Y, Shang S, Guo S et al (2021) Lnc2Cancer 3.0: an updated resource for experimentally supported lncRNA/circRNA cancer associations and web tools based on RNA-seq and scRNA-seq data. Nucleic Acids Res 49:D1251–D1258. https://doi.org/10.1093/nar/gkaa1006

Garant J-M, Perreault J-P, Scott MS (2017) Motif independent identification of potential RNA G-quadruplexes by G4RNA screener. Bioinformatics 33:3532–3537. https://doi.org/10.1093/bioinformatics/btx498

Gawronski KAB, Kim J (2017) Single cell transcriptomics of noncoding <scp>RNAs</scp> and their cell-specificity. WIREs RNA 8:e1433. https://doi.org/10.1002/wrna.1433

Ge M, Li A, Wang M (2016) A bipartite network-based method for prediction of long non-coding RNA–protein interactions. Genomics Proteomics Bioinformatics 14:62–71. https://doi.org/10.1016/j.gpb.2016.01.004

Gong T, Ju F, Bu D (2024) Accurate prediction of RNA secondary structure including pseudoknots through solving minimum-cost flow with learned potentials. Commun Biol 7:297. https://doi.org/10.1038/s42003-024-05952-w

Grabherr MG, Haas BJ, Yassour M et al (2011) Full-length transcriptome assembly from RNA-Seq data without a reference genome. Nat Biotechnol 29:644–652. https://doi.org/10.1038/nbt.1883

Griffiths-Jones S (2004) Rfam: annotating non-coding RNAs in complete genomes. Nucleic Acids Res 33:D121–D124. https://doi.org/10.1093/nar/gki081

Gruber AR, Bernhart SH, Lorenz R (2015) The ViennaRNA web services. Springer, pp 307–326

Gupta RA, Shah N, Wang KC et al (2010) Long non-coding RNA HOTAIR reprograms chromatin state to promote cancer metastasis. Nature 464:1071–1076. https://doi.org/10.1038/nature08975

Hamada M, Kiryu H, Sato K et al (2009) Prediction of RNA secondary structure using generalized centroid estimators. Bioinformatics 25:465–473. https://doi.org/10.1093/bioinformatics/btn601

Hamada M, Sato K, Asai K (2011) Improving the accuracy of predicting secondary structure for aligned RNA sequences. Nucleic Acids Res 39:393–402. https://doi.org/10.1093/nar/gkq792

Han S, Liang Y, Ma Q et al (2019) LncFinder: an integrated platform for long non-coding RNA identification utilizing sequence intrinsic composition, structural information and physicochemical property. Brief Bioinform 20:2009–2027. https://doi.org/10.1093/bib/bby065

Hanson RM, Lu X-J (2017) DSSR-enhanced visualization of nucleic acid structures in Jmol. Nucleic Acids Res 45:W528–W533. https://doi.org/10.1093/nar/gkx365

Hofacker IL, Fontana W, Stadler PF et al (1994) Fast folding and comparison of RNA secondary structures. Mon Chem - Chem Mon 125:167–188. https://doi.org/10.1007/BF00818163

Hon C-C, Ramilowski JA, Harshbarger J et al (2017a) An atlas of human long non-coding RNAs with accurate 5′ ends. Nature 543:199–204. https://doi.org/10.1038/nature21374

Hon J, Martínek T, Zendulka J, Lexa M (2017b) Pqsfinder: an exhaustive and imperfection-tolerant search tool for potential quadruplex-forming sequences in R. Bioinformatics 33:3373–3379. https://doi.org/10.1093/bioinformatics/btx413

Hu H, Zhu C, Ai H et al (2017a) LPI-ETSLP: lncRNA–protein interaction prediction using eigenvalue transformation-based semi-supervised link prediction. Mol BioSyst 13:1781–1787. https://doi.org/10.1039/C7MB00290D

Hu L, Xu Z, Hu B, Lu ZJ (2017b) COME: a robust coding potential calculation tool for lncRNA identification and characterization based on multiple features. Nucleic Acids Res 45:e2. https://doi.org/10.1093/nar/gkw798

Hu H, Zhang L, Ai H et al (2018) HLPI-ensemble: prediction of human lncRNA-protein interactions based on ensemble strategy. RNA Biol 1–10. https://doi.org/10.1080/15476286.2018.1457935

Huang Y, Li H, Xiao Y (2018) 3dRPC: a web server for 3D RNA–protein structure prediction. Bioinformatics 34:1238–1240. https://doi.org/10.1093/bioinformatics/btx742

Huang L, Zhang H, Deng D et al (2019a) LinearFold: linear-time approximate RNA folding by 5′-to-3′ dynamic programming and beam search. Bioinformatics 35:i295–i304. https://doi.org/10.1093/bioinformatics/btz375

Huang Y-A, Huang Z-A, You Z-H et al (2019b) Predicting lncRNA-miRNA interaction via graph convolution auto-encoder. Front Genet 10:758. https://doi.org/10.3389/fgene.2019.00758

Huang L, Jiao S, Yang S et al (2021) LGFC-CNN: prediction of lncRNA-protein interactions by using multiple types of features through deep learning. Genes (Basel) 12:1689. https://doi.org/10.3390/genes12111689

Huarte M, Guttman M, Feldser D et al (2010) A large intergenic noncoding RNA induced by p53 mediates global gene repression in the p53 response. Cell 142:409–419. https://doi.org/10.1016/j.cell.2010.06.040

Huppert JL, Balasubramanian S (2005) Prevalence of quadruplexes in the human genome. Nucleic Acids Res 33:2908–2916. https://doi.org/10.1093/nar/gki609

Irfan M, Javed Z, Khan K et al (2022) Apoptosis evasion via long non-coding RNAs in colorectal cancer. Cancer Cell Int 22:280. https://doi.org/10.1186/s12935-022-02695-8

Iwakiri J, Hamada M, Asai K (2016) Bioinformatics tools for lncRNA research. Biochim Biophys Acta Gene Regul Mech 1859:23–30. https://doi.org/10.1016/j.bbagrm.2015.07.014

Jia K, Gao Y, Shi J et al (2020) Annotation and curation of the causality information in LncRNADisease. Database 2020:baz150. https://doi.org/10.1093/database/baz150

Jin K, Wang S, Zhang Y et al (2019) Long non-coding RNA PVT1 interacts with MYC and its downstream molecules to synergistically promote tumorigenesis. Cell Mol Life Sci 76:4275–4289. https://doi.org/10.1007/s00018-019-03222-1

John B, Enright AJ, Aravin A et al (2004) Human MicroRNA targets. PLoS Biol 2:e363. https://doi.org/10.1371/journal.pbio.0020363

Jones AN, Sattler M (2019) Challenges and perspectives for structural biology of lncRNAs—the example of the Xist lncRNA A-repeats. J Mol Cell Biol 11:845–859. https://doi.org/10.1093/jmcb/mjz086

Junge A, Refsgaard JC, Garde C et al (2017) RAIN: RNA–protein association and interaction networks. Database 2017:baw167. https://doi.org/10.1093/database/baw167

Karagkouni D, Paraskevopoulou MD, Tastsoglou S et al (2019) DIANA-LncBase v3: indexing experimentally supported miRNA targets on non-coding transcripts. Nucleic Acids Res. https://doi.org/10.1093/nar/gkz1036

Karam JAQ, Fréreux C, Mohanty BK et al (2024) The RNA-binding protein PCBP1 modulates transcription by recruiting the G-quadruplex-specific helicase DHX9. J Biol Chem 300:107830. https://doi.org/10.1016/j.jbc.2024.107830

Kato Y, Sato K, Hamada M et al (2010) RactIP: fast and accurate prediction of RNA-RNA interaction using integer programming. Bioinformatics 26:i460–i466. https://doi.org/10.1093/bioinformatics/btq372

Kaur B, Sharma P, Arora P, Sood V (2023) QUFIND: tool for comparative prediction and mining of G4 quadruplexes overlapping with CpG islands. Front Genet 14:1265808. https://doi.org/10.3389/fgene.2023.1265808

Kikin O, D'Antonio L, Bagga PS (2006) QGRS mapper: a web-based server for predicting G-quadruplexes in nucleotide sequences. Nucleic Acids Res 34:W676–W682. https://doi.org/10.1093/nar/gkl253

Kim D, Pertea G, Trapnell C et al (2013) TopHat2: accurate alignment of transcriptomes in the presence of insertions, deletions and gene fusions. Genome Biol 14:R36. https://doi.org/10.1186/gb-2013-14-4-r36

Kim D, Paggi JM, Park C et al (2019) Graph-based genome alignment and genotyping with HISAT2 and HISAT-genotype. Nat Biotechnol 37:907–915. https://doi.org/10.1038/s41587-019-0201-4

Knudsen B, Hein J (2003) Pfold: RNA secondary structure prediction using stochastic context-free grammars. Nucleic Acids Res 31:3423–3428. https://doi.org/10.1093/nar/gkg614

Landrum MJ, Lee JM, Benson M et al (2018) ClinVar: improving access to variant interpretations and supporting evidence. Nucleic Acids Res 46:D1062–D1067. https://doi.org/10.1093/nar/gkx1153

Langmead B, Salzberg SL (2012) Fast gapped-read alignment with Bowtie 2. Nat Methods 9:357–359. https://doi.org/10.1038/nmeth.1923

Lee C, Kikyo N (2012) Strategies to identify long noncoding RNAs involved in gene regulation. Cell Biosci 2:37. https://doi.org/10.1186/2045-3701-2-37

Li A, Zhang J, Zhou Z (2014) PLEK: a tool for predicting long non-coding RNAs and messenger RNAs based on an improved k-mer scheme. BMC Bioinform 15:311. https://doi.org/10.1186/1471-2105-15-311

Li A, Ge M, Zhang Y et al (2015a) Predicting long noncoding RNA and protein interactions using heterogeneous network model. Biomed Res Int 2015:1–11. https://doi.org/10.1155/2015/671950

Li A, Zhang J, Zhou Z et al (2015b) ALDB: a domestic-animal long noncoding RNA database. PLoS One 10:e0124003. https://doi.org/10.1371/journal.pone.0124003
Li J, Ma W, Zeng P et al (2015c) LncTar: a tool for predicting the RNA targets of long noncoding RNAs. Brief Bioinform 16:806–812. https://doi.org/10.1093/bib/bbu048
Li X, Zhou B, Chen L et al (2017) GRID-seq reveals the global RNA–chromatin interactome. Nat Biotechnol 35:940–950. https://doi.org/10.1038/nbt.3968
Li J, Zhang X, Liu C (2020) The computational approaches of lncRNA identification based on coding potential: status quo and challenges. Comput Struct Biotechnol J 18:3666–3677. https://doi.org/10.1016/j.csbj.2020.11.030
Li Q, Mo W, Ding Y, Ding X (2021a) Study of lncRNA TPA in promoting invasion and metastasis of breast cancer mediated by TGF-β signaling pathway. Front Cell Dev Biol 9:688751. https://doi.org/10.3389/fcell.2021.688751
Li Y, Sun H, Feng S et al (2021b) Capsule-LPI: a LncRNA–protein interaction predicting tool based on a capsule network. BMC Bioinform 22:246. https://doi.org/10.1186/s12859-021-04171-y
Liang Y, Yin X, Zhang Y et al (2024) Predicting lncRNA–protein interactions through deep learning framework employing multiple features and random forest algorithm. BMC Bioinform 25:108. https://doi.org/10.1186/s12859-024-05727-4
Lin MF, Jungreis I, Kellis M (2011) PhyloCSF: a comparative genomics method to distinguish protein coding and non-coding regions. Bioinformatics 27:i275–i282. https://doi.org/10.1093/bioinformatics/btr209
Lin X, Lu Y, Zhang C et al (2024) LncRNADisease v3.0: an updated database of long non-coding RNA-associated diseases. Nucleic Acids Res 52:D1365–D1369. https://doi.org/10.1093/nar/gkad828
Liu L, Li Z, Liu C et al (2022) LncRNAWiki 2.0: a knowledgebase of human long non-coding RNAs with enhanced curation model and database system. Nucleic Acids Res 50:D190–D195. https://doi.org/10.1093/nar/gkab998
Lu Q, Ren S, Lu M et al (2013) Computational prediction of associations between long non-coding RNAs and proteins. BMC Genomics 14:651. https://doi.org/10.1186/1471-2164-14-651
Lyu K, Chow EY-C, Mou X et al (2021) RNA G-quadruplexes (rG4s): genomics and biological functions. Nucleic Acids Res 49:5426–5450. https://doi.org/10.1093/nar/gkab187
Ma L, Cao J, Liu L et al (2019) LncBook: a curated knowledgebase of human long non-coding RNAs. Nucleic Acids Res 47:D128–D134. https://doi.org/10.1093/nar/gky960
Macosko EZ, Basu A, Satija R et al (2015) Highly parallel genome-wide expression profiling of individual cells using Nanoliter droplets. Cell 161:1202–1214. https://doi.org/10.1016/j.cell.2015.05.002
Makris E, Kolaitis A, Andrikos C et al (2023) Knotify+: toward the prediction of RNA H-type pseudoknots. Including Bulges and Internal Loops Biomolecules 13:308. https://doi.org/10.3390/biom13020308
Mao FY, Tu M-J, Traber GM, Yu A-M (2024) Comparison of three computational tools for the prediction of RNA tertiary structures. Noncoding RNA 10:55. https://doi.org/10.3390/ncrna10060055
Markham NR, Zuker M (2008) UNAFold: software for nucleic acid folding and hybridization. In: Bioinformatics: structure, function and applications. Humana Press, Totowa, pp 3–31
Maroni P, Gomarasca M, Lombardi G (2023) Long non-coding RNAs in bone metastasis: progresses and perspectives as potential diagnostic and prognostic biomarkers. Front Endocrinol (Lausanne) 14:1156494. https://doi.org/10.3389/fendo.2023.1156494
Mas-Ponte D, Carlevaro-Fita J, Palumbo E et al (2017) LncATLAS database for subcellular localization of long noncoding RNAs. RNA 23:1080–1087. https://doi.org/10.1261/rna.060814.117
Mattick JS, Amaral PP, Carninci P et al (2023) Long non-coding RNAs: definitions, functions, challenges and recommendations. Nat Rev Mol Cell Biol 24:430–447. https://doi.org/10.1038/s41580-022-00566-8

McCown PJ, Wang MC, Jaeger L, Brown JA (2019) Secondary structural model of human MALAT1 reveals multiple structure–function relationships. Int J Mol Sci 20:5610. https://doi.org/10.3390/ijms20225610

Moore MJ, Scheel TKH, Luna JM et al (2015) miRNA–target chimeras reveal miRNA 3′-end pairing as a major determinant of Argonaute target specificity. Nat Commun 6:8864. https://doi.org/10.1038/ncomms9864

Mou X, Liew SW, Kwok CK (2022) Identification and targeting of G-quadruplex structures in *MALAT1* long non-coding RNA. Nucleic Acids Res 50:397–410. https://doi.org/10.1093/nar/gkab1208

Mückstein U, Tafer H, Hackermüller J et al (2006) Thermodynamics of RNA–RNA binding. Bioinformatics 22:1177–1182. https://doi.org/10.1093/bioinformatics/btl024

Mukherjee S, Pramanik P, Basuchowdhuri P, Bhattacharya S (2024) G4-attention: deep learning model with attention for predicting DNA G-quadruplexes. bioRxiv

Muppirala UK, Honavar VG, Dobbs D (2011) Predicting RNA-protein interactions using only sequence information. BMC Bioinform 12:489. https://doi.org/10.1186/1471-2105-12-489

Murakami Y, Spriggs RV, Nakamura H, Jones S (2010) PiRaNhA: a server for the computational prediction of RNA-binding residues in protein sequences. Nucleic Acids Res 38:W412–W416. https://doi.org/10.1093/nar/gkq474

O'Leary NA, Wright MW, Brister JR et al (2016) Reference sequence (RefSeq) database at NCBI: current status, taxonomic expansion, and functional annotation. Nucleic Acids Res 44:D733–D745. https://doi.org/10.1093/nar/gkv1189

Ontiveros-Palacios N, Cooke E, Nawrocki EP et al (2025) Rfam 15: RNA families database in 2025. Nucleic Acids Res 53:D258–D267. https://doi.org/10.1093/nar/gkae1023

Ørom UA, Derrien T, Beringer M et al (2010) Long noncoding RNAs with enhancer-like function in human cells. Cell 143:46–58. https://doi.org/10.1016/j.cell.2010.09.001

Pafilis E, Frankild SP, Fanini L et al (2013) The SPECIES and ORGANISMS resources for fast and accurate identification of taxonomic names in text. PLoS One 8:e65390. https://doi.org/10.1371/journal.pone.0065390

Pan X, Fan Y-X, Yan J, Shen H-B (2016) IPMiner: hidden ncRNA-protein interaction sequential pattern mining with stacked autoencoder for accurate computational prediction. BMC Genomics 17:582. https://doi.org/10.1186/s12864-016-2931-8

Paraskevopoulou MD, Karagkouni D, Vlachos IS et al (2018) microCLIP super learning framework uncovers functional transcriptome-wide miRNA interactions. Nat Commun 9:3601. https://doi.org/10.1038/s41467-018-06046-y

Paytuví Gallart A, Hermoso Pulido A, Martínez A, de Lagrán I et al (2016) GREENC: a Wiki-based database of plant lncRNAs. Nucleic Acids Res 44:D1161–D1166. https://doi.org/10.1093/nar/gkv1215

Pedersen JS (2004) A comparative method for finding and folding RNA secondary structures within protein-coding regions. Nucleic Acids Res 32:4925–4936. https://doi.org/10.1093/nar/gkh839

Peng C, Han S, Zhang H, Li Y (2019) RPITER: a hierarchical deep learning framework for ncRNA–protein interaction prediction. Int J Mol Sci 20:1070. https://doi.org/10.3390/ijms20051070

Pertea M, Pertea GM, Antonescu CM et al (2015) StringTie enables improved reconstruction of a transcriptome from RNA-seq reads. Nat Biotechnol 33:290–295. https://doi.org/10.1038/nbt.3122

Pinkney HR, Wright BM, Diermeier SD (2020) The lncRNA toolkit: databases and in silico tools for lncRNA analysis. Noncoding RNA 6:49. https://doi.org/10.3390/ncrna6040049

Puig Lombardi E, Londoño-Vallejo A (2020) A guide to computational methods for G-quadruplex prediction. Nucleic Acids Res 48:1–15. https://doi.org/10.1093/nar/gkz1097

Quek XC, Thomson DW, Maag JLV et al (2015) lncRNAdb v2.0: expanding the reference database for functional long noncoding RNAs. Nucleic Acids Res 43:D168–D173. https://doi.org/10.1093/nar/gku988

Quinn JJ, Chang HY (2016) Unique features of long non-coding RNA biogenesis and function. Nat Rev Genet 17:47–62. https://doi.org/10.1038/nrg.2015.10

Ramakrishnaiah Y, Morris AP, Dhaliwal J et al (2023) Linc2function: a comprehensive pipeline and webserver for long non-coding RNA (lncRNA) identification and functional predictions using deep learning approaches. Epigenomes 7:22. https://doi.org/10.3390/epigenomes7030022

Ramani R, Krumholz K, Huang Y-F, Siepel A (2019) PhastWeb: a web interface for evolutionary conservation scoring of multiple sequence alignments using phastCons and phyloP. Bioinformatics 35:2320–2322. https://doi.org/10.1093/bioinformatics/bty966

Ramírez-Colmenero A, Oktaba K, Fernandez-Valverde SL (2020) Evolution of genome-organizing long non-coding RNAs in metazoans. Front Genet 11:589697. https://doi.org/10.3389/fgene.2020.589697

Reeder J, Giegerich R (2004) Design, implementation and evaluation of a practical pseudoknot folding algorithm based on thermodynamics. BMC Bioinform 5:104. https://doi.org/10.1186/1471-2105-5-104

Rehmsmeier M, Steffen P, Höchsmann M, Giegerich R (2004) Fast and effective prediction of microRNA/target duplexes. RNA 10:1507–1517. https://doi.org/10.1261/rna.5248604

Rinn JL, Chang HY (2012) Genome regulation by long noncoding RNAs. Annu Rev Biochem 81:145–166. https://doi.org/10.1146/annurev-biochem-051410-092902

Rintala-Maki ND, Sutherland LC (2009) Identification and characterisation of a novel antisense non-coding RNA from the RBM5 gene locus. Gene 445:7–16. https://doi.org/10.1016/j.gene.2009.06.009

Rivas E, Eddy SR (1999) A dynamic programming algorithm for RNA structure prediction including pseudoknots 1 1Edited by I. Tinoco. J Mol Biol 285:2053–2068. https://doi.org/10.1006/jmbi.1998.2436

Rivas E, Lang R, Eddy SR (2012) A range of complex probabilistic models for RNA secondary structure prediction that includes the nearest-neighbor model and more. RNA 18:193–212. https://doi.org/10.1261/rna.030049.111

Ronneberger O, Fischer P, Brox T (2015) U-Net: convolutional networks for biomedical image segmentation. Springer, pp 234–241

Rupaimoole R, Lee J, Haemmerle M et al (2015) Long noncoding RNA ceruloplasmin promotes cancer growth by altering glycolysis. Cell Rep 13:2395–2402. https://doi.org/10.1016/j.celrep.2015.11.047

Sanbonmatsu K (2022) Getting to the bottom of lncRNA mechanism: structure–function relationships. Mamm Genome 33:343–353. https://doi.org/10.1007/s00335-021-09924-x

Sato K, Hamada M, Asai K, Mituyama T (2009) CENTROIDFOLD: a web server for RNA secondary structure prediction. Nucleic Acids Res 37:W277–W280. https://doi.org/10.1093/nar/gkp367

Sato K, Kato Y, Hamada M et al (2011) IPknot: fast and accurate prediction of RNA secondary structures with pseudoknots using integer programming. Bioinformatics 27:i85–i93. https://doi.org/10.1093/bioinformatics/btr215

Sato K, Akiyama M, Sakakibara Y (2021) RNA secondary structure prediction using deep learning with thermodynamic integration. Nat Commun 12:941. https://doi.org/10.1038/s41467-021-21194-4

Scaria V, Hariharan M, Arora A, Maiti S (2006) Quadfinder: server for identification and analysis of quadruplex-forming motifs in nucleotide sequences. Nucleic Acids Res 34:W683–W685. https://doi.org/10.1093/nar/gkl299

Schmidt N, Lareau CA, Keshishian H et al (2020) The SARS-CoV-2 RNA–protein interactome in infected human cells. Nat Microbiol 6:339–353. https://doi.org/10.1038/s41564-020-00846-z

Schriml LM, Munro JB, Schor M et al (2022) The human disease ontology 2022 update. Nucleic Acids Res 50:D1255–D1261. https://doi.org/10.1093/nar/gkab1063

Seemann SE, Richter AS, Gesell T et al (2011) PETcofold: predicting conserved interactions and structures of two multiple alignments of RNA sequences. Bioinformatics 27:211–219. https://doi.org/10.1093/bioinformatics/btq634

Seifuddin F, Singh K, Suresh A et al (2020) lncRNAKB, a knowledgebase of tissue-specific functional annotation and trait association of long noncoding RNA. Sci Data 7:326. https://doi.org/10.1038/s41597-020-00659-z

Sharma S, Hassan MY, Barbhuiya NH et al (2025) A dataset curated for the assessment of G4s in the LncRNAs dysregulated in various human cancers. Sci Data 12:849. https://doi.org/10.1038/s41597-025-05176-5

Shen C, Ding Y, Tang J et al (2019) LPI-KTASLP: prediction of LncRNA-protein interaction by semi-supervised link learning with multivariate information. IEEE Access 7:13486–13496. https://doi.org/10.1109/ACCESS.2019.2894225

Sherry ST, Ward MH, Kholodov M et al (2001) dbSNP: the NCBI database of genetic variation. Nucleic Acids Res 29:308–311. https://doi.org/10.1093/nar/29.1.308

Singh U, Khemka N, Rajkumar MS et al (2017) PLncPRO for prediction of long non-coding RNAs (lncRNAs) in plants and its application for discovery of abiotic stress-responsive lncRNAs in rice and chickpea. Nucleic Acids Res 45:e183–e183. https://doi.org/10.1093/nar/gkx866

Singh J, Hanson J, Paliwal K, Zhou Y (2019) RNA secondary structure prediction using an ensemble of two-dimensional deep neural networks and transfer learning. Nat Commun 10:5407. https://doi.org/10.1038/s41467-019-13395-9

Singh J, Paliwal K, Zhang T et al (2021) Improved RNA secondary structure and tertiary base-pairing prediction using evolutionary profile, mutational coupling and two-dimensional transfer learning. Bioinformatics 37:2589–2600. https://doi.org/10.1093/bioinformatics/btab165

Singh D, Assaraf YG, Gacche RN (2022) Long non-coding RNA mediated drug resistance in breast cancer. Drug Resist Updat 63:100851. https://doi.org/10.1016/j.drup.2022.100851

Smith C, Heyne S, Richter AS et al (2010) Freiburg RNA tools: a web server integrating INTARNA, EXPARNA and LOCARNA. Nucleic Acids Res 38:W373–W377. https://doi.org/10.1093/nar/gkq316

Sun L, Luo H, Bu D et al (2013) Utilizing sequence intrinsic composition to classify protein-coding and long non-coding transcripts. Nucleic Acids Res 41:e166–e166. https://doi.org/10.1093/nar/gkt646

Sun L, Liu H, Zhang L, Meng J (2015) lncRScan-SVM: a tool for predicting long non-coding RNAs using support vector machine. PLoS One 10:e0139654. https://doi.org/10.1371/journal.pone.0139654

Suresh V, Liu L, Adjeroh D, Zhou X (2015) RPI-Pred: predicting ncRNA-protein interaction using sequence and structural information. Nucleic Acids Res 43:1370–1379. https://doi.org/10.1093/nar/gkv020

Sweeney BA, Petrov AI, Ribas CE et al (2021) RNAcentral 2021: secondary structure integration, improved sequence search and new member databases. Nucleic Acids Res 49:D212–D220. https://doi.org/10.1093/nar/gkaa921

Szcześniak MW, Bryzghalov O, Ciomborowska-Basheer J, Makałowska I (2019) CANTATAdb 2.0: expanding the collection of plant long noncoding RNAs. Springer, pp 415–429

Szklarczyk D, Kirsch R, Koutrouli M et al (2023) The STRING database in 2023: protein–protein association networks and functional enrichment analyses for any sequenced genome of interest. Nucleic Acids Res 51:D638–D646. https://doi.org/10.1093/nar/gkac1000

Tafer H, Hofacker IL (2008) RNAplex: a fast tool for RNA–RNA interaction search. Bioinformatics 24:2657–2663. https://doi.org/10.1093/bioinformatics/btn193

Tassinari M, Richter SN, Gandellini P (2021) Biological relevance and therapeutic potential of G-quadruplex structures in the human noncoding transcriptome. Nucleic Acids Res 49:3617–3633. https://doi.org/10.1093/nar/gkab127

Tate JG, Bamford S, Jubb HC et al (2019) COSMIC: the catalogue of somatic mutations in cancer. Nucleic Acids Res 47:D941–D947. https://doi.org/10.1093/nar/gky1015

Terribilini M, Sander JD, Lee J-H et al (2007) RNABindR: a server for analyzing and predicting RNA-binding sites in proteins. Nucleic Acids Res 35:W578–W584. https://doi.org/10.1093/nar/gkm294

Tian X, Shen L, Wang Z et al (2021) A novel lncRNA–protein interaction prediction method based on deep forest with cascade forest structure. Sci Rep 11:18881. https://doi.org/10.1038/s41598-021-98277-1

Trapnell C, Williams BA, Pertea G et al (2010) Transcript assembly and quantification by RNA-Seq reveals unannotated transcripts and isoform switching during cell differentiation. Nat Biotechnol 28:511–515. https://doi.org/10.1038/nbt.1621

Tripathi R, Patel S, Kumari V et al (2016) DeepLNC, a long non-coding RNA prediction tool using deep neural network. Network Model Anal Health Inf Bioinf 5:21. https://doi.org/10.1007/s13721-016-0129-2

Turner M, Danino YM, Barshai M et al (2022) rG4detector, a novel RNA G-quadruplex predictor, uncovers their impact on stress granule formation. Nucleic Acids Res 50:11426–11441. https://doi.org/10.1093/nar/gkac950

Uszczynska-Ratajczak B, Lagarde J, Frankish A et al (2018) Towards a complete map of the human long non-coding RNA transcriptome. Nat Rev Genet 19:535–548. https://doi.org/10.1038/s41576-018-0017-y

Van Nostrand EL, Freese P, Pratt GA et al (2020) A large-scale binding and functional map of human RNA-binding proteins. Nature 583:711–719. https://doi.org/10.1038/s41586-020-2077-3

Varizhuk A, Ischenko D, Smirnov I et al (2014) An improved search algorithm to find G-quadruplexes in genome sequences. bioRxiv. https://doi.org/10.1101/001990

Varizhuk A, Ischenko D, Tsvetkov V et al (2017) The expanding repertoire of G4 DNA structures. Biochimie 135:54–62. https://doi.org/10.1016/j.biochi.2017.01.003

Vignesh K, Thangamalar P (2024) Regulatory role of lncRNA MALAT1 and miR-150-3p interaction in breast cancer progression and therapeutic implications. Cancer Plus 6:2633. https://doi.org/10.36922/cp.2633

Volders P-J, Anckaert J, Verheggen K et al (2019) LNCipedia 5: towards a reference set of human long non-coding RNAs. Nucleic Acids Res 47:D135–D139. https://doi.org/10.1093/nar/gky1031

Wan Y, Kertesz M, Spitale RC et al (2011) Understanding the transcriptome through RNA structure. Nat Rev Genet 12:641–655. https://doi.org/10.1038/nrg3049

Wang L, Park HJ, Dasari S et al (2013) CPAT: coding-potential assessment tool using an alignment-free logistic regression model. Nucleic Acids Res 41:e74–e74. https://doi.org/10.1093/nar/gkt006

Wang J, Wang J, Huang Y, Xiao Y (2019) 3dRNA v2.0: an updated web server for RNA 3D structure prediction. Int J Mol Sci 20:4116. https://doi.org/10.3390/ijms20174116

Wang Y, Liu Y, Wang S et al (2020) ATTfold: RNA secondary structure prediction with pseudoknots based on attention mechanism. Front Genet 11:612086. https://doi.org/10.3389/fgene.2020.612086

Wang J, Zhao Y, Gong W et al (2021) EDLMFC: an ensemble deep learning framework with multi-scale features combination for ncRNA–protein interaction prediction. BMC Bioinform 22:133. https://doi.org/10.1186/s12859-021-04069-9

Wang W, Wang Y, Sun B et al (2023) LPLSG: prediction of lncRNA-protein interaction based on local network structure. Curr Bioinform 18:276–284. https://doi.org/10.2174/1574893618666230223143914

Watkins AM, Rangan R, Das R (2020) FARFAR2: improved de novo Rosetta prediction of complex global RNA folds. Structure 28:963–976.e6. https://doi.org/10.1016/j.str.2020.05.011

Wekesa JS, Luan Y, Chen M, Meng J (2019) A hybrid prediction method for plant lncRNA-protein interaction. Cells 8:521. https://doi.org/10.3390/cells8060521

Wekesa JS, Luan Y, Meng J (2020a) LPI-DL: a recurrent deep learning model for plant lncRNA-protein interaction and function prediction with feature optimization. In: 2020 IEEE international conference on bioinformatics and biomedicine (BIBM). IEEE, pp 499–502

Wekesa JS, Meng J, Luan Y (2020b) Multi-feature fusion for deep learning to predict plant lncRNA-protein interaction. Genomics 112:2928–2936. https://doi.org/10.1016/j.ygeno.2020.05.005

Wekesa JS, Meng J, Luan Y (2020c) A deep learning model for plant lncRNA-protein interaction prediction with graph attention. Mol Gen Genomics 295:1091–1102. https://doi.org/10.1007/s00438-020-01682-w

Wenzel A, Akbaşli E, Gorodkin J (2012) RIsearch: fast RNA–RNA interaction search using a simplified nearest-neighbor energy model. Bioinformatics 28:2738–2746. https://doi.org/10.1093/bioinformatics/bts519

Willmott D, Murrugarra D, Ye Q (2020) Improving RNA secondary structure prediction via state inference with deep recurrent neural networks. Comput Math Biophys 8:36–50. https://doi.org/10.1515/cmb-2020-0002

Wingett SW, Andrews S (2018) FastQ screen: a tool for multi-genome mapping and quality control. F1000Res 7:1338. https://doi.org/10.12688/f1000research.15931.2

Wood DE, Lu J, Langmead B (2019) Improved metagenomic analysis with kraken 2. Genome Biol 20:257. https://doi.org/10.1186/s13059-019-1891-0

Wu W, Yan Z, Nguyen TC et al (2019) Mapping RNA–chromatin interactions by sequencing with iMARGI. Nat Protoc 14:3243–3272. https://doi.org/10.1038/s41596-019-0229-4

Wucher V, Legeai F, Hédan B et al (2017) FEELnc: a tool for long non-coding RNA annotation and its application to the dog transcriptome. Nucleic Acids Res 45:e57. https://doi.org/10.1093/nar/gkw1306

Xiao Y, Zhang J, Deng L (2017) Prediction of lncRNA-protein interactions using HeteSim scores based on heterogeneous networks. Sci Rep 7:3664. https://doi.org/10.1038/s41598-017-03986-1

Xiao J, Lin L, Luo D et al (2020) Long noncoding RNA TRPM2-AS acts as a microRNA sponge of miR-612 to promote gastric cancer progression and radioresistance. Oncogene 9:29. https://doi.org/10.1038/s41389-020-0215-2

Xu D, Xu H, Zhang Y et al (2021) LncRNA-protein interaction prediction based on regularized nonnegative matrix factorization and sequence information. MATCH Commun Math Comput Chem 85:555–574

Xu D, Wang W, Wang D et al (2024) Long noncoding RNA MALAT-1: a versatile regulator in cancer progression, metastasis, immunity, and therapeutic resistance. Noncoding RNA Res 9:388–406. https://doi.org/10.1016/j.ncrna.2024.01.015

Yang C, Yang L, Zhou M et al (2018) LncADeep: an *ab initio* lncRNA identification and functional annotation tool based on deep learning. Bioinformatics 34:3825–3834. https://doi.org/10.1093/bioinformatics/bty428

Yang S, Wang Y, Lin Y et al (2020) LncMirNet: predicting LncRNA–miRNA interaction based on deep learning of ribonucleic acid sequences. Molecules 25:4372. https://doi.org/10.3390/molecules25194372

Yang T, He Y, Wang Y (2024) Introducing TEC-LncMir for prediction of lncRNA-miRNA interactions through deep learning of RNA sequences. Brief Bioinform 26:bbaf046. https://doi.org/10.1093/bib/bbaf046

Yang S, Pham NT, Li Z et al (2025) Advances in RNA secondary structure prediction and RNA modifications: methods, data, and applications. arXiv

Yates B, Braschi B, Gray KA et al (2017) Genenames.org: the HGNC and VGNC resources in 2017. Nucleic Acids Res 45:D619–D625. https://doi.org/10.1093/nar/gkw1033

Yi H-C, You Z-H, Cheng L et al (2020a) Learning distributed representations of RNA and protein sequences and its application for predicting lncRNA-protein interactions. Comput Struct Biotechnol J 18:20–26. https://doi.org/10.1016/j.csbj.2019.11.004

Yi H-C, You Z-H, Wang M-N et al (2020b) RPI-SE: a stacking ensemble learning framework for ncRNA-protein interactions prediction using sequence information. BMC Bioinform 21:60. https://doi.org/10.1186/s12859-020-3406-0

Yu H, Qi Y, Yang B et al (2023) G4Atlas: a comprehensive transcriptome-wide G-quadruplex database. Nucleic Acids Res 51:D126–D134. https://doi.org/10.1093/nar/gkac896

Yuan Y, Yang E, Zhang R (2024) Wfold: a new method for predicting RNA secondary structure with deep learning. Comput Biol Med 182:109207. https://doi.org/10.1016/j.compbiomed.2024.109207

Zakov S, Goldberg Y, Elhadad M, Ziv-ukelson M (2011) Rich parameterization improves RNA structure prediction. J Comput Biol 18:1525–1542. https://doi.org/10.1089/cmb.2011.0184

Zangouei AS, Zangoue M, Taghehchian N et al (2023) Cell cycle related long non-coding RNAs as the critical regulators of breast cancer progression and metastasis. Biol Res 56:1. https://doi.org/10.1186/s40659-022-00411-4

Zerbino DR, Achuthan P, Akanni W et al (2018) Ensembl 2018. Nucleic Acids Res 46:D754–D761. https://doi.org/10.1093/nar/gkx1098

Zhan Z-H, Jia L-N, Zhou Y et al (2019) BGFE: a deep learning model for ncRNA-protein interaction predictions based on improved sequence information. Int J Mol Sci 20:978. https://doi.org/10.3390/ijms20040978

Zhang W, Yue X, Tang G et al (2018) SFPEL-LPI: sequence-based feature projection ensemble learning for predicting LncRNA-protein interactions. PLoS Comput Biol 14:e1006616. https://doi.org/10.1371/journal.pcbi.1006616

Zhang H, Zhang C, Li Z et al (2019) A new method of RNA secondary structure prediction based on convolutional neural network and dynamic programming. Front Genet 10. https://doi.org/10.3389/fgene.2019.00467

Zhang S-W, Zhang X-X, Fan X-N, Li W-N (2020a) LPI-CNNCP: prediction of lncRNA-protein interactions by using convolutional neural network with the copy-padding trick. Anal Biochem 601:113767. https://doi.org/10.1016/j.ab.2020.113767

Zhang T, Wang M, Xi J, Li A (2020b) LPGNMF: predicting long non-coding RNA and protein interaction using graph regularized nonnegative matrix factorization. IEEE/ACM Trans Comput Biol Bioinform 17:189–197. https://doi.org/10.1109/TCBB.2018.2861009

Zhang Y, Wang J, Xiao Y (2022) 3dRNA: 3D structure prediction from linear to circular RNAs. J Mol Biol 434:167452. https://doi.org/10.1016/j.jmb.2022.167452

Zhao J, Song X, Wang K (2016) lncScore: alignment-free identification of long noncoding RNA from assembled novel transcripts. Sci Rep 6:34838. https://doi.org/10.1038/srep34838

Zhao Q, Zhang Y, Hu H et al (2018) IRWNRLPI: integrating random walk and Neighborhood regularized logistic matrix factorization for lncRNA-protein interaction prediction. Front Genet 9:239. https://doi.org/10.3389/fgene.2018.00239

Zhao L, Wang J, Li Y et al (2021a) NONCODEV6: an updated database dedicated to long non-coding RNA annotation in both animals and plants. Nucleic Acids Res 49:D165–D171. https://doi.org/10.1093/nar/gkaa1046

Zhao S, Zhang X, Chen S, Zhang S (2021b) Long noncoding RNAs: fine-tuners hidden in the cancer signaling network. Cell Death Discov 7:283. https://doi.org/10.1038/s41420-021-00678-8

Zhao G, Li P, Qiao X et al (2022) Predicting lncRNA–protein interactions by heterogenous network embedding. Front Genet 12:814073. https://doi.org/10.3389/fgene.2021.814073

Zheng S, Yang L, Zou Y et al (2020) Correction to: long non-coding RNA HUMT hypomethylation promotes lymphangiogenesis and metastasis via activating FOXK1 transcription in triple-negative breast cancer. J Hematol Oncol 13:23. https://doi.org/10.1186/s13045-020-00861-x

Zheng Y, Luo H, Teng X et al (2023) NPInter v5.0: ncRNA interaction database in a new era. Nucleic Acids Res 51:D232–D239. https://doi.org/10.1093/nar/gkac1002

Zhou Y-K, Hu J, Shen Z-A et al (2020) LPI-SKF: predicting lncRNA-protein interactions using similarity kernel fusions. Front Genet 11:615144. https://doi.org/10.3389/fgene.2020.615144

Zhuo L, Song B, Liu Y et al (2022) Predicting ncRNA–protein interactions based on dual graph convolutional network and pairwise learning. Brief Bioinform 23:bbac339. https://doi.org/10.1093/bib/bbac339

Zuker M, Mathews DH, Turner DH (1999) Algorithms and thermodynamics for RNA secondary structure prediction: a practical guide. In: RNA biochemistry and biotechnology. Springer, Dordrecht, pp 11–43

Chapter 3
Long Non-coding RNAs in Stem Cells and Regeneration

Eduardo Pérez-Campos, Daniel Scott-Algara, María Teresa Hernández-Huerta, Laura Pérez-Campos Mayoral, Socorro Pina-Canseco, Carlos Romero-Díaz, Holanda Isabel Cruz-Luis, Gloria Rubí Sánchez Hernández, Miriam Emily Avendaño-Villegas, José Luis Hernández-Morales, Héctor Alejandro Cabrera-Fuentes, Edgar Ramos-Martínez, Iván Ramos-Martínez, Eduardo Pérez-Campos Mayoral, Margarito Martínez-Cruz, Eli Cruz-Parada, and Carlos Lastre-Domínguez

Abstract Long non-coding RNAs (lncRNAs) orchestrate stem cell self-renewal, differentiation, and pluripotency through chromatin remodeling and transcriptional control. These regulatory transcripts govern the fate of both pluripotent populations—including embryonic, epiblast, and induced pluripotent stem cells—and tissue-specific lineages such as hematopoietic, neuronal, and mesenchymal progenitors. Mechanistically, lncRNAs operate as molecular signals, decoys, scaffolds, or competing endogenous RNAs. By sequestering microRNAs, they exert precise spatiotemporal control over stem cell fate across diverse physiological landscapes,

---

E. Pérez-Campos (✉)
National Institute of Technology of Mexico/IT.(TecNM) Oaxaca, Oaxaca, Mexico

"Dr. Eduardo Pérez Ortega" Clinical Pathology Laboratory, Oaxaca, Mexico
e-mail: pcampos@itoaxaca.edu.mx

D. Scott-Algara
Unité de Biologie Cellulaire des Lymphocytes and Direction of International Affairs, Institut Pasteur, Paris, France

M. T. Hernández-Huerta · H. I. Cruz-Luis · G. R. Sánchez Hernández · J. L. Hernández-Morales
SECIHTI, Faculty of Medicine and Surgery, Benito Juárez Autonomous University of Oaxaca, Oaxaca, Mexico

L. Pérez-Campos Mayoral · S. Pina-Canseco · E. Ramos-Martínez · E. Pérez-Campos Mayoral
UNAM-UABJO Faculty of Medicine Research Centre, Faculty of Medicine and Surgery, Benito Juárez Autonomous University of Oaxaca, Oaxaca, Mexico

L. M. Vaschetto (ed.), *Long Non-Coding RNAs, Human Health, and Diseases*, Epigenetics and Human Health 16, https://doi.org/10.1007/978-3-032-19668-2_3

including the musculoskeletal, nervous, and cardiovascular systems. Dysregulation of these non-coding networks drives a spectrum of pathologies. In cardiovascular disease, aberrant lncRNA expression modulates angiogenesis and plaque stability, while in type 2 diabetes, it exacerbates insulin resistance and oxidative stress. Furthermore, these RNAs are implicated in neurodegenerative conditions like Alzheimer's and Parkinson's disease, where they influence neuroinflammation and protein homeostasis. Beyond their roles in degeneration, lncRNAs serve as diagnostic biomarkers for bone disorders and critical drivers of malignancy. In cancer stem cells, they sustain "stemness" and therapy resistance by modulating key signaling cascades, including the Wnt/β-catenin and Notch pathways. Consequently, lncRNAs represent promising targets for novel diagnostic and therapeutic interventions spanning regenerative medicine to precision oncology.

Keywords Stem cells · Long non-coding RNAs · Tissue regeneration · Gene regulation · Differentiation · Pluripotency · Cancer stem cells · Cardiovascular diseases · Neurodegenerative diseases · Therapeutic targets

Abbreviations

AD	Alzheimer's disease
ADD3	γ-Adducin
ADSCs	Adipose-derived mesenchymal stem cells
AFP	Alpha-fetoprotein
ALS	Amyotrophic lateral sclerosis
ANCR	Anti-differentiation non-coding RNA
ANRIL	Antisense non-coding RNA in the INK4 locus
APC	Adenomatous polyposis coli
ASOs	Antisense oligonucleotides

C. Romero-Díaz · M. E. Avendaño-Villegas · M. Martínez-Cruz · E. Cruz-Parada · C. Lastre-Domínguez
National Institute of Technology of Mexico/IT.(TecNM) Oaxaca, Oaxaca, Mexico

H. A. Cabrera-Fuentes
National Institute of Technology of Mexico/IT.(TecNM) Oaxaca, Oaxaca, Mexico

UNAM-UABJO Faculty of Medicine Research Centre, Faculty of Medicine and Surgery, Benito Juárez Autonomous University of Oaxaca, Oaxaca, Mexico

R&D Group, Vice Presidency for Scientific Research and Innovation Imam Abdulrahman Bin Faisal University Dammam, Dammam, Saudi Arabia

National Institute of Technology of Mexico/IT.(TecNM), Tijuana, Baja California, Mexico

I. Ramos-Martínez
Mycology Unit, Department of Microbiology and Parasitology, Faculty of Medicine, National Autonomous University of Mexico, Mexico City, Mexico

BCR	B-cell receptor
B-III-TUB	Beta-III-tubulin
BM-MSCs	Bone marrow-derived mesenchymal stem cells
BMP	Bone morphogenetic proteins
ceRNA	Competing endogenous RNA
circRNAs	Circular RNAs
CMPs	Common myeloid progenitor
CRC	Colorectal cancer
CSCs	Cancer stem cells
CUDR	Cancer-upregulated drug-resistant
CVDs	Cardiovascular diseases
DECs	Definitive endoderm cells
DLL4	Delta-like ligand 4
DOS	Delta and OSM-11-like proteins
DSL	Delta/Serrate/LAG-2
EMT	Epithelial-mesenchymal transition
EPB41L4A-AS1	Erythrocyte membrane protein band 4.1 like 4A antisense 1
EpiSCs	Epiblast stem cells
ESCs	Embryonic stem cells
Gas5	Growth-arrest-specific transcript 5
GBM	Glioblastoma
GCP	Cerebellar granule progenitor cells
GSK-3β	Glycogen synthase kinase-3β
hAMSCs	Human adipose-derived mesenchymal stem cells
HCC	Hepatocellular carcinoma
HD	Huntington's disease
HOTAIR	HOX transcript antisense RNA
HOTAIRM1	Myeloid HOX antisense intergenic RNA 1
HPSC	Human pluripotent stem cells
HSCs	Hematopoietic stem cells
HSPC	Hematopoietic stem cells and progenitor cells
ICM	Inner cell mass
ID2	Inhibitor of DNA binding 2
IL-6	Interleukin 6
iPSCs	Induced pluripotent stem cells
lncRNAs	Long non-coding RNAs
LT-HSCs	Long-term hematopoietic stem cells
MALAT1	Metastasis-associated lung adenocarcinoma transcript 1
MEPs	Megakaryocyte-erythroid progenitor
mESC	Mouse embryonic stem cell
MI	Myocardial infarction
MIAT	Myocardial infarction-associated transcript
miRNAs	microRNAs
MK	Megakaryocyte
MKPs	Megakaryocyte progenitor

MPP	Multipotent progenitor
MSCs	Mesenchymal stem cells
ncRNAs	Non-coding RNAs
NEAT1	Nuclear paraspeckle assembly transcript 1
Nrf2	Nuclear factor-erythroid 2 p45-related factor 2
OA	Osteoarthritis
OCT4	Octamer binding transcription factor 4
OP	Osteoporosis
OS	Osteosarcoma
PD	Parkinson's disease
PRRs	Pattern-recognition receptors
PSCs	Pluripotent stem cells
RBPs	RNA-binding proteins
RMST	Rhabdomyosarcoma 2-associated transcript
RNP	Ribonucleoprotein
ROS	Reactive oxygen species
Shh	Sonic hedgehog
SMA	Smooth muscle actin
SNCA-AS1	Synuclein alpha antisense 1
ST-HSCs	Short-term HSC
T2DM	Type-2 diabetes mellitus
TCR	T-cell receptor
TLRs	Toll-like receptors
TNFR	NF receptor
TSSCs	Tissue-specific stem cells
TUG1	Taurine-upregulated gene 1

3.1 Introduction

Stem cells define tissue homeostasis through two intrinsic properties: Self-renewal maintains the stem cell pool throughout the organism's lifespan. This process occurs via two distinct mechanisms: symmetric division, which generates two identical daughter stem cells, and asymmetric division, which produces one stem cell and one differentiated progenitor (Shahriyari and Komarova 2013). However, the extent of this regenerative potential is lineage-dependent. For example, long-term hematopoietic stem cells (LT-HSCs) renew themselves strongly and can rebuild blood and immune systems even after several transplants (Sakurai et al. 2024).

While self-renewal maintains the stem cell pool, differentiation enables unspecialized progenitors to acquire distinct phenotypes and specialized functions. However, the capacity for differentiation varies significantly across the stem cell hierarchy, necessitating a classification based on developmental potential. Consequently, stem cells are categorized into two primary classes: pluripotent stem cells (PSCs) and tissue-specific stem cells (TSSCs). PSCs, characterized by their ability to generate any somatic cell lineage, include embryonic stem cells (ESCs) derived from the inner cell mass of the blastocyst, epiblast stem cells (EpiSCs), and induced pluripotent stem cells (iPSCs) (Takahashi and Yamanaka 2006; Nichols and Smith 2009; Iltis 2023). In contrast, TSSCs are multipotent and can only form cell types from their tissue of origin. A prime example is the bone marrow mesenchymal stem cell (BM-MSC), which differentiates into osteoblasts, chondrocytes, and adipocytes but lacks the capacity to generate unrelated lineages such as neurons (Kondoh 2024). Due to their high proliferation rates, both human pluripotent stem cells (HPSCs) and mesenchymal stem cells (MSCs) are critical for regenerative medicine applications ranging from cardiovascular disease to arthritis (Hoang et al. 2022).

Stem cell function is strictly governed by a specialized microenvironment known as the niche. Within this specialized microenvironment, exosomes facilitate intercellular communication, regulating critical processes such as inflammation, apoptosis, and fibrosis (Scadden 2014; Imam et al. 2024). Integral to this signaling network are telocytes—interstitial cells characterized by long, narrow processes. These cells maintain homeostasis and promote tissue repair by establishing direct contact with neighboring cells and delivering signaling vesicles containing microRNAs (miRNAs), lncRNAs, and genomic DNA (Babadag and Celebi-Saltik 2023). Dysregulation of this niche homeostasis is inextricably linked to pathologies such as cancer (Sonawala et al. 2022). Consequently, this review examines how lncRNA-stem cell interactions within these niches drive disease progression and evaluates their potential as therapeutic targets.

3.2 lncRNAs in the Regulation of Stem Cell Functions

Non-coding RNAs (ncRNAs) play a key role in deciding cell lineages and intercellular communication, often mediating their effects via extracellular vesicles (Fatima et al. 2017). Among them, lncRNAs are critical for keeping pluripotency and guiding differentiation (Aich and Chakraborty 2020). LncRNAs modulate cellular reprogramming through diverse mechanisms, ranging from direct transcriptional control to epigenetic remodeling. For instance, RoR is critical for the induction of pluripotency, acting as a regulator of the core transcription factors. In contrast, Gas5 governs self-renewal in mouse embryonic stem cells through an epigenetic mechanism, specifically by inducing DNA demethylation to reshape the chromatin landscape (Tu et al. 2018). Many transcripts help maintain pluripotency. In mouse models, lncRNAs such as *AK028326* and *AK141205* have been identified as key pluripotency factors (Sheik Mohamed et al. 2010). In the context of iPSCs, epigenetic modulation is paramount; transcripts like *Osilr9* and *Oplr16* function as chromatin modulators to sustain the pluripotent state (Zhu et al. 2023b; Jia et al. 2020). Furthermore, a coordinated network including *FAST*, *Cyrano*, *linc1614*, and *LincU* ensures the maintenance of naive pluripotency, preventing premature differentiation (Jiapaer et al. 2018; Guo et al. 2018b).

In TSSCs, lncRNAs drive lineage commitment through precise molecular interactions. In the neural niche, for example, the lncRNAs *ES1* and *ES2* interact with the transcription factor *SOX2* and the polycomb repressive complex 2 (*SUZ12*) to regulate neuronal differentiation (Ng et al. 2012). SOX2 activity is further modulated by RMST, a transcriptional co-regulator required for SOX2 recruitment to the promoters of neurogenic effectors (Ng et al. 2012). In MSCs, Bvht promotes differentiation into cardiovascular progenitor cells through the transcription factor *Mesp1* (Hou et al. 2017), while OG promotes osteogenesis by interacting with heterogeneous nuclear ribonucleoprotein K (hnRNPK) and regulating bone morphogenetic protein (BMP) expression (Tang et al. 2019). The lncRNA *MALAT1* regulates osteogenic differentiation in bone MSCs by promoting special AT-rich sequence-binding protein 2 (SATB2) expression through the miR-34c/SATB2 axis (Yang et al. 2019a) and acting as a competing endogenous RNA for *miR-129-5p* (Lee and Hong 2020). This demonstrates that lncRNAs regulate stem cell fate decisions in different tissue contexts through diverse molecular mechanisms.

3.2.1 miRNA-lncRNA Interactions in Hematopoiesis

Hematopoiesis is the ongoing process of blood cell production from HSCs. This group is very diverse. LT-HSCs reside at the apex of the hierarchy, possessing the capacity for indefinite self-renewal while producing all blood lineages (Lee and Hong 2020). Recent lineage-tracing studies have further stratified this pool, identifying variants such as Vwf-tdTomato+ HSCs, which exhibit a bias toward platelet generation, contrasting with Vwf-tdTomato− HSCs that replenish all myeloid and lymphoid lineages in a balanced manner (Carrelha et al. 2024). Hematopoiesis precursors include LT-HSCs, short-term HSC (ST-HSCs), multipotent progenitor (MPP)-1, MPP (mainly MPP2), common myeloid progenitor (CMPs), megakaryocyte-erythroid progenitor (MEPs), and megakaryocyte progenitor (MKPs) (Noetzli et al. 2019).

miRNAs are small RNA molecules that lack protein-coding potential. They control gene expression by binding to mRNA, blocking translation, or causing mRNA breakdown. In hematopoiesis, miRNAs function as master regulators, controlling the commitment, proliferation, and differentiation of HSCs into all mature blood cells. Their expression levels can increase or decrease at specific stages, targeting proto-oncogenes to either inhibit or stimulate different blood cell lineages such as myeloid and erythroid cells (Nassiri et al. 2023).

Essential miRNAs in hematopoiesis include *miR-223*, a positive modulator of granulocytic differentiation whose expression increases significantly during this process, regulating a feedback loop with transcription factors *NFI-A* and *C/EBPalpha* to promote granulocyte differentiation (Fazi et al. 2005). The miRNA *miR-150* determines megakaryocyte-erythrocyte progenitor (MEP) fate by targeting the *Myb* proto-oncogene; decreased *miR-150* levels, accompanied by increased Myb, promote erythroid cell differentiation (Lu et al. 2008). The *miR-144/451* cluster directly inhibits the proto-oncogenic protein Myc to promote erythroid differentiation and enucleation while protecting erythroid cells from oxidative stress by repressing the Ywhaz/14-3-3 zeta protein (Yu et al. 2010). Other miRNAs participate in hematopoietic lineage precursors, including the *miR-23a/27A/24-2* cluster and *miR-155* in HSCs (Emmrich et al. 2012; Lu et al. 2008), and *miR-26b-5p* in mature platelets (Pedersen et al. 2024). Platelet function involves *miR-223*, *miR-126*, *miR-96*, *miR-200b*, *miR-495*, and *miR-107* (Parra-Izquierdo et al. 2020; Li et al. 2023b).

LncRNAs introduce a complex level of regulation to this network, primarily by modulating miRNA bioavailability. They function through two principal mechanisms: the direct suppression of miRNA processing or the competitive sequestration of mature miRNAs (the ceRNA effect). In the former scenario, transcripts like lncRNA-PFAR bind to the precursor of *miR-141*, inhibiting its maturation and reducing downstream *miR-141* levels (Zhang et al. 2021d). In the latter, lncRNAs act as molecular sponges, harboring binding sites for specific miRNAs and thereby preventing them from repressing their mRNA targets (Xiao et al. 2018; Li et al. 2024a).

lncRNAs in HSCs include HOTAIR, which acts through inhibition of p15 expression (Gao et al. 2018), *HOXa11-AS* (Horvat-Switzer and Thompson 2005), and *NORAD* (Wang et al. 2024). Members of the HOXA group have been linked to the emergence and maintenance of long-term repopulating hematopoietic stem cells, and dysregulation of HOXA genes—either directly or through genetic lesions to upstream regulators—is associated with malignant phenotypes (Abuhantash et al. 2021). lncRNAs also act on HSPCs in the bone marrow niche and regulate the hematopoietic stem cell pool through trogocytosis, a process in which a cell physically extracts and ingests cellular material from another cell (Gao et al. 2024b). In MSCs, osteoblasts, and endothelial cells, lncRNAs can act as surrogate niche components in the bone marrow (Tanaka et al. 2023). Different lncRNAs participate at each stage of cell proliferation and maturation, with specific examples associated with megakaryocytopoiesis precursors detailed in Table 3.1.

3.3 Mechanisms of Tissue Repair and Regeneration

3.3.1 General Mechanisms

Tissue regeneration is the natural process of repairing or replacing damaged tissue through stem cell activation, cell proliferation, differentiation, and migration. In this process, lncRNAs contribute by recruiting RNA-binding proteins (RBPs) to modulate mRNA stability, regulating epigenetic modifications, controlling signaling pathways, influencing splicing, and directing subcellular localization. LncRNAs function through diverse molecular modalities. In the nucleus, they act as epigenetic scaffolds; for instance, *Oct4P4* recruits the methyltransferase SUV39H1 to silence the *Oct4* promoter. In the cytoplasm, they interact with RBPs to modulate mRNA stability, as seen when ANCR binds PTBP1 to stabilize *ID2* mRNA.

Table 3.1 lncRNAs involved in megakaryocytopoiesis

Cells	lncRNA	Main function
Hematopoietic stem cells (HSCs). HSCs are multipotent cells capable of self-renewal and differentiation into all blood cells	lncRNA Xist	Studied in hematologic malignancies (Noetzli et al. 2019)
	lncRNA H19	Participates in the embryonic development of HSCs (Zhou et al. 2019)
	lncRNA myeloid HOX antisense intergenic RNA 1 (HOTAIRM1)	Is one of the lncRNAs highly upregulated in hiPSC-HSC-derived MKs (Dahariya et al. 2023)
	lncRNA EGO	Participates in the differentiation of hematopoietic stem cells (Ghafouri-Fard et al. 2021b)
	lincRNA-EPS	Participates in the suppression of apoptotic/inflammatory response genes (Alvarez-Dominguez and Lodish 2017)
	HoxBlinc	Involved in the regulation of HOXA/B gene transcription (Zhu et al. 2021)
	lncRNA Spehd	Involved in the regulation of hematopoietic stem and progenitor cells (Delas et al. 2019)

(continued)

Table 3.1 (continued)

Cells	lncRNA	Main function
2. *HSC variants*, with a tendency toward MK (Vwf-tdTomato+ HSC)	–	–
3. *Megakaryocyte (MK) progenitor cells*. These specialized cells undergo endomitosis, a DNA replication process that occurs without cell division, leading to the formation of polyploid cells with multiple copies of their chromosomes	lncRNA AS-RBM15	Participates in MK differentiation (Noetzli et al. 2019, Tran et al. 2016)
	DANCR	Is highly expressed in the MK-erythroid precursor (Ren et al. 2021)
4. *Immature MKs*. These cells continue to mature. They develop a system of internal membranes and granules for platelet function	Nuclear paraspeckle assembly transcript 1 (lncRNA NEAT1)	Has multiple functions, such as participating in MK differentiation and thrombopoiesis (Bian et al. 2020)
	AS-RBM15	Participates in MK differentiation (Tran et al. 2016)
5. *Mature MKs*. These are large polyploid cells that extend proplatelet extensions into the bone marrow sinusoids	JPX lncRNA	Participates in differentiation or maturation and can act as an miRNA sponge (Dahariya et al. 2021)

3.3.2 Bone and Cartilage Regeneration

In the context of osteochondral regeneration, lncRNAs act as binary switches, either promoting or inhibiting differentiation. Some lncRNAs like *SNHG1* help cartilage repair in jawbone marrow mesenchymal stem cells by regulating signal transducer and activator of transcription 3 (STAT3) phosphorylation and suppressing oxidative stress (Liu et al. 2024). Similarly, *UCA1* and *DANCR* have been identified as pro-chondrogenic factors in synovium-derived MSCs, driving proliferation and matrix synthesis (Zhang et al. 2017b; Shu et al. 2019). Conversely, transcripts such as *Meg3* suppress this process, inhibiting chondrogenic differentiation. Maintaining this equilibrium is critical for tissue homeostasis.

In bone regeneration, *LINC00665* inhibits proliferation and promotes apoptosis of bone marrow mesenchymal cells; reducing its expression ameliorates joint damage and attenuates osteoarthritis progression, suggesting therapeutic potential (Chen et al. 2024). *LINC00707* promotes bone regeneration in human bone marrow-derived mesenchymal stem cells by binding *miR-370-3p* and upregulating Wnt family member 2B (*WNT2B*) (Jia et al. 2019). *HRL-SC*, a hypoxia-sensitive lncRNA, promotes proliferation and migration of human dental pulp stem cells via the PI3K/AKT signaling pathway, with applications in pulp regeneration (Zeng et al. 2022). In contrast, *HOXA10* inhibits osteogenesis in human periodontal ligament stem cells by downregulating β-catenin and increasing Dickkopf Wnt signaling pathway inhibitor 1 (DKK1) expression (Wang 2021a).

Several lncRNAs demonstrate potential as biomarkers for osteoarthritis; for example, serum expression of *LINC00305* and its variant *rs2850711* (A/T) increases in osteoarthritis patients and correlates with pathology severity (Wahba et al. 2020). *MIR22HG* and *ENST00000619282* show significantly higher expression in osteoarthritis patients compared to healthy controls, while *DSCR9*, *LINC01189*, and *MAPKAPK5-AS1* exhibit lower expression (Wen et al. 2020). *ITSN1-2* is upregulated in plasma and synovial tissue in osteoarthritis and positively correlates with nucleotide-binding oligomerization domain-containing protein 2 (*NOD2*) mRNA, both associated with osteoarthritis inflammation (Yue et al. 2019). *TSPEAR-AS2* is significantly downregulated in osteoarthritis patient plasma and inversely correlates with *miR-212-3p* expression (Lv et al. 2022). *AL928768.3* and *AC091493.1* show high synovial expression in osteoarthritis patients, serving as potential biomarkers for disease risk and activity (Sun et al. 2020). *LINK-A* positively regulates aggression and inflammation through two independent pathways, with higher expression in osteoarthritis synovial tissue (Wang et al. 2021b).

Other lncRNAs demonstrate functional complexity across cellular niches through intricate regulatory networks with miRNAs. *MCM3AP-AS1* exhibits dual functionality, mediating cartilage repair in osteoarthritis and regulating osteogenic differentiation of dental pulp stem cells through the miR-143-3p/IGFBP5 axis (Yang et al. 2022a). ZBED3-AS1 activates Wnt/β-catenin signaling to promote chondrogenesis in human synovial fluid-derived MSCs, highlighting therapeutic relevance for osteoarthritis (Ou et al. 2017).

3.3.3 *Nervous System Regeneration*

Nervous system regeneration requires stem cells to become neural precursors, oligodendrocytes, astrocytes, and neurons. A specific group of lncRNAs controls this process, including *Neat1*, *Gas5*, and *TUG1* (Tavares et al. 2024). Functionally, these transcripts modulate the complex signaling environments required for axonal regeneration and the bridging of lesion sites. For instance, interferon gamma antisense 1 (*IFNG-AS1*) promotes neurogenesis via the *miR-21a*-3p/PI3K/AKT signaling axis (Fu et al. 2024), while the downregulation of *H19* in induced neural stem cells has been shown to accelerate neuronal differentiation (Gao et al. 2024a). Beyond lncRNAs, other regulatory mechanisms drive regeneration. In axolotls, spinal cord regeneration after injury involves developmental signaling pathways where inhibiting *miR-200a* upregulates the mesodermal marker brachyury in spinal cord cells while maintaining neural stem cell marker sox2 expression, with in vivo tracking revealing subsequent differentiation into neural and mesodermal lineages (Walker et al. 2022).

3.3.4 Cardiovascular Regeneration

LncRNAs also control heart development, health, and disease remodeling through different mechanisms including epigenetic remodeling, transcriptional regulation, and post-transcriptional processing (Wang et al. 2021c; Chini et al. 2023). In the injured heart, lncRNAs reactivate developmental programs to support myocardial repair, influence inflammatory responses, mediate apoptosis, and control fibrosis. Specific lncRNAs work with chromatin-modifying complexes to regulate cardiac gene transcription (Dahariya et al. 2021) or function as molecular sponges for miRNAs, thereby fine-tuning proteins crucial for cell survival and stress responses (Ratti et al. 2020). During cardiogenesis, lncRNAs act as molecular scaffolds or decoys, directing chromatin modifiers to specific genomic loci to either activate or repress cardiac-specific genes (Alexanian and Ounzain 2020; Zhang et al. 2020). Furthermore, certain lncRNAs interact directly with transcription factors such as GATA binding protein 4 (GATA4) and NK2 homeobox 5 (NKX2-5), promoting progenitor cell differentiation into mature cardiomyocytes (Huang et al. 2023). Representative examples are summarized in Table 3.2.

LncRNAs derived from MSCs have demonstrated significant cardioprotective effects. For instance, high expression of *HAND2-AS1* in bone marrow MSCs protects cardiomyoblasts (Li et al. 2023a), while *UCA1* from human umbilical cord MSCs protects cardiac microvascular endothelial cells from hypoxia/reoxygenation injury (Diao and Zhang 2021). In the context of myocardial infarction (MI), MSC-derived extracellular vesicles (EVs) play a crucial role. Specifically, *A2M-AS1* carried within MSC vesicles attenuates myocardial ischemia-reperfusion-induced

Table 3.2 lncRNAs involved in the regulation of cardiovascular function

Mechanism	lncRNA	Role in cardiovascular regulation
Epigenetic modulation	HOTAIR (HOX transcript antisense RNA)	Recruits PRC2 to modify the chromatin state of cardiac genes, resulting in context-dependent gene repression or activation (Mattick et al. 2023)
Transcriptional regulation	MIAT (myocardial infarction-associated transcript)	Interacts with transcription factors to modulate the expression of genes critical for cardiac function, particularly under conditions of stress or injury (Xie et al. 2025)
Post-transcriptional control	MALAT1 (metastasis-associated lung adenocarcinoma transcript 1)	Functions as a competing endogenous RNA to sequester miRNAs, regulating mRNA stability and translation in cardiac cells (Yoon et al. 2013)

cardiomyocyte apoptosis via the miR-556-5p/XIAP axis (Yu et al. 2023). Furthermore, upregulation of transcription factor 21 (Tcf21) achieved through MSC-derived EVs and TARID-loaded lipid nanoparticles promotes cardiac repair (Zhu et al. 2023a). Other prominent lncRNAs also confer cardioprotection: *NEAT1* protects cardiomyocytes against apoptosis; *MALAT1* from human umbilical cord MSCs blocks aging-induced cardiac dysfunction; *H19* overexpression is associated with reduced cardiomyocyte apoptosis, decreased IL-6 and TNF-α levels, and improved cardiac function and reduced infarct size; and Kruppel-like factor 3 antisense RNA 1 (*KLF3-AS1*) overexpression mitigates cell apoptosis, pyroptosis, and MI size (Huang 2020).

3.3.5 Skin and Wound Healing

The MSC-derived exosomal lncRNA *H19* promotes healing across multiple tissue contexts. In diabetic foot ulcers, *H19* increases phosphatase and tensin homolog (PTEN) expression by binding *miR-152-3p* (Li et al. 2020a). Similarly, in diabetic mouse skin, *H19* derived from hair follicle mesenchymal stem cells accelerates wound healing (Yang et al. 2023). In the context of cartilage repair, *H19* from umbilical cord MSCs acts as a ceRNA against *miR-29b-3p*, upregulating Forkhead box O3 (FoxO3) to promote post-traumatic focal repair (Yan et al. 2021). Regarding other transcripts, *HOTAIR* can improve burn wound healing by modulating epidermal stem cells, with increased levels observed in wound tissue during repair (Shi et al. 2020). Finally, *AK015322* induces *miR-21* expression to promote the proliferation and differentiation of hair follicle stem cells via the PI3K/AKT signaling pathway, demonstrating potential for tissue-engineering applications in injured skin (Cai et al. 2018).

3.4 lncRNAs in Pathology

While lncRNAs drive homeostatic regeneration, their dysregulation underpins the pathogenesis of complex diseases. In conditions such as myocardial infarction and diabetes, the regenerative capacity of stem cells is impaired or overwhelmed. Conversely, in cancer, the self-renewal pathways essential for regeneration are co-opted to drive malignancy.

3.4.1 Myocardial Infarction and Ischemia-Reperfusion Injury

ncRNAs exert stage-specific regulatory roles in myocardial infarction recovery, influencing angiogenesis, tissue repair, and immunomodulation. During the acute phase, rapid activation of reparative mechanisms limits myocardial damage (Ong et al. 2018), with lncRNAs contributing to both inflammatory and reparative processes (Noetzli et al. 2019). Key reparative functions include promoting angiogenesis to restore cardiac perfusion. For example, *miR-21-5p* enhances vascular formation by upregulating vascular endothelial growth factor A (VEGFA) and platelet-derived growth factor subunit B (PDGF-BB) in endothelial cells, while *MALAT1* stimulates endothelial cell proliferation and angiogenesis via increased S-phase cyclin expression. ncRNA-modulated stem cells lead to angiogenesis at the infarct site, improving cardiac oxygenation. Preclinical studies demonstrate that therapeutic modulation of these lncRNAs can mitigate ischemia-reperfusion injury, potentially reducing infarct size and preserving cardiac function (Niu et al. 2024).

During the subacute phase, ncRNAs facilitate cardiac tissue repair (Zhuang et al. 2024), enhance differentiation of transplanted stem cells into cardiomyocytes, and provide cardioprotection. For instance, *miR-210* reduces mitochondrial oxygen consumption and increases glycolytic activity, while heart-specific lncRNA *NR_045363* promotes cardiomyocyte proliferation and improves cardiac function. In the chronic phase, where irreversible damage often occurs, ncRNAs induce stem cells to exert immunomodulatory effects, mitigating post-transplant inflammation. Specifically, anti-miRNA-204-3p upregulates C-X-C motif chemokine receptor 4 (CXCR4) expression in mesenchymal stem cells, enhancing their proliferation and migration while promoting an anti-inflammatory macrophage phenotype in cardiac tissue (Kaikkonen et al. 2011).

3.4.2 *Atherosclerosis*

Atherosclerosis is a chronic inflammatory condition marked by lipid accumulation, endothelial dysfunction, and atherosclerotic plaque formation (Zhao et al. 2012). lncRNAs control several key steps in this disease; for instance, *H19* modulates lipid metabolism, inflammatory signaling, and smooth muscle cell behavior, with therapeutic targeting potentially stabilizing plaques and mitigating adverse cardiovascular outcomes (Sopic et al. 2025). Dysregulated expression of *Gas5*, *MAARS*, *MALAT1*, *MIAT*, and *ZNF800* is associated with plaque instability and increased risk of cardiovascular events. Aberrant *MALAT1* and *MIAT* expression correlates with endothelial dysfunction and vascular remodeling, while *Gas5* participates in pathways influencing cell apoptosis and inflammation—critical factors in plaque destabilization (Petkovic et al. 2023; Huang et al. 2021).

3.4.3 *Heart Failure*

In heart failure, dysregulated lncRNA expression contributes to maladaptive cardiac remodeling, fibrosis, and impaired myocardial contractility (He et al. 2022; Mably and Wang 2024). lncRNAs modulate critical signaling pathways, including transforming growth factor beta (TGF-β), Wnt, and mitogen-activated protein kinase (MAPK), which are central to the fibrotic response and cellular stress adaptation (Shin et al. 2024; Han et al. 2020; Zhang et al. 2023).

3.4.4 Type-2 Diabetes Mellitus

LncRNAs play a multifaceted role in the pathogenesis of type-2 diabetes mellitus (T2DM) by modulating insulin resistance, oxidative stress, inflammation, cell death, fibrosis, and vascular proliferation (Geng et al. 2024). Clinically, T2DM patients exhibit distinct lncRNA profiles; for instance, elevated *Meg3* expression and reduced *H19* expression are linked to complications such as retinopathy and neuropathy, as well as renal indicators like urea, creatinine, and the urine albumin-creatinine ratio (Alrefai et al. 2023). Genetic variations also lead to disease risk. Mice expressing the mutant lncRNA *Mut-Reg1cp*, for example, develop insulin resistance and beta-cell dysfunction. Mechanistically, *Mut-Reg1cp* is transported via pancreatic islet-derived exosomes to peripheral tissues, where it inhibits adiponectin receptor 1 (*AdipoR1*) translation, thereby promoting insulin resistance (Clevers et al. 2014).

Oxidative stress is a critical driver of T2DM, contributing to insulin resistance and beta-cell dysfunction by increasing reactive oxygen species (ROS) generation, which in turn promotes inflammation and vascular damage (Eguchi et al. 2021). LncRNAs are key regulators of this process. *MALAT1*, for example, modulates oxidative stress by interacting with nuclear factor-erythroid 2 p45-related factor 2 (Nrf2), preventing it from upregulating antioxidant genes such as *Nqo1* and *Cat*; consequently, *MALAT1* ablation has been shown to enhance insulin secretion and insulin receptor substrate 1 (IRS-1) activation (Tran et al. 2016). Furthermore, mitochondria-associated lncRNAs influence T2DM pathogenesis by governing ROS generation, ATP production, and mitochondrial dynamics (Durr et al. 2022). Specific interactions include *Kcnq1ot1* binding with *miR-378a* in mitochondria to increase ATP synthase expression; blocking this interaction ameliorates cardiac dysfunction in murine T2DM models (Shu et al. 2019). Similarly, *Meg3* overexpression has been shown to alleviate cognitive dysfunction and reduce neuroinflammation in diabetic rats by inhibiting ras-related C3 botulinum toxin substrate 1 (*Rac1*) expression and mitigating mitochondrial-associated apoptosis (Wang et al. 2021d).

3.4.5 Neurodegenerative Diseases

lncRNAs regulate gene expression, inflammatory response, oxidative stress, protein homeostasis, and miRNA interactions in neurodegenerative diseases, including Alzheimer's disease (AD), Parkinson's disease (PD) (Zhao et al. 2024), amyotrophic lateral sclerosis (ALS) (Li and Wang 2023; Rajabi et al. 2024), and

Huntington's disease (HD) (Qian et al. 2019; Wang et al. 2021a). In AD, *BACE1-AS* stabilizes BACE1 gene transcription, increasing beta-amyloid production (Li et al. 2019). The lncRNAs *p21* and *Meg3* reduce neuroinflammation, while *SOX21-AS1* promotes oxidative stress (Canseco-Rodriguez et al. 2022). *MALAT1* plays a neuroprotective role by regulating pathways associated with synapse and neuronal survival (Li et al. 2020b). In PD, *SNHG1* regulates the degradation of misfolded proteins (Li et al. 2020b), while *HOTAIR* modulates neuroinflammation and susceptibility to oxidative stress in dopaminergic degeneration (Zhang et al. 2021c). In ALS, *NEAT1* modulates miRNAs related to oxidative stress and inflammation (Nishimoto et al. 2021), and *lnc-ABCA12-3* is associated with faster ALS progression and shorter survival (Yu et al. 2024b). In HD, *HTT-AS* regulates mutant HTT expression (Tsagakis et al. 2020).

3.4.6 Senescence, lncRNAs, and Associated Diseases

Aging affects stem cell homeostasis, reducing renewal and proliferation capacity (Jamieson and Weissman 2023). During aging, decreased expression of certain pluripotency-associated lncRNAs contributes to progressive loss of regenerative capacity and tissue deterioration. Some lncRNAs activate pathways related to cellular senescence, including proinflammatory processes and genome destabilization (Cabrera-Fuentes and Boisvert 2025). *ANRIL* (antisense ncRNA in the INK4 locus) regulates genes related to cellular senescence, including those of the INK4/ARF families, whose deregulation contributes to stem cell functional impairment (Simboeck et al. 2011). The roles of lncRNAs in aging and cellular senescence are discussed in Chap. 5 by de Jesus's research team. *ANRIL* polymorphisms are associated with many age-related diseases, including AD, PD, atherosclerosis, and ischemic heart disease (Shou et al. 2024). *LYPLAL1-AS1* binds to the *MIRLET7B* promoter region and inhibits *miR-let-7b* transcription; reduced *miR-let-7b* expression reduces senescence, while increased expression of *miR-let-7b* mimetics promotes it (Yang et al. 2022b). The lncRNA *SAN* mediates adipose-derived stem cell senescence by binding miR-143-3p and regulating γ-adducin (ADD3) expression. SAN suppression in adipose-derived stem cells improved cell function and inhibited cellular senescence. When these cells were applied to skin wounds, enhanced re-epithelialization, collagen deposition, and neovascularization were observed compared with adipose-derived stem cells expressing lncRNA *SAN* (Xiong et al. 2023).

3.4.7 Osteoporosis

Osteoporosis is a bone disease characterized by bone fragility and fracture risk, influenced by genetic, environmental, and epigenetic factors including DNA methylation, histone modification, and ncRNAs (miRNAs, circRNAs, and lncRNAs).

These ncRNAs regulate gene expression, affecting bone metabolism and disease progression (Chen et al. 2023).

Expression of *DANCR* is elevated in the blood mononuclear cells of patients with osteoporosis, where it contributes to disease progression by stimulating bone resorption through IL6 and TNFRI secretion (Tong et al. 2015). In contrast, *NEF* is downregulated in plasma, effectively distinguishing patients from healthy controls. Because *NEF* levels are inversely correlated with IL-6, analyzing these markers together offers significant diagnostic value (Ma et al. 2019). *SNHG1* shows low postmenopausal plasma expression sufficient to distinguish osteoporosis patients (Huang et al. 2019). *MIAT* is positively regulated in women with osteoporosis and proposed as a high-accuracy serum diagnostic candidate (Yu et al. 2024a). *ZNF529* is upregulated in mononuclear cells of postmenopausal osteoporosis patients and considered a biomarker (Yihua et al. 2024). Altered lncRNA expression during osteoporosis development represents potential diagnostic biomarkers. The role of lncRNAs in bone diseases is detailed in Chap. 10 by Selvamurugan's research group.

3.4.8 Osteosarcoma

Advanced osteosarcoma is characterized by unbalanced lncRNA expression, suggesting their potential utility as biomarkers. Prominent diagnostic candidates include *HOTTIP*, whose overexpression correlates with advanced clinical stage, distant metastasis, and short survival (Li et al. 2015), and *HOST2*, which is linked to tumor stage and metastasis via B-cell lymphoma 2 (Bcl-2) downregulation and Bcl-2-associated X protein (Bax) upregulation (Qi et al. 2018). Furthermore, *GAPLINC* positively regulates CD44 expression and shows elevated levels in metastatic tissues and during disease progression (Liao et al. 2018). *TUG1* is also overexpressed in osteosarcoma, acting as a sponge for the tumor suppressor *miR-143-5p* (Yu et al. 2019) and associating with larger tumor size. Mechanistically, *DLX6-AS1* regulates proliferation, invasion, and migration by functioning as a ceRNA for *miR-641/HOXA9* signaling (Zhang et al. 2019). Regarding prognosis, *GNAS-AS1* is associated with unfavorable pathologic features and shorter overall survival (Mi et al. 2021). Finally, TP73-AS1 exhibits high expression in advanced osteosarcoma, correlating with large tumor size, distant metastasis, and poor survival rates (Chen et al. 2019).

Liquid biopsy biomarkers offer less invasive alternatives to tissue biopsies. *UCA1* is upregulated in both tissue and serum of osteosarcoma patients, relating to tumor status, metastasis, and worse prognosis (Wen et al. 2017). *HNF1A-AS1* is upregulated in osteosarcoma tissues and cell lines, with serum expression distinguishing patients from healthy individuals (Cai et al. 2017). *FAL1* presents differential serum expression levels between osteosarcoma patients and healthy individuals, relating to patient status (Wang et al. 2018).

Prognostic biomarkers include *BC050642*, positively expressed in osteosarcoma tissue and cell lines, promoting proliferation and inhibiting apoptosis while discouraging favorable clinical predictions (Yang et al. 2019b). Other prognostic biomarkers include *FGFR3-AS1* (Sun et al. 2016), *ZEB1-AS1*, *CCAL* (Zhou et al. 2017), *ATB* (Han et al. 2017), *SOX2-OT* (Wang et al. 2017), *EWSAT1* (Zhang et al. 2017a), *MEG3* (Shen et al. 2019), *TTN-AS1* (Fu et al. 2019), *MALAT1* (Liu et al. 2019b), and *LINC01614* (Cai et al. 2021).

3.4.9 Cancer Stem Cells and lncRNAs as Targets

CSCs represent small, long-lived subpopulations of tumor cells characterized by their capacity for differentiation and excessive self-renewal potential, traits driven by numerous intracellular signal transduction pathways (Table 3.3) (Bisht et al. 2022). These cells originate from normal stem cells transformed by oncogenic mutations and microenvironmental factors (Venkatesh et al. 2018); the same factors keep CSC traits alive, drive tumor metastasis, and foster therapy resistance (Zeng et al. 2023). While CSCs are often identified by surface markers such as CD24, CD44, CD49f, CD66, CD90, CD117/c-kit, CD133, CD326, and CD126, as well as the pluripotency markers OCT4, SOX2, and NANOG/ALDH1A1/ABCG2/LGR5, the clinical utility of these markers in solid tumors is complicated by their expression on non-cancerous cells (Murai and Matsuda 2023). Consequently, lncRNAs, which regulate the expression of CSC-related transcription factors, are emerging as valuable alternative markers (Luo et al. 2016).

Aberrant expression of lncRNAs is a hallmark of various malignancies (Yue et al. 2021). Prominent lncRNAs such as AK126698 (Fu et al. 2016), *HOTAIR* (Liu et al. 2019a), *TPT1-AS1* (Zhang et al. 2021; Wu et al. 2019), *MALAT1* (Hou et al. 2023), *HSPA1A* (Hsp70) (Zhang 2020), *Meg3*, *NEAT1*, *BLACAT1* (Jin 2020), and *Gas5* play critical roles in tumorigenesis. These transcripts regulate key signaling pathways—including Wnt/β-catenin, NF-κB, JAK-STAT (Ghasemian et al. 2023), PI3K/AKT, Sonic hedgehog, and Notch—at both transcriptional and post-transcriptional levels (Guo et al. 2018a). Because of their specific localization and proximity to relevant genes, these lncRNAs offer significant potential as diagnostic, prognostic, and therapeutic targets. The roles of lncRNAs in cancer are detailed in Chap. 6 by Aoki et al.

MALAT1 serves as an independent prognostic marker for hepatocellular carcinoma recurrence and is associated with reduced overall survival and advanced tumor stages in cervical cancer, breast cancer, and multiple myeloma (Schwerdtfeger et al. 2021). Mechanistically, *MALAT1* overexpression in liver CSCs enhances stemness properties by upregulating *YAP1* expression via the "sponging" of *miR-375*. *YAP1*, a key effector of the Hippo pathway, is crucial for regulating cell proliferation, apoptosis, and organ growth (Hansen et al. 2015). Inhibiting *MALAT1* reverses this sponging effect, liberating *miR-375* to exert its antitumor activity through post-transcriptional repression of oncogenes (Zhao et al. 2020). Similarly, *HOTAIR* can predict recurrence in small cell lung cancer, urothelial carcinoma, colon cancer, and cervical cancer. Reducing *HOTAIR* expression inhibits human

Table 3.3 Signaling pathways in the regulation of cancer stem cells (CSCs)

Signaling pathways	Molecules involved	Data of interest	Examples of CSC markers	Associated lncRNAs
Wnt/β--catenin	Involves the binding of Wnt ligands to the Frizzled and LRP receptors, facilitated by the accumulation of inactive β-catenin, regulated by GSK-3β, Axin, APC, and Disheveled (Ratti et al. 2020)	Abnormal activation of this pathway causes CSC stemness, promotes the sphere-forming ability of CSCs, and contributes to resistance to CSC-associated therapy (Manni and Min 2022)	OCT-4 in lung cancer (A549 cells) (Teng et al. 2010); $CD44^{+}/CD24^{low}$ in breast cancer (Jang et al. 2015) and CRC (Sousa-Franco et al. 2019); and CD45 in CRC (Park et al. 2021)	AK126698 (Tolue Ghasaban and Moghbeli 2024), HOTAIR (Hu et al. 2018), and SNHG5 (Li et al. 2022) promotes proliferation and cancer stem cell-like properties of HCC
Notch	Activated when a ligand (e.g., DSL/DOS, DSL-only, DOS co-ligands, and noncanonical) binds to a Notch receptor (types 1, 2, 3, and 4) on the cell membrane. Ligand binding triggers Notch cleavage by the ADAM metalloprotease at site 2 (Kopan and Ilagan 2009)	Plays a critical role in connecting angiogenesis and CSC self-renewal; it is increasingly recognized as a target for CSC elimination (Chini et al. 2023)	DLL-4 (Capaccione and Pine 2013) and CD90 in HCC-CSCs (Zolfaghari-Dehkharghani et al. 2024)	HOTAIR and MALAT1
NF-κB	The canonical NF-κB pathway is activated by TLRs and proinflammatory cytokines, responding to a broad range of stimuli, including ligands of cytokine receptors, PRRs, TNFR superfamily members, TCR, and BCR. In contrast, the noncanonical NF-κB pathway exhibits selective activation by a subset of TNFR superfamily members such as LTβR, BAFFR, CD40, and RANK (Liu et al. 2017)	Tumorsphere-forming cells exhibit enhanced p65 phosphorylation and heightened NF-κB signaling. Notably, preferential NF-κB activation is observed in cancer stem cell (CSC) tumor subsets (Rinkenbaugh and Baldwin 2016). In ovarian cancer, CSCs can demonstrate CCL5-dependent NF-κB p65 activation, which drives MMP9 expression, cell migration, angiogenesis, and epithelial differentiation (Kaltschmidt et al. 2019). Consequently, NF-κB inhibition presents a potential strategy to reduce CSC self-renewal within the tumor microenvironment, thereby potentially slowing cancer progression, relapse, and chemotherapy resistance (Zakaria et al. 2018)	CD44, SOX2, *OCT4*, SCA-1, KLF4, and NANOG in ovarian CSCs and lung CSCs (Chefetz et al. 2013)	FOXD2-AS1, KRT19P3, and NKILA (Ahmad et al. 2022)

JAK-STAT	Composed of ligand-receptor complexes, JAKs (JAK1, JAK2, JAK3, and TYK2), and STATs (STAT1, STAT2, STAT3, STAT4, STAT5a, STAT5b, and STAT6), which cause receptor dimerization. Activated JAK phosphorylates STAT, which dissociates from the receptor and forms homodimers or heterodimers through the SH2-domain–phosphotyrosine interactions, thereby regulating the transcription of the target genes (Hu et al. 2021)	JAK-STAT pathways promote the survival of CSCs in breast or colon cancer. Also, it leads to increased tumorigenic and metastatic ability, CSC transition, and drug resistance (Mattick et al. 2023)	CD44–STAT3 complex (Mattick et al. 2023); c-Myc, SOX2, and *OCT4* in colon CSCs (Shu et al. 2019)	ARSR and CUDR (Chen et al. 2017)
PI3K/AKT	Ras, RTKs, and GPCRs initiate PI3K activation, resulting in the phosphorylation of PIP2 to PIP3. This facilitates the recruitment of PDK1 and AKT to the plasma membrane, where PDK1 and mTORC2 phosphorylate AKT. Activated AKT subsequently phosphorylates downstream effectors that govern cell growth, survival, angiogenesis, and migration/invasion. Conversely, PTEN dephosphorylates PIP3 back to PIP2, thereby inactivating AKT (Vidal et al. 2022)	Activation of the PI3K/Akt/mTOR pathway is implicated in the acquisition of epithelial-mesenchymal transition (EMT) and cancer stem cell (CSC) phenotypes, contributing to prostate cancer radioresistance (Chang et al. 2013). The PTEN/PI3K/AKT pathway plays a crucial role in the maintenance of prostate CSCs, suggesting that targeting PI3K signaling could offer a therapeutic strategy for prostate cancer by eliminating these cells (Dubrovska et al. 2009)	CD133 and CD44 on prostate CSCs; CD44 and ALDH1A1 (Zhang et al. 2016)	HOTAIR, Gas5, FER1L4, LINC00628, PICART1, LOC101928316, ADAMTS9-AS2, SLC25A5-AS1, Meg3, AB073614, and SNHG6 (Ghafouri-Fard et al. 2021a)
Sonic hedgehog (shh)	Shh signal activation requires the Shh ligand to bind to the Patched (Ptc)-Smoothened (Smo) receptor complex. This initiates the downstream Shh signaling cascade by encoding membrane proteins that generate intracellular signals to regulate several protein kinases (Choudhry et al. 2014)	Downregulation of Shh and Wnt/β-catenin activity resulted in the inhibition of breast CSCs (Li et al. 2018). The Shh pathway has been implicated in the regulation of CSC self-renewal and is also required for the maintenance of CSC self-renewal of embryonal rhabdomyosarcoma (Xu et al. 2017)	CD44, CD133, SOX2, *OCT4*, Shh, and Gli1 proteins (Berrino and Omar 2024)	lncRNA-Hh

(continued)

Table 3.3 (continued)

Signaling pathways	Molecules involved	Data of interest	Examples of CSC markers	Associated lncRNAs
Hsp27-AKT-HK2	Hsp27 directly interacts and activates AKT, increasing HK2 expression through mTOR (Liu 2019a)	Elevated expression of phospho-Hsp27 or HK2 in esophageal tumor samples correlates with poor patient prognosis, with concurrent overexpression of both markers indicating a worse outcome. Additionally, Hsp27-AKT-HK2 pathway activation drives chemoresistance, increased glycolysis, and oxidative phosphorylation in esophageal cancer stem cells (CSCs) (Maehara et al. 2021)	HK2 and Hsp27 in esophageal cancer (Li and Wang 2023); $CD133^+$ and $CD49f^+$ in HCC (Chen et al. 2016)	DLEU2, HK2 et FAK (Dong et al. 2021)

APC adenomatous polyposis coli, *GSK-3β* glycogen synthase kinase-3β, *CRC* colorectal cancer, *DSL* Delta/Serrate/LAG-2, *DOS* Delta and OSM-11-like proteins, *DLL4* Delta-like ligand 4, *HCC* hepatocellular carcinoma, *PRRs* pattern-recognition receptors, *TNFR* TNF receptor, *TCR* T-cell receptor, *BCR* B-cell receptor, *TLRs* toll-like receptors, *EMT* epithelial-mesenchymal transition, *Shh* sonic hedgehog, *CUDR* cancer-upregulated drug-resistant

colorectal CSCs, suggesting a potential regimen to curb invasiveness and metastasis (Dou et al. 2016). In age-related cancers, *HOTAIR* further promotes metastasis and chemoresistance by modulating chromatin structure and gene expression (Xin et al. 2021).

In the context of glioblastoma, E2F1-mediated transcription leads to the overexpression of *TALNEC2*. Inhibiting this lncRNA reduces cell proliferation, induces G1/S-phase cell cycle arrest, and decreases both self-renewal and mesenchymal transformation of glioblastoma stem cells. Furthermore, *TALNEC2* inhibition increases radiation sensitivity and prolongs survival in tumor-bearing mice, effects mediated in part by restoring *miR-21* and *miR-11* expression levels (Alexanian and Ounzain 2020). In colorectal cancer, the lncRNA *cCSC1* induces stemness by activating Hedgehog signaling; its inhibition attenuates CSC self-renewal and enhances sensitivity to 5-fluorouracil (Zhou et al. 2020). These epigenetically active lncRNAs can be targeted using siRNAs, antisense oligonucleotides, and CRISPR/Cas9, providing novel and promising approaches for cancer treatment (Srinivas et al. 2024).

3.5 Study Limitations

This work has several limitations. Publication bias may exist because studies with significant results get published more often in top journals, which can skew the evidence. LncRNA databases evolve rapidly, which complicates the work, as different resources provide inconsistent gene names, isoform counts, and functional annotations. Low expression levels and the tissue-specific nature of many lncRNAs may result in their underrepresentation in RNA sequencing datasets, leading to an incomplete picture of lncRNA functions in stem cell-specific contexts.

3.6 Conclusion

LncRNAs regulate stem cell biology by controlling self-renewal, differentiation, and interactions within the cellular niche. These molecules are essential for tissue regeneration in the nervous, skeletal, and cardiovascular systems, while their dysfunction contributes to diseases ranging from metabolic disorders to cancer. We should focus on two goals: validating lncRNAs as biomarkers for non-invasive diagnosis, such as liquid biopsies, and developing targeted therapies using antisense oligonucleotides and CRISPR/Cas9. Advancing our understanding of these mechanisms will directly improve regenerative medicine and cancer treatment.

Acknowledgments The authors thank the National Technological Institute of Mexico for funding (23623.25-P). The authors also wish to thank Eli Cruz Parada for his support throughout this work. This chapter was further refined and revised with the help of the artificial intelligence (AI) tools Gemini 3 Pro and Perplexity Pro. Each section was carefully reviewed and approved by the authors to ensure the text is appropriate in terms of content and scientific accuracy and to maintain the integrity of the original content. EPC and HACF are members of the Comité Científico de Salud of the Servicios de Salud de Oaxaca (SSO), Oaxaca, Mexico. EPC, MTHH, LPCM and HACF are members of the Comité Oaxaqueño de Trombosis, Hemostasia y Endotelio (COTHE) of the SSO, Oaxaca, Mexico.

Funding Information This work received no specific funding.

Compliance with Ethical Standards

Conflict of Interest The authors declare that they have no conflict of interest.

Ethical Approval This chapter contains no studies with human participants or animals performed by any of the authors.

References

Ahmad S, Abbas M, Ullah MF, Aziz MH, Beylerli O, Alam MA, Syed MA, Uddin S, Ahmad A (2022) Long non-coding RNAs regulated NF-kappaB signaling in cancer metastasis: micromanaging by not so small non-coding RNAs. Semin Cancer Biol 85:155–163

Aich M, Chakraborty D (2020) Role of lncRNAs in stem cell maintenance and differentiation. Curr Top Dev Biol 138:73–112. https://doi.org/10.1016/bs.ctdb.2019.11.003

Alexanian M, Ounzain S (2020) Long noncoding RNAs in cardiac development. Cold Spring Harb Perspect Biol 12. https://doi.org/10.1101/cshperspect.a037374

Alrefai AA, Khader HF, Elbasuony HA, Elzorkany KM, Saleh AA (2023) Evaluation of the expression levels of lncRNAs H19 and MEG3 in patients with type 2 diabetes mellitus. Mol Biol Rep 50:6075–6085. https://doi.org/10.1007/s11033-023-08569-0

Alvarez-Dominguez JR, Lodish HF (2017) Emerging mechanisms of long noncoding RNA function during normal and malignant hematopoiesis. Blood 130:1965–1975

Babadag S, Celebi-Saltik B (2023) A cellular regulator of the niche: telocyte. Tissue Barriers 11:2131955. https://doi.org/10.1080/21688370.2022.2131955

Berrino C, Omar A (2024) Unravelling the mysteries of the sonic hedgehog pathway in cancer stem cells: activity, crosstalk and regulation. Curr Issues Mol Biol 46:5397–5419

Bian W, Chen W, Jiang X, Qu H, Jiang J, Yang J, Liang X, Zhao B, Sun Y, Zhang C (2020) Downregulation of long non-coding RNA nuclear paraspeckle assembly transcript 1 inhibits MEG-01 differentiation and platelet-like particles activity. Front Genet 11:571467

Bisht S, Nigam M, Kunjwal SS, Sergey P, Mishra AP, Sharifi-Rad J (2022) Cancer stem cells: from an insight into the basics to recent advances and therapeutic targeting. Stem Cells Int 2022:9653244

Cabrera-Fuentes HA, Boisvert WA (2025) Refining the impact of early intermittent hyperlipidemia on atherosclerosis: unveiling the role of neutrophil reprogramming, sex differences, gut microbiota, and maternal hypercholesterolemia. FASEB J 39:e70425. https://doi.org/10.1096/fj.202403281R

Cai L, Lv J, Zhang Y, Li J, Wang Y, Yang H (2017) The lncRNA HNF1A-AS1 is a negative prognostic factor and promotes tumorigenesis in osteosarcoma. J Cell Mol Med 21:2654–2662. https://doi.org/10.1111/jcmm.12944

Cai B, Zheng Y, Ma S, Xing Q, Wang X, Yang B, Yin G, Guan F (2018) Long non-coding RNA regulates hair follicle stem cell proliferation and differentiation through PI3K/AKT signal pathway. Mol Med Rep 17:5477–5483. https://doi.org/10.3892/mmr.2018.8546

Cai Q, Zhao X, Wang Y, Li S, Wang J, Xin Z, Li F (2021) LINC01614 promotes osteosarcoma progression via miR-520a-3p/SNX3 axis. Cell Signal 83:109985. https://doi.org/10.1016/j.cellsig.2021.109985

Canseco-Rodriguez A, Masola V, Aliperti V, Meseguer-Beltran M, Donizetti A, Sanchez-Perez AM (2022) Long non-coding RNAs, extracellular vesicles and inflammation in Alzheimer's disease. Int J Mol Sci 23. https://doi.org/10.3390/ijms232113171

Capaccione KM, Pine SR (2013) The notch signaling pathway as a mediator of tumor survival. Carcinogenesis 34:1420–1430

Carrelha J, Mazzi S, Winroth A, Hagemann-Jensen M, Ziegenhain C, Hogstrand K, Seki M, Brennan MS, Lehander M, Wu B, Meng Y, Markljung E, Norfo R, Ishida H, Belander Stralin K, Grasso F, Simoglou Karali C, Aliouat A, Hillen A, Chari E, Siletti K, Thongjuea S, Mead AJ, Linnarsson S, Nerlov C, Sandberg R, Yoshizato T, Woll PS, Jacobsen SEW (2024) Alternative platelet differentiation pathways initiated by nonhierarchically related hematopoietic stem cells. Nat Immunol 25:1007–1019. https://doi.org/10.1038/s41590-024-01845-6

Chang L, Graham PH, Hao J, Ni J, Bucci J, Cozzi PJ, Kearsley JH, Li Y (2013) Acquisition of epithelial-mesenchymal transition and cancer stem cell phenotypes is associated with activation of the PI3K/Akt/mTOR pathway in prostate cancer radioresistance. Cell Death Dis 4:e875

Chefetz I, Alvero AB, Holmberg JC, Lebowitz N, Craveiro V, Yang-Hartwich Y, Yin G, Squillace L, Gurrea Soteras M, Aldo P, Mor G (2013) TLR2 enhances ovarian cancer stem cell self-renewal and promotes tumor repair and recurrence. Cell Cycle 12:511–521

Chen S, Liu H, Wang Y, Wang S, Yang B, Sun D, Sun P (2024) Overexpression of lncRNA LINC00665 inhibits the proliferation and chondroblast differentiation of bone marrow mesenchymal stem cells by targeting miR-214-3p. J Orthop Surg Res 19:2. https://doi.org/10.1186/s13018-023-04475-0

Chen Y, Sun Y, Xue X, Ma H (2023) Comprehensive analysis of epigenetics mechanisms in osteoporosis. Front Genet 14:1153585. https://doi.org/10.3389/fgene.2023.1153585

Chen CL, Uthaya Kumar DB, Punj V, Xu J, Sher L, Tahara SM, Hess S, Machida K (2016) NANOG metabolically reprograms tumor-initiating stem-like cells through tumorigenic changes in oxidative phosphorylation and fatty acid metabolism. Cell Metab 23:206–219

Chen S, Zhu J, Wang F, Guan Z, Ge Y, Yang X, Cai J (2017) LncRNAs and their role in cancer stem cells. Oncotarget 8:110685–110692

Chen X, Zhou Y, Liu S, Zhang D, Yang X, Zhou Q, Song Y, Liu Y (2019) LncRNA TP73-AS1 predicts poor prognosis and functions as oncogenic lncRNA in osteosarcoma. J Cell Biochem 120:2569–2575

Chini A, Guha P, Malladi VS, Guo Z, Mandal SS (2023) Novel long non-coding RNAs associated with inflammation and macrophage activation in human. Sci Rep 13:4036

Choudhry Z, Rikani AA, Choudhry AM, Tariq S, Zakaria F, Asghar MW, Sarfraz MK, Haider K, Shafiq AA, Mobassarah NJ (2014) Sonic hedgehog signalling pathway: a complex network. Ann Neurosci 21:28–31

Clevers H, Loh KM, Nusse R (2014) Stem cell signaling. An integral program for tissue renewal and regeneration: Wnt signaling and stem cell control. Science 346:1248012. https://doi.org/10.1126/science.1248012

Dahariya S, Raghuwanshi S, Sangeeth A, Malleswarapu M, Kandi R, Gutti RK (2021) Megakaryoblastic leukemia: a study on novel role of clinically significant long non-coding RNA signatures in megakaryocyte development during treatment with phorbol ester. Cancer Immunol Immunother 70:3477–3488

Dahariya S, Raghuwanshi S, Thamodaran V, Velayudhan SR, Gutti RK (2023) Role of long non-coding RNAs in human-induced pluripotent stem cells derived megakaryocytes: A p53, HOX antisense intergenic RNA myeloid 1, and miR-125b interaction study. J Pharmacol Exp Ther 384:92–101

Delas MJ, Jackson BT, Kovacevic T, Vangelisti S, Munera Maravilla E, Wild SA, Stork EM, Erard N, Knott SRV, Hannon GJ (2019) lncRNA Spehd regulates hematopoietic stem and progenitor cells and is required for multilineage differentiation. Cell Rep 27:719–729 e6

Diao L, Zhang Q (2021) Transfer of lncRNA UCA1 by hUCMSCs-derived exosomes protects against hypoxia/reoxygenation injury through impairing miR-143-targeted degradation of Bcl-2. Aging (Albany NY) 13:5967–5985. https://doi.org/10.18632/aging.202520

Dong P, Xiong Y, Konno Y, Ihira K, Kobayashi N, Yue J, Watari H (2021) Long non-coding RNA DLEU2 drives EMT and glycolysis in endometrial cancer through HK2 by competitively binding with miR-455 and by modulating the EZH2/miR-181a pathway. J Exp Clin Cancer Res 40:216

Dou J, Ni Y, He X, Wu D, Li M, Wu S, Zhang R, Guo M, Zhao F (2016) Decreasing lncRNA HOTAIR expression inhibits human colorectal cancer stem cells. Am J Transl Res 8:98–108

Dubrovska A, Kim S, Salamone RJ, Walker JR, Maira SM, Garcia-Echeverria C, Schultz PG, Reddy VA (2009) The role of PTEN/Akt/PI3K signaling in the maintenance and viability of prostate cancer stem-like cell populations. Proc Natl Acad Sci USA 106:268–273

Durr AJ, Hathaway QA, Kunovac A, Taylor AD, Pinti MV, Rizwan S, Shepherd DL, Cook CC, Fink GK, Hollander JM (2022) Manipulation of the miR-378a/mt-ATP6 regulatory axis rescues ATP synthase in the diabetic heart and offers a novel role for lncRNA Kcnq1ot1. Am J Physiol Cell Physiol 322:C482–C495. https://doi.org/10.1152/ajpcell.00446.2021

Eguchi N, Vaziri ND, Dafoe DC, Ichii H (2021) The role of oxidative stress in pancreatic beta cell dysfunction in diabetes. Int J Mol Sci 22:1509. https://doi.org/10.3390/ijms22041509

Emmrich S, Henke K, Hegermann J, Ochs M, Reinhardt D, Klusmann JH (2012) miRNAs can increase the efficiency of ex vivo platelet generation. Ann Hematol 91:1673–1684. https://doi.org/10.1007/s00277-012-1517-z

Fatima F, Ekstrom K, Nazarenko I, Maugeri M, Valadi H, Hill AF, Camussi G, Nawaz M (2017) Non-coding RNAs in mesenchymal stem cell-derived extracellular vesicles: deciphering regulatory roles in stem cell potency, inflammatory resolve, and tissue regeneration. Front Genet 8:161. https://doi.org/10.3389/fgene.2017.00161

Fazi F, Rosa A, Fatica A, Gelmetti V, De Marchis ML, Nervi C, Bozzoni I (2005) A minicircuitry comprised of microRNA-223 and transcription factors NFI-A and C/EBPalpha regulates human granulopoiesis. Cell 123:819–831. https://doi.org/10.1016/j.cell.2005.09.023

Fu X, Li H, Liu C, Hu B, Li T, Wang Y (2016) Long noncoding RNA AK126698 inhibits proliferation and migration of non-small cell lung cancer cells by targeting Frizzled-8 and suppressing Wnt/β-catenin signaling pathway. Onco Targets Ther 9:3815–3827

Fu D, Lu C, Qu X, Li P, Chen K, Shan L, Zhu X (2019) LncRNA TTN-AS1 regulates osteosarcoma cell apoptosis and drug resistance via the miR-134-5p/MBTD1 axis. Aging (Albany NY) 11:8374–8385. https://doi.org/10.18632/aging.102325

Fu Y, Zhang YL, Liu RQ, Xu MM, Xie JL, Zhang XL, Xie GM, Han YT, Zhang XM, Zhang WT, Zhang J, Zhang J (2024) Exosome lncRNA IFNG-AS1 derived from mesenchymal stem cells of human adipose ameliorates neurogenesis and ASD-like behavior in BTBR mice. J Nanobiotechnol 22:66. https://doi.org/10.1186/s12951-024-02338-2

Gao S, Zhou B, Li H, Huang X, Wu Y, Xing C, Yu X, Ji Y (2018) Long noncoding RNA HOTAIR promotes the selfrenewal of leukemia stem cells through epigenetic silencing of p15. Exp Hematol 67:32–40.e3. https://doi.org/10.1016/j.exphem.2018.08.005

Gao M, Dong Q, Yang Z, Zou D, Han Y, Chen Z, Xu R (2024a) Long non-coding RNA H19 regulates neurogenesis of induced neural stem cells in a mouse model of closed head injury. Neural Regen Res 19:872–880. https://doi.org/10.4103/1673-5374.382255

Gao X, Carpenter RS, Boulais PE, Zhang D, Marlein CR, Li H, Smith M, Chung DJ, Maryanovich M, Will B, Steidl U, Frenette PS (2024b) Regulation of the hematopoietic stem cell pool by C-Kit-associated trogocytosis. Science 385:eadp2065. https://doi.org/10.1126/science.adp2065

Geng M, Liu W, Li J, Yang G, Tian Y, Jiang X, Xin Y (2024) LncRNA as a regulator in the development of diabetic complications. Front Endocrinol (Lausanne) 15:1324393. https://doi.org/10.3389/fendo.2024.1324393

Ghafouri-Fard S, Abak A, Tondro Anamag F, Shoorei H, Majidpoor J, Taheri M (2021a) The emerging role of non-coding RNAs in the regulation of PI3K/AKT pathway in the carcinogenesis process. Biomed Pharmacother 137:111279

Ghafouri-Fard S, Niazi V, Taheri M (2021b) Role of miRNAs and lncRNAs in hematopoietic stem cell differentiation. Noncoding RNA Res 6:8–14
Ghasemian A, Omear HA, Mansoori Y, Mansouri P, Deng X, Darbeheshti F, Zarenezhad E, Kohansal M, Pezeshki B, Wang Z, Tang H (2023) Long non-coding RNAs and JAK/STAT signaling pathway regulation in colorectal cancer development. Front Genet 14:1297093. https://doi.org/10.3389/fgene.2023.1297093
Guo F, Cao Z, Guo H, Li S (2018a) The action mechanism of lncRNA-HOTAIR on the drug resistance of non-small cell lung cancer by regulating Wnt signaling pathway. Exp Ther Med 15:4885–4889. https://doi.org/10.3892/etm.2018.6052
Guo X, Wang Z, Lu C, Hong W, Wang G, Xu Y, Liu Z, Kang J (2018b) LincRNA-1614 coordinates Sox2/PRC2-mediated repression of developmental genes in pluripotency maintenance. J Mol Cell Biol 10:118–129. https://doi.org/10.1093/jmcb/mjx041
Han F, Wang C, Wang Y, Zhang L (2017) Long noncoding RNA ATB promotes osteosarcoma cell proliferation, migration and invasion by suppressing miR-200s. Am J Cancer Res 7:770–783
Han M, Liao Z, Liu F, Chen X, Zhang B (2020) Modulation of the TGF-beta signaling pathway by long noncoding RNA in hepatocellular carcinoma. Biomark Res 8:70. https://doi.org/10.1186/s40364-020-00252-x
Hansen CG, Moroishi T, Guan KL (2015) YAP and TAZ: a nexus for Hippo signaling and beyond. Trends Cell Biol 25:499–513. https://doi.org/10.1016/j.tcb.2015.05.002
He X, Du T, Long T, Liao X, Dong Y, Huang ZP (2022) Signaling cascades in the failing heart and emerging therapeutic strategies. Signal Transduct Target Ther 7:134. https://doi.org/10.1038/s41392-022-00972-6
Hoang DM, Pham PT, Bach TQ, Ngo ATL, Nguyen QT, Phan TTK, Nguyen GH, Le PTT, Hoang VT, Forsyth NR, Heke M, Nguyen LT (2022) Stem cell-based therapy for human diseases. Signal Transduct Target Ther 7:272. https://doi.org/10.1038/s41392-022-01134-4
Horvat-Switzer RD, Thompson AA (2005) HoxA11 is expressed in the developing hemangioblast as well as early hematopoietic precursor stem cells. Blood 106:4206–4206. https://doi.org/10.1182/blood.V106.11.4206.4206
Hou J, Long H, Zhou C, Zheng S, Wu H, Guo T, Wu Q, Zhong T, Wang T (2017) Long noncoding RNA Braveheart promotes cardiogenic differentiation of mesenchymal stem cells in vitro. Stem Cell Res Ther 8:4. https://doi.org/10.1186/s13287-016-0454-5
Hou J, Zhang G, Wang X, Wang Y, Wang K (2023) Functions and mechanisms of lncRNA MALAT1 in cancer chemotherapy resistance. Biomark Res 11:23
Hu XY, Hou PF, Li TT, Quan HY, Li ML, Lin T, Liu JJ, Bai J, Zheng JN (2018) The roles of Wnt/beta-catenin signaling pathway related lncRNAs in cancer. Int J Biol Sci 14:2003–2011
Hu X, Li J, Fu M, Zhao X, Wang W (2021) The JAK/STAT signaling pathway: from bench to clinic. Signal Transduct Target Ther 6:402
Huang S, Zhu X, Xiao D, Zhuang J, Liang G, Liang C, Zheng X, Ke Y, Chang Y (2019) LncRNA SNHG1 was downregulated after menopause and participates in postmenopausal osteoporosis. Biosci Rep 39:BSR20190445. https://doi.org/10.1042/BSR20190445
Huang SF, Zhao G, Peng XF, Ye WC (2021) The pathogenic role of Long non-coding RNA H19 in atherosclerosis via the miR-146a-5p/ANGPTL4 pathway. Front Cardiovasc Med 8:770163. https://doi.org/10.3389/fcvm.2021.770163
Huang X, Jie S, Li W, Liu C (2023) GATA4-activated lncRNA MALAT1 promotes osteogenic differentiation through inhibiting NEDD4-mediated RUNX1 degradation. Cell Death Discov 9:150. https://doi.org/10.1038/s41420-023-01422-0
Huang Y (2020) Exosomal lncRNAs from mesenchymal stem cells as the novel modulators to cardiovascular disease. Stem Cell Res Ther 11:315. https://doi.org/10.1186/s13287-020-01812-6
Iltis AS, Koster G, Reeves E, Matthews KRW (2023) Ethical, legal, regulatory, and policy issues concerning embryoids: a systematic review of the literature. Stem Cell Res Ther 14:209. https://doi.org/10.1186/s13287-023-03448-8
Imam R, Aboulhoda BE, Amer MM, Hassan FE, Alghamdi MA, Abdel-Hamed MR (2024) Role of mesenchymal stem cells-derived exosomes on inflammation, apoptosis, fibrosis and telocyte

modulation in doxorubicin-induced cardiotoxicity: a closer look at the structural level. Microsc Res Tech 87:1598–1614. https://doi.org/10.1002/jemt.24544

Jamieson CHM, Weissman IL (2023) Stem-cell aging and pathways to precancer evolution. N Engl J Med 389:1310–1319. https://doi.org/10.1056/NEJMra2304431

Jang GB, Kim JY, Cho SD, Park KS, Jung JY, Lee HY, Hong IS, Nam JS (2015) Blockade of Wnt/beta-catenin signaling suppresses breast cancer metastasis by inhibiting CSC-like phenotype. Sci Rep 5:12465

Jia B, Wang Z, Sun X, Chen J, Zhao J, Qiu X (2019) Long noncoding RNA LINC00707 sponges miR-370-3p to promote osteogenesis of human bone marrow-derived mesenchymal stem cells through upregulating WNT2B. Stem Cell Res Ther 10:67. https://doi.org/10.1186/s13287-019-1161-9

Jia L, Wang Y, Wang C, Du Z, Zhang S, Wen X, Zhou L, Li H, Chen H, Li D, Zhang S, Li W, Xu W, Hoffman AR, Cui J, Hu JF (2020) Oplr16 serves as a novel chromatin factor to control stem cell fate by modulating pluripotencyspecific chromosomal looping and TET2-mediated DNA demethylation. Nucleic Acids Res 48:3935–3948. https://doi.org/10.1093/nar/gkaa097

Jiapaer Z, Li G, Ye D, Bai M, Li J, Guo X, Du Y, Su D, Jia W, Chen W, Wang G, Yu Y, Zhu F, Wan X, Kang J (2018) LincU preserves naive pluripotency by restricting ERK activity in embryonic stem cells. Stem Cell Rep 11:395–409. https://doi.org/10.1016/j.stemcr.2018.06.010

Jin W (2020) Role of JAK/STAT3 signaling in the regulation of metastasis, the transition of cancer stem cells, and chemoresistance of cancer by epithelial-mesenchymal transition. Cells 9:217. https://doi.org/10.3390/cells9010217

Kaikkonen MU, Lam MT, Glass CK (2011) Non-coding RNAs as regulators of gene expression and epigenetics. Cardiovasc Res 90:430–440. https://doi.org/10.1093/cvr/cvr097

Kaltschmidt C, Banz-Jansen C, Benhidjeb T, Beshay M, Forster C, Greiner J, Hamelmann E, Jorch N, Mertzlufft F, Pfitzenmaier J, Simon M, Schulte am Esch J, Vordemvenne T, Wahnert D, Weissinger F, Wilkens L, Kaltschmidt B (2019) A role for NF-kappaB in organ specific cancer and cancer stem cells. Cancer 11:655

Kondoh H (2024) Different types of pluripotent stem cells represent different developmental stages. Results Probl Cell Differ 72:11–25. https://doi.org/10.1007/978-3-031-39027-2_2

Kopan R, Ilagan MX (2009) The canonical notch signaling pathway: unfolding the activation mechanism. Cell 137:216–233

Lee JY, Hong SH (2020) Hematopoietic stem cells and their roles in tissue regeneration. Int J Stem Cells 13:1–12. https://doi.org/10.15283/ijsc19127

Li Q, Bu Y, Shao H, Li W, Zhao D, Wang J (2023a) Protective effect of bone marrow mesenchymal stem cellderived exosomes on cardiomyoblast hypoxia-reperfusion injury through the HAND2-AS1/miR-17-5p/Mfn2 axis. BMC Cardiovasc Disord 23:114. https://doi.org/10.1186/s12872-023-03148-4

Li F, Cao L, Hang D, Wang F, Wang Q (2015) Long non-coding RNA HOTTIP is up-regulated and associated with poor prognosis in patients with osteosarcoma. Int J Clin Exp Pathol 8:11414–11420

Li Y, Hu J, Guo D, Ma W, Zhang X, Zhang Z, Lu G, He S (2022) LncRNA SNHG5 promotes the proliferation and cancer stem cell-like properties of HCC by regulating UPF1 and Wnt-signaling pathway. Cancer Gene Ther 29:1373–1383

Li J, Huang S, Shi L, Chen G, Liu X, Liu M, Guo G (2024) Interaction between long noncoding RNA and microRNA in lung inflammatory diseases. Immun Inflamm Dis 12:e1129. https://doi.org/10.1002/iid3.1129

Li B, Luan S, Chen J, Zhou Y, Wang T, Li Z, Fu Y, Zhai A, Bi C (2020a) The MSC-derived exosomal lncRNA H19 promotes wound healing in diabetic foot ulcers by upregulating PTEN via microRNA-152-3p. Mol Ther Nucleic Acids 19:814–826. https://doi.org/10.1016/j.omtn.2019.11.034

Li F, Wang Y, Yang H, Xu Y, Zhou X, Zhang X, Xie Z, Bi J (2019) The effect of BACE1-AS on beta-amyloid generation by regulating BACE1 mRNA expression. BMC Mol Biol 20:23. https://doi.org/10.1186/s12867-019-0140-0

Li K, Wang Z (2023) lncRNA NEAT1: key player in neurodegenerative diseases. Ageing Res Rev 86:101878

Li X, Wang X, Xie C, Zhu J, Meng Y, Chen Y, Li Y, Jiang Y, Yang X, Wang S, Chen J, Zhang Q, Geng S, Wu J, Zhong C, Zhao Y (2018) Sonic hedgehog and Wnt/beta-catenin pathways mediate curcumin inhibition of breast cancer stem cells. Anti-Cancer Drugs 29:208–215

Li L, Xu Y, Zhao M, Gao Z (2020b) Neuro-protective roles of long non-coding RNA MALAT1 in Alzheimer's disease with the involvement of the microRNA-30b/CNR1 network and the following PI3K/AKT activation. Exp Mol Pathol 117:104545. https://doi.org/10.1016/j.yexmp.2020.104545

Li W, Lv Y, Sun Y (2023b) Roles of non-coding RNA in megakaryocytopoiesis and thrombopoiesis: new target therapies in ITP. Platelets 34:2157382. https://doi.org/10.1080/09537104.2022.2157382

Liao S, Zhou S, Wang C (2018) GAPLINC is a predictor of poor prognosis and regulates cell migration and invasion in osteosarcoma. Biosci Rep 38:BSR20181171. https://doi.org/10.1042/BSR20181171

Liu T, Zhang L, Joo D, Sun SC (2017) NF-kappaB signaling in inflammation. Signal Transduct Target Ther 2:17023

Liu CC, Chou KT, Hsu JW, Lin JH, Hsu TW, Yen DH, Hung SC, Hsu HS (2019a) High metabolic rate and stem cell characteristics of esophageal cancer stem-like cells depend on the Hsp27-AKT-HK2 pathway. Int J Cancer 145:2144–2156

Liu CC, Chou KT, Hsu JW, Lin JH, Hsu TW, Yen DH, Hung SC, Hsu HS (2019b) High metabolic rate and stem cell characteristics of esophageal cancer stem-like cells depend on the Hsp27-AKT-HK2 pathway. Int J Cancer 145:2144–2156. https://doi.org/10.1002/ijc.32301

Liu H, Liu H, Yang Q, Fan Z (2024) LncRNA SNHG1 enhances cartilage regeneration by modulating chondrogenic differentiation and angiogenesis potentials of JBMMSCs via mitochondrial function regulation. Stem Cell Res Ther 15:177. https://doi.org/10.1186/s13287-024-03793-2

Liu B, Liu Q, Pan S, Huang Y, Qi Y, Li S, Xiao Y, Jia L (2019a) The HOTAIR/miR-214/ST6GAL1 crosstalk modulates colorectal cancer procession through mediating sialylated c-Met via JAK2/STAT3 cascade. J Exp Clin Cancer Res 38:455. https://doi.org/10.1186/s13046-019-1468-5

Liu M, Yang P, Mao G, Deng J, Peng G, Ning X, Yang H, Sun H (2019b) Long non-coding RNA MALAT1 as a valuable biomarker for prognosis in osteosarcoma: A systematic review and meta-analysis. Int J Surg 72:206–213

Lu J, Guo S, Ebert BL, Zhang H, Peng X, Bosco J, Pretz J, Schlanger R, Wang JY, Mak RH, Dombkowski DM, Preffer FI, Scadden DT, Golub TR (2008) MicroRNA-mediated control of cell fate in megakaryocyte-erythrocyte progenitors. Dev Cell 14:843–853. https://doi.org/10.1016/j.devcel.2008.03.012

Luo J, Wang P, Wang R, Wang J, Liu M, Xiong S, Li Y, Cheng B (2016) The Notch pathway promotes the cancer stem cell characteristics of CD90+ cells in hepatocellular carcinoma. Oncotarget 7:9525–9537. https://doi.org/10.18632/oncotarget.6672

Ma X, Guo Z, Gao W, Wang J, Liu Y, Gao F, Sun S, Zhou X, Yang Z, Zheng W (2019) LncRNA-NEF is downregulated in postmenopausal osteoporosis and is related to course of treatment and recurrence. J Int Med Res 47:3299–3306. https://doi.org/10.1177/0300060519847854

Mably JD, Wang DZ (2024) Long non-coding RNAs in cardiac hypertrophy and heart failure: functions, mechanisms and clinical prospects. Nat Rev Cardiol 21:326–345. https://doi.org/10.1038/s41569-023-00952-5

Maehara O, Suda G, Natsuizaka M, Shigesawa T, Kanbe G, Kimura M, Sugiyama M, Mizokami M, Nakai M, Sho T, Morikawa K, Ogawa K, Ohashi S, Kagawa S, Kinugasa H, Naganuma S, Okubo N, Ohnishi S, Takeda H, Sakamoto N (2021) FGFR2 maintains cancer cell differentiation via AKT signaling in esophageal squamous cell carcinoma. Cancer Biol Ther 22:372–380

Manni W, Min W (2022) Signaling pathways in the regulation of cancer stem cells and associated targeted therapy. MedComm 3:e176

Mattick JS, Amaral PP, Carninci P, Carpenter S, Chang HY, Chen LL, Chen R, Dean C, Dinger ME, Fitzgerald KA, Gingeras TR, Guttman M, Hirose T, Huarte M, Johnson R, Kanduri C,

Kapranov P, Lawrence JB, Lee JT, Mendell JT, Mercer TR, Moore KJ, Nakagawa S, Rinn JL, Spector DL, Ulitsky I, Wan Y, Wilusz JE, Wu M (2023) Long non-coding RNAs: definitions, functions, challenges and recommendations. Nat Rev Mol Cell Biol 24:430–447

Mi Z, Dong Y, Wang Z, Ye P (2021) Biomarker potential of lncRNA GNAS-AS1 in osteosarcoma prognosis and effect on cellular function. J Orthop Surg Res 16:470. https://doi.org/10.1186/s13018-021-02611-2

Murai T, Matsuda S (2023) Targeting the PI3K-Akt-mTOR signaling pathway involved in vasculogenic mimicry promoted by cancer stem cells. Am J Cancer Res 13:5039–5046

Nassiri SM, Ahmadi Afshar N, Almasi P (2023) Insight into microRNAs' involvement in hematopoiesis: current standing point of findings. Stem Cell Res Ther 14:282. https://doi.org/10.1186/s13287-023-03504-3

Ng SY, Johnson R, Stanton LW (2012) Human long non-coding RNAs promote pluripotency and neuronal differentiation by association with chromatin modifiers and transcription factors. EMBO J 31:522–533. https://doi.org/10.1038/emboj.2011.459

Nichols J, Smith A (2009) Naive and primed pluripotent states. Cell Stem Cell 4:487–492

Nishimoto Y, Nakagawa S, Okano H (2021) NEAT1 lncRNA and amyotrophic lateral sclerosis. Neurochem Int 150:105175. https://doi.org/10.1016/j.neuint.2021.105175

Niu X, Zhang J, Hu S, Dang W, Wang K, Bai M (2024) lncRNA Oip5-as1 inhibits excessive mitochondrial fission in myocardial ischemia/reperfusion injury by modulating DRP1 phosphorylation. Cell Mol Biol Lett 29:72. https://doi.org/10.1186/s11658-024-00588-4

Noetzli LJ, French SL, Machlus KR (2019) New insights into the differentiation of megakaryocytes from hematopoietic progenitors. Arterioscler Thromb Vasc Biol 39:1288–1300

Ong SB, Hernandez-Resendiz S, Crespo-Avilan GE, Mukhametshina RT, Kwek XY, Cabrera-Fuentes HA, Hausenloy DJ (2018) Inflammation following acute myocardial infarction: multiple players, dynamic roles, and novel therapeutic opportunities. Pharmacol Ther 186:73–87. https://doi.org/10.1016/j.pharmthera.2018.01.001

Ou F, Su K, Sun J, Liao W, Yao Y, Zheng Y, Zhang Z (2017) The LncRNA ZBED3-AS1 induces chondrogenesis of human synovial fluid mesenchymal stem cells. Biochem Biophys Res Commun 487:457–463. https://doi.org/10.1016/j.bbrc.2017.04.090

Park SY, Kim JY, Jang GB, Choi JH, Kim JH, Lee CJ, Lee S, Baek JH, Park KK, Kim JM, Chang HJ, Cho NC, Nam JS (2021) Aberrant activation of the CD45-Wnt signaling axis promotes stemness and therapy resistance in colorectal cancer cells. Theranostics 11:8755–8770

Parra-Izquierdo I, McCarty OJT, Aslan JE (2020) Platelet miR-223 delivery rescues vascular cells in Kawasaki disease. Circ Res 127:874–876. https://doi.org/10.1161/CIRCRESAHA.120.317796

Pedersen OB, Hvas AM, Pasalic L, Kristensen SD, Grove EL, Nissen PH (2024) Platelet function and maturity and related microRNA expression in whole blood in patients with ST-segment elevation myocardial infarction. Thromb Haemost 124:192–202. https://doi.org/10.1055/s-0043-1776305

Petkovic A, Erceg S, Munjas J, Ninic A, Vladimirov S, Davidovic A, Vukmirovic L, Milanov M, Cvijanovic D, Mitic T, Sopic M (2023) LncRNAs as regulators of atherosclerotic plaque stability. Cells 12:1832. https://doi.org/10.3390/cells12141832

Qi XD, Xu SY, Song Y (2018) Prognostic value of long non-coding RNA HOST2 expression and its tumor-promotive function in human osteosarcoma. Eur Rev Med Pharmacol Sci 22:921–927. https://doi.org/10.26355/eurrev_201802_14371

Qian C, Ye Y, Mao H, Yao L, Sun X, Wang B, Zhang H, Xie L, Zhang H, Zhang Y, Zhang S, He X (2019) Downregulated lncRNA-SNHG1 enhances autophagy and prevents cell death through the miR-221/222/p27/mTOR pathway in Parkinson's disease. Exp Cell Res 384:111614. https://doi.org/10.1016/j.yexcr.2019.111614

Rajabi D, Khanmohammadi S, Rezaei N (2024) The role of long noncoding RNAs in amyotrophic lateral sclerosis. Rev Neurosci 35:533–547. https://doi.org/10.1515/revneuro-2023-0155

Ratti M, Lampis A, Ghidini M, Salati M, Mirchev MB, Valeri N, Hahne JC (2020) MicroRNAs (miRNAs) and long non-coding RNAs (lncRNAs) as new tools for cancer therapy: first steps from bench to bedside. Target Oncol 15:261–278

Ren Y, Zhu J, Han Y, Li P, Wu J, Qu H, Zhang Z, Fang X (2021) Regulatory association of long noncoding RNAs and chromatin accessibility facilitates erythroid differentiation. Blood Adv 5:5396–5409

Rinkenbaugh AL, Baldwin AS (2016) The NF-kappaB pathway and cancer stem cells. Cells 5:16

Sakurai M, Ishitsuka K, Becker HJ, Yamazaki S (2024) Ex vivo expansion of human hematopoietic stem cells and clinical applications. Cancer Sci 115:698–705. https://doi.org/10.1111/cas.16066

Scadden DT (2014) Nice neighborhood: emerging concepts of the stem cell niche. Cell. 157:41–50

Schwerdtfeger M, Desiderio V, Kobold S, Regad T, Zappavigna S, Caraglia M (2021) Long non-coding RNAs in cancer stem cells. Transl Oncol 14:101134

Shahriyari L, Komarova NL (2013) Symmetric vs. asymmetric stem cell divisions: an adaptation against cancer? PLoS One 8:e76195. https://doi.org/10.1371/journal.pone.0076195

Sheik Mohamed J, Gaughwin PM, Lim B, Robson P, Lipovich L (2010) Conserved long noncoding RNAs transcriptionally regulated by Oct4 and Nanog modulate pluripotency in mouse embryonic stem cells. RNA 16:324–337. https://doi.org/10.1261/rna.1441510

Shen B, Zhou N, Hu T, Zhao W, Wu D, Wang S (2019) LncRNA MEG3 negatively modified osteosarcoma development through regulation of miR-361-5p and FoxM1. J Cell Physiol 234:13464–13480. https://doi.org/10.1002/jcp.28026

Shin JJ, Park J, Shin HS, Arab I, Suk K, Lee WH (2024) Roles of lncRNAs in NF-kappaB-mediated macrophage inflammation and their implications in the pathogenesis of human diseases. Int J Mol Sci 25:2670. https://doi.org/10.3390/ijms25052670

Shou F, Li G, Morshedi M (2024) Long non-coding RNA ANRIL and its role in the development of age-related diseases. Mol Neurobiol 61:7919–7929. https://doi.org/10.1007/s12035-024-04074-y

Shu T, He L, Wang X, Pang M, Yang B, Feng F, Wu Z, Liu C, Zhang S, Liu B, Wang Q, Rong L (2019) Long noncoding RNA UCA1 promotes chondrogenic differentiation of human bone marrow mesenchymal stem cells via miRNA-145-5p/SMAD5 and miRNA-124-3p/SMAD4 axis. Biochem Biophys Res Commun 514:316–322

Simboeck E, Ribeiro JD, Teichmann S, Di Croce L (2011) Epigenetics and senescence: learning from the INK4-ARF locus. Biochem Pharmacol 82:1361–1370. https://doi.org/10.1016/j.bcp.2011.07.084

Sonawala K, Ramalingam S, Sellamuthu I (2022) Influence of long non-coding RNA in the regulation of cancer stem cell Signaling pathways. Cells 11:3492. https://doi.org/10.3390/cells11213492

Sopic M, Vladimirov S, Munjas J, Mitic T, Hall IF, Jusic A, Ruzic D, Devaux Y, Athero NETCACA (2025) Targeting noncoding RNAs to treat atherosclerosis. Br J Pharmacol 182:220–245. https://doi.org/10.1111/bph.16412

Sousa-Franco A, Rebelo K, Da Rocha ST, Bernardes de Jesus B (2019) LncRNAs regulating stemness in aging. Aging Cell 18:e12870

Srinivas T, Siqueira E, Guil S (2024) Techniques for investigating lncRNA transcript functions in neurodevelopment. Mol Psychiatry 29:874–890. https://doi.org/10.1038/s41380-023-02377-5

Sun J, Wang X, Fu C, Wang X, Zou J, Hua H, Bi Z (2016) Long noncoding RNA FGFR3-AS1 promotes osteosarcoma growth through regulating its natural antisense transcript FGFR3. Mol Biol Rep 43:427–436. https://doi.org/10.1007/s11033-016-3975-1

Sun L, Tu J, Liu C, Pan A, Xia X, Chen X (2020) Analysis of lncRNA expression profiles by sequencing reveals that lnc-AL928768.3 and lnc-AC091493.1 are novel biomarkers for disease risk and activity of rheumatoid arthritis. Inflammopharmacology 28:437–450. https://doi.org/10.1007/s10787-019-00666-6

Takahashi K, Yamanaka S (2006) Induction of pluripotent stem cells from mouse embryonic and adult fibroblast cultures by defined factors. Cell 126:663–676

Tang S, Xie Z, Wang P, Li J, Wang S, Liu W, Li M, Wu X, Su H, Cen S, Ye G, Zheng G, Wu Y, Shen H (2019) LncRNAOG promotes the osteogenic differentiation of bone marrow-derived mesenchymal stem cells under the regulation of hnRNPK. Stem Cells 37:270–283. https://doi.org/10.1002/stem.2937

Tavares ESJ, Pessoa J, Nobrega-Pereira S, Bernardes de Jesus B (2024) The impact of long non-coding RNAs in tissue regeneration and senescence. Cells 13:119. https://doi.org/10.3390/cells13020119

Teng Y, Wang X, Wang Y, Ma D (2010) Wnt/beta-catenin signaling regulates cancer stem cells in lung cancer A549 cells. Biochem Biophys Res Commun 392:373–379

Tolue Ghasaban F, Moghbeli M (2024) Long non-coding RNAs as the pivotal regulators of epithelial mesenchymal transition through WNT/beta-catenin signaling pathway in tumor cells. Pathol Res Pract 263:155683

Tong X, Gu PC, Xu SZ, Lin XJ (2015) Long non-coding RNA-DANCR in human circulating monocytes: a potential biomarker associated with postmenopausal osteoporosis. Biosci Biotechnol Biochem 79:732–737. https://doi.org/10.1080/09168451.2014.998617

Tran NT, Su H, Khodadadi-Jamayran A, Lin S, Zhang L, Zhou D, Pawlik KM, Townes TM, Chen Y, Mulloy JC, Zhao X (2016) The AS-RBM15 lncRNA enhances RBM15 protein translation during megakaryocyte differentiation. EMBO Rep 17:887–900

Tsagakis I, Douka K, Birds I, Aspden JL (2020) Long non-coding RNAs in development and disease: conservation to mechanisms. J Pathol 250:480–495. https://doi.org/10.1002/path.5405

Tu J, Tian G, Cheung HH, Wei W, Lee TL (2018) Gas5 is an essential lncRNA regulator for self-renewal and pluripotency of mouse embryonic stem cells and induced pluripotent stem cells. Stem Cell Res Ther 9:71. https://doi.org/10.1186/s13287-018-0813-5

Venkatesh V, Nataraj R, Thangaraj GS, Karthikeyan M, Gnanasekaran A, Kaginelli SB, Kuppanna G, Kallappa CG, Basalingappa KM (2018) Targeting notch signalling pathway of cancer stem cells. Stem Cell Investig 5:5. https://doi.org/10.21037/sci.2018.02.02

Vidal S, Bouzaher YH, El Motiam A, Seoane R, Rivas C (2022) Overview of the regulation of the class IA PI3K/AKT pathway by SUMO. Semin Cell Dev Biol 132:51–61

Wahba AS, Ibrahim ME, Mesbah NM, Saleh SM, Abo-Elmatty DM, Mehanna ET (2020) Serum LINC00305 expression and its genetic variant rs2850711 are associated with clinical and laboratory features of rheumatoid arthritis. Br J Biomed Sci 77:142–147. https://doi.org/10.1080/09674845.2020.1744942

Walker SE, Sabin KZ, Gearhart MD, Yamamoto K, Echeverri K (2022) Regulation of stem cell identity by miR-200a during spinal cord regeneration. Development 149:dev200033. https://doi.org/10.1242/dev.200033

Wang C, Li Y, Yu K, Jiang Z, Wang Y, Yang G (2021a) HOXA10 inhibit the osteogenic differentiation of periodontal ligament stem cells by regulating β-catenin localization and DKK1 expression. Connect Tissue Res 62:393–401

Wang J, Shen C, Li R, Wang C, Xiao Y, Kuang Y, Lao M, Xu S, Shi M, Cai X, Liang L, Xu H (2021b) Increased long noncoding RNA LINK-A contributes to rheumatoid synovial inflammation and aggression. JCI Insight 6:e146757

Wang Z, Tan M, Chen G, Li Z, Lu X (2017) LncRNA SOX2-OT is a novel prognostic biomarker for osteosarcoma patients and regulates osteosarcoma cells proliferation and motility through modulating SOX2. IUBMB Life 69:867–876. https://doi.org/10.1002/iub.1681

Wang Y, Chen J, Cowan DB, Wang DZ (2021b) Non-coding RNAs in cardiac regeneration: mechanism of action and therapeutic potential. Semin Cell Dev Biol 118:150–162. https://doi.org/10.1016/j.semcdb.2021.07.007

Wang Y, Lv Y, Jiang X, Yu X, Wang D, Liu D, Liu X, Sun Y (2024) Long non-coding RNA NORAD regulates megakaryocyte differentiation and proplatelet formation via the DUSP6/ERK signaling pathway. Biochem Biophys Res Commun 715:150004

Wang C, Li Y, Yu K, Jiang Z, Wang Y, Yang G (2021a) HOXA10 inhibit the osteogenic differentiation of periodontal ligament stem cells by regulating beta-catenin localization and DKK1 expression. Connect Tissue Res 62:393–401. https://doi.org/10.1080/03008207.2020.1756271

Wang Z, Xia P, Hu J, Huang Y, Zhang F, Li L, Wang E, Guo Q, Ye Z (2021c) LncRNA MEG3 alleviates diabetic cognitive impairments by reducing mitochondrial-derived apoptosis through promotion of FUNDC1-related mitophagy via Rac1-ROS axis. ACS Chem Neurosci 12:2280–2307. https://doi.org/10.1021/acschemneuro.0c00682

Wang Y, Zhao Z, Zhang S, Li Z, Li D, Yang S, Zhang H, Zeng X, Liu J (2018) LncRNA FAL1 is a negative prognostic biomarker and exhibits pro-oncogenic function in osteosarcoma. J Cell Biochem 119:8481–8489. https://doi.org/10.1002/jcb.27074

Wen JJ, Ma YD, Yang GS, Wang GM (2017) Analysis of circulating long non-coding RNA UCA1 as potential biomarkers for diagnosis and prognosis of osteosarcoma. Eur Rev Med Pharmacol Sci 21:498–503

Wen J, Liu J, Jiang H, Wan L, Xin L, Sun Y, Zhang P, Sun Y, Zhang Y, Du X, Wang X, Wang J (2020) lncRNA expression profiles related to apoptosis and autophagy in peripheral blood mononuclear cells of patients with rheumatoid arthritis. FEBS Open Bio 10:1642–1654. https://doi.org/10.1002/2211-5463.12913

Wu W, Gao H, Li X, Zhu Y, Peng S, Yu J, Zhan G, Wang J, Liu N, Guo X (2019) LncRNA TPT1-AS1 promotes tumorigenesis and metastasis in epithelial ovarian cancer by inducing TPT1 expression. Cancer Sci 110:1587–1598

Xiao L, Wu J, Wang JY, Chung HK, Kalakonda S, Rao JN, Gorospe M, Wang JY (2018) Long noncoding RNA uc.173 promotes renewal of the intestinal mucosa by inducing degradation of microRNA 195. Gastroenterology 154:599–611. https://doi.org/10.1053/j.gastro.2017.10.009

Xie X, Huang M, Ma S, Xin Q, Wang Y, Hu L, Zhao H, Li P, Liu M, Yuan R, Miao Y, Zhu Y, Cong W (2025) The role of long non-coding RNAs in cardiovascular diseases: a comprehensive review. Noncoding RNA Res 11:158–187

Xin X, Li Q, Fang J, Zhao T (2021) LncRNA HOTAIR: a potential prognostic factor and therapeutic target in human cancers. Front Oncol 11:679244. https://doi.org/10.3389/fonc.2021.679244

Xiong H, Ren S, Chen J, Yang X, Liu Y, Xu Z, Guo J, Jiang T, Yuan M, Liu Y, Zhang G, Li W, Machens HG, Chen Z (2023) Knockdown of long noncoding RNA SAN rejuvenates aged adipose-derived stem cells via miR-143-3p/ADD3 axis. Stem Cell Res Ther 14:213. https://doi.org/10.1186/s13287-023-03441-1

Xu X, Lu Y, Li Y, Prinz RA (2017) Sonic hedgehog Signaling in thyroid cancer. Front Endocrinol 8:284

Yan L, Liu G, Wu X (2021) The umbilical cord mesenchymal stem cell-derived exosomal lncRNA H19 improves osteochondral activity through miR-29b-3p/FoxO3 axis. Clin Transl Med 11:e255. https://doi.org/10.1002/ctm2.255

Yang X, Yang J, Lei P, Wen T (2019a) LncRNA MALAT1 shuttled by bone marrow-derived mesenchymal stem cellssecreted exosomes alleviates osteoporosis through mediating microRNA-34c/SATB2 axis. Aging (Albany NY) 11:8777–8791. https://doi.org/10.18632/aging.102264

Yang Y, Fei M, Zhou X, Li Y, Jin D (2019b) The potential value of lncRNA-BC050642 in osteosarcoma origination and outcomes. Artif Cells Nanomed Biotechnol 47:1859–1866. https://doi.org/10.1080/21691401.2019.1611593

Yang C, Xu X, Lin P, Luo B, Luo S, Huang H, Zhu J, Huang M, Peng S, Wu Q, Yin L (2022a) Overexpression of long noncoding RNA MCM3AP-AS1 promotes osteogenic differentiation of dental pulp stem cells via miR-143-3p/IGFBP5 axis. Hum Cell 35:150–162. https://doi.org/10.1007/s13577-021-00648-3

Yang Y, Liu S, He C, Lv T, Zeng L, Zhang F, Chen H, Zhao RC (2022b) LncRNA LYPLAL1-AS1 rejuvenates human adipose-derived mesenchymal stem cell senescence via transcriptional MIRLET7B inactivation. Cell Biosci 12:45. https://doi.org/10.1186/s13578-022-00782-x

Yang H, Zhang Y, Du Z, Wu T, Yang C (2023) Hair follicle mesenchymal stem cell exosomal lncRNA H19 inhibited NLRP3 pyroptosis to promote diabetic mouse skin wound healing. Aging (Albany NY) 15:791–809. https://doi.org/10.18632/aging.204513

Yihua Z, Chenjian P, Lining W, Shujun J, Haomiao Y, Kaixuan Z, Fenglei D, Yong M, Xudong C, Chunlei Z, Haitao S (2024) Identification of long non-coding RNA as potential biomarkers for the diagnosis of postmenopausal osteoporosis. Int J Endocrinol 2024:8820697. https://doi.org/10.1155/2024/8820697

Yoon JH, Abdelmohsen K, Gorospe M (2013) Posttranscriptional gene regulation by long noncoding RNA. J Mol Biol 425:3723–3730

Yu D, Dos Santos CO, Zhao G, Jiang J, Amigo JD, Khandros E, Dore LC, Yao Y, D'Souza J, Zhang Z, Ghaffari S, Choi J, Friend S, Tong W, Orange JS, Paw BH, Weiss MJ (2010) miR-451 protects against erythroid oxidant stress by repressing 14-3-3zeta. Genes Dev 24:1620–1633. https://doi.org/10.1101/gad.1942110

Yu X, Hu L, Li S, Shen J, Wang D, Xu R, Yang H (2019) Long non-coding RNA Taurine upregulated gene 1 promotes osteosarcoma cell metastasis by mediating HIF-1alpha via miR-143-5p. Cell Death Dis 10:280. https://doi.org/10.1038/s41419-019-1509-1

Yu H, Pan Y, Dai M, Wang X, Chen H (2023) Mesenchymal stem cell-originated exosomal Lnc A2M-AS1 alleviates hypoxia/reperfusion-induced apoptosis and oxidative stress in cardiomyocytes. Cardiovasc Drugs Ther 37:891–904. https://doi.org/10.1007/s10557-022-07339-7

Yu C, Chen B, Su H, Yang Y (2024a) Long non-coding RNA MIAT serves as a biomarker of fragility fracture and promotes fracture healing. J Orthop Surg Res 19:343. https://doi.org/10.1186/s13018-024-04824-7

Yu Y, Pang D, Huang J, Li C, Cui Y, Shang H (2024b) Downregulation of Lnc-ABCA12-3 modulates UBQLN1 expression and protein homeostasis pathways in amyotrophic lateral sclerosis. Sci Rep 14:21383. https://doi.org/10.1038/s41598-024-72666-8

Yue T, Fan X, Zhang Z, Liu Z, Guo M, Bai F, Gong X, Gao C, Xiao L (2019) Downregulation of lncRNA ITSN1-2 correlates with decreased disease risk and activity of rheumatoid arthritis (RA), and reduces RA fibroblast-like synoviocytes proliferation and inflammation via inhibiting NOD2/RIP2 signaling pathway. Am J Transl Res 11:4650–4666

Yue J, Wu Y, Qiu L, Zhao R, Jiang M, Zhang H (2021) LncRNAs link cancer stemness to therapy resistance. Am J Cancer Res 11:1051–1068

Zakaria N, Mohd Yusoff N, Zakaria Z, Widera D, Yahaya BH (2018) Inhibition of NF-kappaB Signaling reduces the stemness characteristics of lung cancer stem cells. Front Oncol 8:166

Zeng J, Chen M, Yang Y, Wu B (2022) A novel hypoxic lncRNA, HRL-SC, promotes the proliferation and migration of human dental pulp stem cells through the PI3K/AKT signaling pathway. Stem Cell Res Ther 13:286. https://doi.org/10.1186/s13287-022-02970-5

Zeng Z, Fu M, Hu Y, Wei Y, Wei X, Luo M (2023) Regulation and signaling pathways in cancer stem cells: implications for targeted therapy for cancer. Mol Cancer 22:172. https://doi.org/10.1186/s12943-023-01877-w

Zhang C, Han B, Xu T, Li D (2020) The biological function and potential mechanism of long non-coding RNAs in cardiovascular disease. J Cell Mol Med 24:12900–12909. https://doi.org/10.1111/jcmm.15968

Zhang GY, Zhang JF, Hu XM, Luo ZP, Ma YZ (2017a) Clinical significance of long non-coding RNA EWSAT1 as a novel prognostic biomarker in osteosarcoma. Eur Rev Med Pharmacol Sci 21:5337–5341. https://doi.org/10.26355/eurrev_201712_13918

Zhang L, Yang C, Chen S, Wang G, Shi B, Tao X, Zhou L, Zhao J (2017b) Long noncoding RNA DANCR is a positive regulator of proliferation and chondrogenic differentiation in human synovium-derived stem cells. DNA Cell Biol 36:136–142. https://doi.org/10.1089/dna.2016.3544

Zhang N, Meng X, Mei L, Zhao C, Chen W (2019) LncRNA DLX6-AS1 promotes tumor proliferation and metastasis in osteosarcoma through modulating miR-641/HOXA9 signaling pathway. J Cell Biochem 120:11478–11489. https://doi.org/10.1002/jcb.28426

Zhang Q, Huang XM, Liao JX, Dong YK, Zhu JL, He CC, Huang J, Tang YW, Wu D, Tian JY (2021a) LncRNA HOTAIR promotes neuronal damage through facilitating NLRP3 mediated-pyroptosis activation in Parkinson's disease via regulation of miR-326/ELAVL1 axis. Cell Mol Neurobiol 41:1773–1786. https://doi.org/10.1007/s10571-020-00946-8

Zhang T, Zhang G, Yang W, Chen H, Hu J, Zhao Z, Cheng C, Li G, Xie Y, Li Y, Kong R, Wang Y, Wang G, Chen H, Bai XW, Pan S, Sun B, Li L (2021b) Lnc-PFAR facilitates autophagy and exacerbates pancreatic fibrosis by reducing pre-miR-141 maturation in chronic pancreatitis. Cell Death Dis 12:996. https://doi.org/10.1038/s41419-021-04236-z

Zhang HF, Wu C, Alshareef A, Gupta N, Zhao Q, Xu XE, Jiao JW, Li EM, Xu LY, Lai R (2016) The PI3K/AKT/c-MYC Axis promotes the acquisition of cancer stem-like features in esophageal squamous cell carcinoma. Stem Cells 34:2040–2051

Zhang S, Wang B, Xiao H, Dong J, Li Y, Zhu C, Jin Y, Li H, Cui M, Fan S (2020) 2020. LncRNA HOTAIR enhances breast cancer radioresistance through facilitating HSPA1A expression via sequestering miR-449b-5p. Thoracic Cancer 11:1801–1816

Zhang L, Ye F, Zuo Z, Cao D, Peng Y, Li Z, Huang J, Duan L (2021) Long noncoding RNA TPT1-AS1 promotes the progression and metastasis of colorectal cancer by upregulating the TPT1-mediated FAK and JAK-STAT3 signalling pathways. Aging 13:3779–3797

Zhang H, Zhou Y, Wen D, Wang J (2023) Noncoding RNAs: master regulator of fibroblast to myofibroblast transition in fibrosis. Int J Mol Sci 24:1801

Zhao W, Ji X, Zhang F, Li L, Ma L (2012) Embryonic stem cell markers. Molecules 17:6196–6236. https://doi.org/10.3390/molecules17066196

Zhao L, Lou G, Li A, Liu Y (2020) lncRNA MALAT1 modulates cancer stem cell properties of liver cancer cells by regulating YAP1 expression via miR-375 sponging. Mol Med Rep 22:1449–1457. https://doi.org/10.3892/mmr.2020.11196

Zhao N, Wang J, Huang S, Zhang J, Bao J, Ni H, Gao X, Zhang C (2024) The landscape of programmed cell deathrelated lncRNAs in Alzheimer's disease and Parkinson's disease. Apoptosis 29:1584–1599. https://doi.org/10.1007/s10495-024-01984-z

Zhou DK, Yang XW, Li H, Yang Y, Zhu ZJ, Wu N (2017) Up-regulation of long noncoding RNA CCAL predicts poor patient prognosis and promotes tumor metastasis in osteosarcoma. Int J Biol Markers 32:e108–e112. https://doi.org/10.5301/jbm.5000240

Zhou J, Xu J, Zhang L, Liu S, Ma Y, Wen X, Hao J, Li Z, Ni Y, Li X, Zhou F, LI Q, Wang F, Wang X, Si Y, Zhang P, Liu C, Bartolomei M, Tang F, Liu B, Yu J, Lan Y (2019) Combined single-cell profiling of lncRNAs and functional screening reveals that H19 is pivotal for embryonic hematopoietic stem cell development. Cell Stem Cell 24:285–298.e5

Zhou H, Xiong Y, Peng L, Wang R, Zhang H, Fu Z (2020) LncRNA-cCSC1 modulates cancer stem cell properties in colorectal cancer via activation of the Hedgehog signaling pathway. J Cell Biochem 121:2510–2524. https://doi.org/10.1002/jcb.29473

Zhu G, Luo H, Feng Y, Guryanova OA, Xu J, Chen S, Lai Q, Sharma A, Xu B, Zhao Z, Feng R, Ni H, Claxton D, Guo Y, Mesa RA, Qiu Y, Yang FC, Li W, Nimer SD, Huang S, Xu M (2021) HOXBLINC long non-coding RNA activation promotes leukemogenesis in NPM1-mutant acute myeloid leukemia. Nat Commun 12:1956

Zhu D, Liu S, Huang K, Li J, Mei X, Li Z, Cheng K (2023a) Intrapericardial long non-coding RNA-Tcf21 antisense RNA inducing demethylation administration promotes cardiac repair. Eur Heart J 44:1748–1760. https://doi.org/10.1093/eurheartj/ehad114

Zhu Y, Yan Z, Fu C, Wen X, Jia L, Zhou L, Du Z, Wang C, Wang Y, Chen J, Nie Y, Wang W, Cui J, Wang G, Hoffman AR, Hu JF, Li W (2023b) LncRNA Osilr9 coordinates promoter DNA demethylation and the intrachromosomal loop structure required for maintaining stem cell pluripotency. Mol Ther 31:1791–1806. https://doi.org/10.1016/j.ymthe.2022.12.010

Zhuang B, Zhong C, Ma Y, Wang A, Quan H, Hong L (2024) Innovative therapeutic strategies for myocardial infarction across various stages: non-coding RNA and stem cells. Int J Mol Sci 26:231. https://doi.org/10.3390/ijms26010231

Zolfaghari-Dehkharghani M, Mousavi S, Kianifard N, Fazlzadeh A, Parsa H, Tavakoli Pirzaman A, Fazlollahpour-Naghibi A (2024) Importance of long non-coding RNAs in the pathogenesis, diagnosis, and treatment of myocardial infarction. Int J Cardiol Heart Vasc 55:101529

Chapter 4
Long Noncoding RNAs in Nutrient Sensing and Metabolic Regulation

Letícia Ferreira Alves and Sonia Guil

Abstract Long noncoding RNAs (*lncRNAs*) are now recognized as key regulators of gene expression and cellular function. Over the past decades, these molecules have been shown to participate in a wide range of regulatory mechanisms, from chromatin remodeling and transcriptional regulation in the nucleus to mRNA stability and translation in the cytoplasm. Despite their diverse molecular roles, many *lncRNAs* show context-dependent activity, with levels, localization, and interactors varying depending on cell type, environmental signals, or developmental stage. In this chapter, we review recent advances in understanding how *lncRNAs* contribute to metabolic regulation. We highlight their roles in sensing and responding to nutrient availability—such as glucose, amino acids, and oxygen—as well as in modulating lipid metabolism and hormonal signaling. By integrating mechanistic studies across metabolic pathways, we aim to provide a clearer view of how *lncRNAs* help maintain metabolic balance or contribute to metabolic disorders. This growing body of knowledge reinforces the view that *lncRNAs* are not peripheral regulators but central players in the dynamic control of cellular metabolism.

Keywords *lncRNAs* · Metabolism · Epigenetics · Cell signaling · Cancer

Abbreviations

ALDOA	Aldolase A
ALDOC	Aldolase C
ASCT2	Alanine, serine, and cysteine transporter 2
ATP	Adenosine triphosphate
BAT	Brown adipose tissue

L. F. Alves · S. Guil (✉)
Josep Carreras Leukaemia Research Institute (IJC), Badalona, Barcelona, Catalonia, Spain
e-mail: lferreira@carrerasresearch.org; sguil@carrerasresearch.org

L. M. Vaschetto (ed.), *Long Non-Coding RNAs, Human Health, and Diseases*, Epigenetics and Human Health 16, https://doi.org/10.1007/978-3-032-19668-2_4

BCAR4	Breast cancer anti-estrogen resistance 4
Blnc1	Brown fat lncRNA 1
BRAF	B-Raf proto-oncogene
cAMP	Cyclic adenosine monophosphate
CARF	Chromatin accessibility and remodeling factor
CCAR2	Cell cycle and apoptosis regulator 2
CDH1	Anaphase-promoting complex or cyclosome activator protein CDH1, APC/C^{CDH1}
ceRNA	Competing endogenous RNA
ChIP	Chromatin immunoprecipitation
CircMYH9	Circular RNA MYH9
Cyp7a1	Cholesterol 7 alpha-hydroxylase
EBF2	Early B cell factor 2
FADH2	Flavin adenine dinucleotide
FGFR1	Fibroblast growth factor receptor type 1
Foxo1	Forkhead box protein O1
Gas5	Growth arrest-specific 5
GDH	Glutamate dehydrogenase
GLCC1	Glycolysis-associated LncRNA of colorectal cancer
GLI2	GLI family zinc finger 2
Glut1	Glucose transporter 1
GS	Glutamine synthetase
H19	H19 imprinted maternally expressed transcript
HBV+ HCC	Hepatitis B virus-positive hepatocellular carcinoma
HCC	Hepatocellular carcinoma
HIF	Hypoxia-inducible factor
HILPS	Hypoxia-Induced LncRNA for Polo-like Kinase 1 Stabilization
HISLA	HIF-1a-stabilizing long noncoding RNA
HK2	Hexokinase 2
HMGB1	High mobility group box 1
HOTAIR	HOX transcript antisense intergenic RNA
Hoxd	Homeobox D gene
HULC	Highly upregulated in liver cancer
IRAlu	Intronic retrotransposon Alu
LD	Lipid droplet
LDHA	Lactate dehydrogenase A
let-7	Lethal-7
lincRNA	Long intergenic noncoding RNA
LIPTER	Lipid droplet-associated LncRNA
lncRNA	Long noncoding RNA
m^6A	N6-methyladenosine
MALAT1	Metastasis associated lung adenocarcinoma transcript 1
mRNA	Messenger RNA
mTORC1	Mammalian target of rapamycin complex 1
mTORC2	Mammalian target of rapamycin complex 2

MYH10	Myosin heavy chain 10
NADH	Nicotinamide adenine dinucleotide
Neat1	Nuclear paraspeckle assembly transcript 1
NKX2.5	NK2 homeobox 5
NONO	Non-POU domain-containing octamer-binding protein
NRCP	LncRNA ceruloplasmin
p32	(gC1q-R) complement C1q binding protein
p53	Tumor protein p53
PA	Phosphatidic acid
PAH	Phenylalanine hydroxylase
PCAT1	Prostate cancer-associated transcript 1
PCGEM1	Prostate cancer gene expression marker 1
PDH2	Pyruvate dehydrogenase kinase 2
PD-L1	Programmed death-ligand 1
PFKFB3	6-phosphofructo-2-kinase/fructose-2,6-biphosphatase 3
PGI	Phosphoglucose isomerase
Phe	Phenylalanine
PI4P	Phosphatidylinositol 4-phosphate
Pinc	Pregnancy-induced noncoding RNA
PKM2	Pyruvate kinase M2
PLK1	Polo-like kinase 1
Ppargc1a	Peroxisome proliferator-activated receptor gamma coactivator 1-alpha
PRC2	Polycomb repressive complex 2
PRNCR1	Prostate cancer non-protein coding RNA 1
pVHL	Von Hippel–Lindau protein
RBPs	RNA-binding proteins
RICTOR	Rapamycin-insensitive companion of mTOR
RNA pol II	RNA polymerase II
RNA	Ribonucleic acid
ROS	Reactive oxygen species
rRNA	Ribosomal RNA
SAMMSON	Survival associated mitochondrial melanoma specific oncogenic noncoding RNA
SCN	Suprachiasmatic nucleus
SG	Serine and glycine
SIRT1	Sirtuin 1
SLC1A5	Solute carrier family 1 member 5
Snhg9	Small nucleolar RNA host gene 9
STAT1	Signal transducer and activator of transcription 1
SYTL5-OT4	Synaptotagmin-like 5, opposite transcript 4
TAM	Tumor-associated macrophage
TAZ	Tafazzin
TCA	Tricarboxylic acid
USP22	Ubiquitin specific peptidase 22

UTR	Untranslated region
Xist	X-inactive specific transcript
YAP	Yes-associated protein 1
Zfas1	Zinc finger antisense 1

4.1 Introduction

LncRNA research began by overturning a long-held belief in biology: that genes only code for proteins. Previously dismissed as "transcription noise" or "junk RNA," non-translated RNA gained significance with Brannan's 1990 discovery of *H19*, the first identified lncRNA, which demonstrated that RNA molecules could have regulatory roles independent of protein-coding capabilities (Brannan et al. 1990).

Following *H19*, *Xist* (X-inactive-specific transcript) emerged, playing such a crucial biological role that it served as a foundation for ncRNA research (Brockdorff et al. 1992; Brown et al. 1992). Only one X chromosome expresses this lncRNA, which works in dosage compensation by covering the chromosome and recruiting polycomb repressive complex 2 (PRC2) to add silencing histone marks, resulting in the inactivation of the whole chromosome.

Following these two major findings, the study of lncRNA has grown rapidly, yielding thousands of identified molecules and functions. Many lncRNA features make it difficult to determine their functions, since their levels, subcellular localization, and molecular interactions have been shown to differ greatly depending on temporal and spatial contexts. LncRNAs have been reported to act as decoys, guides, scaffolds, and effector molecules (Wang and Chang 2011). This variability in the type of regulation that lncRNA molecules can perform makes them great cell signaling regulators.

Based on their genomic distribution, positional relationships with coding genes, and transcriptional orientations, *lncRNAs* can be categorized into four main groups: Long intervening/intergenic noncoding RNAs (lincRNAs)—located between coding genes; intronic *lncRNAs*—transcribed from intronic regions; sense *lncRNAs*—overlap coding genes on the same strand; and antisense *lncRNAs*—transcribed from the opposite strand of coding genes (Ma et al. 2013).

Other key features of *lncRNAs* contribute to their regulatory roles, including their modular structure. Interestingly, *lncRNAs* often exhibit conservation at the secondary structure level rather than the sequence level (Pang et al. 2006; Kutter et al. 2012). However, their length and complementary-natured regulation make studying their functions particularly challenging. Further complicating this landscape, the human proteome includes hundreds of RNA-binding proteins (RBPs), each capable of binding regions encompassing only a few RNA nucleotides. This leads to numerous interactions between *lncRNAs* and RBPs, forming an intricate and dynamic network. These interactions are highly context-dependent, varying over time and across different cellular environments, which poses an ongoing technological challenge for the field (Pang et al. 2006).

LncRNAs' cellular localization often correlates with their function. While the prevailing view has emphasized their nuclear presence, where they participate in transcriptional regulation and mRNA processing, many *lncRNAs* are also found in the cytoplasm or shuttle between the nucleus and cytosol (Zhang et al. 2014). This dual localization underscores their functional versatility.

In the nucleus, numerous studies have highlighted the role of *lncRNAs* in transcriptional regulation and the multiple contexts in which this regulation can occur. LncRNAs can accumulate at their site of transcription and neighboring regions and regulate transcription by recruiting regulatory protein complexes, such as chromatin modifiers, chromatin organizers, or transcription factors, and act either locally (*in cis*) or at a distance (*in trans*). Additionally, their regulatory activity may involve translocation from their site of synthesis to influence gene expression either globally or in a gene-specific manner. In this context, *lncRNAs* exert their effects through molecular interactions, such as base-pairing with RNAs, binding to DNA, or participating in RNA-protein interactions. Furthermore, in another regulatory context, the act of transcribing an lncRNA locus, rather than the lncRNA transcript itself, can influence the expression of neighboring genes. In such cases, the transcript may play no direct role in regulation, but the process of its transcription can positively or negatively affect the transcription of nearby genes (Marchese et al. 2017).

In the cytoplasm, *lncRNAs* perform a range of critical molecular functions. They regulate mRNA stability, translation, and protein modifications and can act as competing endogenous RNAs (ceRNAs) or precursors of microRNAs (Bridges et al. 2021).

Although *lncRNAs* exhibit extensive functional roles in cultured cells, their contribution to organism phenotypes is more uncertain. Experimental studies involving the knockout of specific *lncRNAs* in model organisms, such as *C. elegans* (Wei et al. 2019) and mice (Amândio et al. 2016), have shown minimal impact on development and overall phenotype. For example, deletion of *HOTAIR* or *MALAT1* in mice resulted in no detectable effects on organismal development, Hoxd gene expression, or global gene expression patterns (Amândio et al. 2016). These findings suggest that, despite their cellular importance, the physiological relevance of certain *lncRNAs* may be context-dependent or redundant in vivo.

Metabolism comprises the essential chemical processes in cells that convert or utilize energy. It is traditionally divided into anabolism, in which energy is consumed to synthesize macromolecules and biomolecular polymers, and catabolism, in which energy is released by breaking down macromolecules into simpler molecules. The proper functioning of metabolism heavily depends on classic signaling pathways, which, as discussed further in this chapter, can be redundant across different metabolic processes. Recent studies have highlighted the role of *lncRNAs* in regulating normal cellular metabolism, as well as their involvement in disease development and regulation (Zong et al. 2023).

This chapter explores the state-of-the-art of research on lncRNA involvement in cellular metabolism, highlighting key scientific findings that mechanistically elucidate the interplay between these molecules and essential signaling pathways in regulating or altering metabolic outcomes.

4.2 Glucose Availability and Glycolysis Regulation

Glucose is an essential molecule in metabolism, as it serves as the primary energy source for cellular activity. The breakdown of carbohydrates, lipids, and proteins ultimately results in the production of energy, and a significant amount of this energy is stored in the form of glucose, the primary metabolic fuel for mammals. At the cellular level, glucose is metabolized to produce adenosine triphosphate (ATP), the cell's energy currency. Glycolysis, the first metabolic pathway to be characterized, is a 10-step cytosolic process in which one glucose molecule is broken down into two pyruvate molecules, generating two ATP molecules independently of oxygen. This pathway is the only energy-producing pathway that can operate under anaerobic conditions or in cells without mitochondria. In the absence of oxygen, pyruvate is converted to lactate. In contrast, when oxygen is available, the pyruvate generated through glycolysis enters the tricarboxylic acid (TCA) cycle, within mitochondria, to drive ATP production. Apart from energy generation, intermediates from glycolysis and the TCA cycle serve as precursors for anabolic processes, contributing to the synthesis of glycogen, lipids, nucleotides, and proteins (Chandel 2021).

4.2.1 Neat1 in Hepatic Gluconeogenesis

Dysregulated hepatic glucose production is a hallmark of type 2 diabetes, leading to chronic hyperglycemia. This metabolic dysfunction is driven in part by the hepatic overproduction of glucose through gluconeogenesis. Zhao et al. (2025) have recently uncovered the mechanism through which paraspeckles modulate this phenomenon (Fig. 4.1). Paraspeckles are nuclear bodies scaffolded by the long isoform of the lncRNA *Neat1* together with multiple proteins. These nuclear aggregates were shown to retain the *Ppargc1a* (a master regulator of gluconeogenesis) and *Foxo1* (a transcription factor that stimulates the expression of gluconeogenic enzymes) mRNAs within the nucleus, preventing their translation and thereby impairing gluconeogenesis. However, in type 2 diabetes, elevated levels of proline (a non-essential amino acid) disrupt this regulatory mechanism. Proline has been observed to bind to the RNA-binding protein NONO, impairing its capacity to interact with *Neat1* and contributing to the destabilization of paraspeckle formation. The loss of paraspeckles, in turn, results in the release of *Ppargc1a* and *Foxo1* mRNAs into the cytoplasm, where they are translated into proteins that drive excessive glucose production. In vivo, restoring *Neat1* expression rescues paraspeckle formation, reduces hepatic glucose output, and ameliorates hyperglycemia in ob/ob mice, an animal model of type 2 diabetes (Zhao et al. 2025).

As with most areas of molecular biology, our current knowledge on the contributions of lncRNA to cell metabolism is, for the vast majority, the result of investigation in a disease-oriented context, especially cancer. While here we will provide

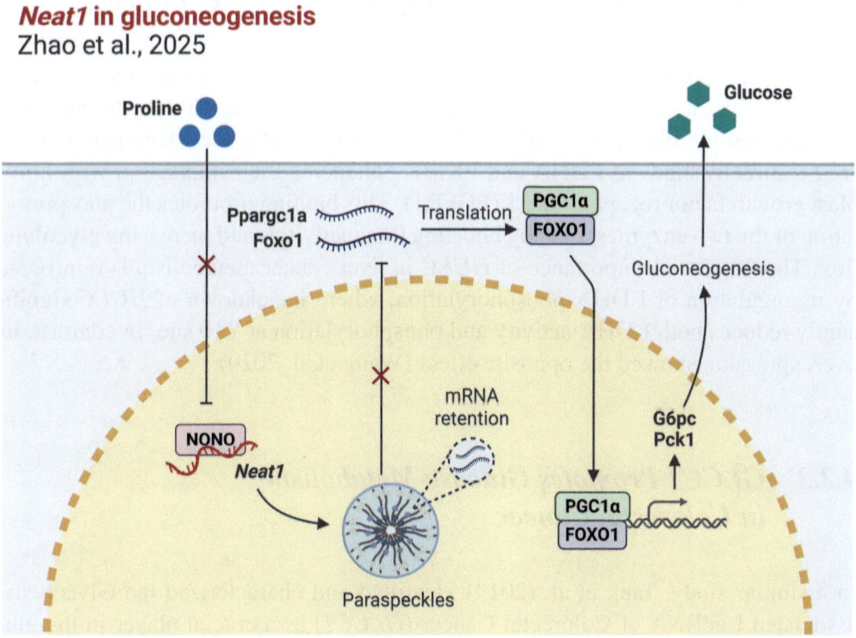

Fig. 4.1 *Neat1* in gluconeogenesis. In healthy contexts, paraspeckles—scaffolded by *Neat1*—retain the *Ppargc1a* and *Foxo1* mRNAs within the nucleus, preventing their translation and thereby impairing gluconeogenesis. In type 2 diabetes, where proline levels are high, this molecule binds to NONO, another component of paraspeckles, preventing it from binding to *Neat1* and contributing to the destabilization of paraspeckle formation. With paraspeckle disassembly, *Ppargc1a* and *Foxo1* mRNAs are released and translated into proteins that drive excessive glucose production. (Adapted from Zhao et al. (2025). Created in BioRender. Guil, S. (2025) https://BioRender.com/rrwuamc)

insights into their roles in normal metabolic processes, our focus will be mainly on cancer metabolism due to the extent of available research in this area.

A well-known feature of cancer cells is their ability to adapt and survive in challenging microenvironments, including nutrient-scarce conditions. Unlike healthy tissues, tumors often experience glucose deprivation, requiring metabolic reprogramming to sustain proliferation and survival. This metabolic switch, known as the Warburg effect (Warburg et al. 1927) and first described by Otto Warburg in 1923, increases the rate of glucose consumption and lactate production regardless of oxygen availability, enabling cancer cells to limit mitochondrial oxidative metabolism and avoid excess reactive oxygen species (ROS) generation, providing a survival advantage during metastasis.

4.2.2 *HULC* Promotes Aerobic Glycolysis in Liver Cancer

In a 2020 study, Highly Upregulated in Liver Cancer (*HULC*) was shown to regulate the enzymatic activities of two key glycolytic enzymes: lactate dehydrogenase A (LDHA) and pyruvate kinase M2 (PKM2). Mechanistically, they demonstrated that *HULC* directly binds to LDHA and PKM2, enhancing their interaction with fibroblast growth factor receptor type 1 (FGFR1). This binding promotes the phosphorylation of the two enzymes, thereby boosting their activities and increasing glycolytic flux. The functional importance of *HULC* in liver cancer metabolism is reinforced by its regulation of LDHA phosphorylation, where knockdown of *HULC* significantly reduces both LDHA activity and phosphorylation at Y10 site. In contrast, its overexpression showed the opposite effect (Wang et al. 2020).

4.2.3 *GLCC1* Promotes Glucose Metabolism in Colorectal Cancer

In a similar study, Tang et al. (2019) identified and characterized the Glycolysis-associated LncRNA of Colorectal Cancer (*GLCC1)* as a crucial player in the glucose deprivation adaptation process in colorectal cancer. They showed that *GLCC1* expression is upregulated in this malignancy and associated with poorer prognosis, larger tumor sizes, and greater invasion depth. The exposure of colorectal cancer cells to glucose-limited conditions led to increased *GLCC1* levels, and *GLCC1* knockdown dramatically impaired cell proliferation in vitro and reduced tumor growth in mouse xenograft models. Further, the silencing of *GLCC1* led to decreased expression of glycolysis-related genes, and functional assays confirmed this metabolic shift, with *GLCC1* knockdown reducing the extracellular acidification rate, a hallmark of glycolysis. Notably, inhibiting glycolysis blocked *GLCC1*-induced cell proliferation, reinforcing its role in metabolic regulation.

Mechanistically, they showed that *GLCC1* interacts with HSP90, improving the stability of c-Myc, a key regulator of glycolysis and carcinogenesis. Under normal conditions, c-Myc undergoes ubiquitin-mediated degradation; however, *GLCC1* facilitates the interaction between HSP90 and USP22, a deubiquitinase, preventing c-Myc degradation. This stabilization promotes LDHA expression, maintaining glycolytic flow and promoting tumor growth. A direct connection between lncRNA activity and tumor metabolism is established by *GLCC1*, which protects c-Myc from degradation and drives metabolic reprogramming in colorectal cancer (Tang et al. 2019).

4.2.4 *BCAR4 Reprograms Glucose Metabolism in Breast Cancer*

Zheng et al. (2017) build upon the complex interactions between *lncRNAs* and cancer metabolism by describing the Breast Cancer Anti-Estrogen Resistance 4 (*BCAR4*), an lncRNA that regulates glucose metabolism in triple-negative breast cancer. This lncRNA, previously identified by the same research group in an earlier study (Xing et al. 2014), was found to enhance invasiveness through the Hedgehog/GLI2 signaling pathway. The Hedgehog signaling pathway is a vital signaling cascade during development that controls the patterning of embryos and the regulation of adult stem cells. It also interacts with the Hippo pathway, another important regulator of organ size, cell proliferation, and apoptosis, via the transcriptional coactivators YAP/TAZ, to determine cell fate. The study showed that YAP is suppressed by Hippo signaling and is activated on glucose stimulation, with decreased phosphorylation and increased *BCAR4* expression through direct promoter binding. Then, *BCAR4* interacts with GLI2 to increase the expression of glycolysis activators HK2 and PFKFB3. When *BCAR4* or its downstream targets were blocked, there was a metabolic switch away from glycolysis, with a corresponding decrease in glucose uptake and tumor proliferation, similar to the Warburg effect. Clinically, elevated *BCAR4* and YAP levels correlated with poor prognosis in breast cancer patients, underscoring the therapeutic potential of targeting lncRNA-driven metabolic pathways (Zheng et al. 2017).

4.2.5 *NRCP Promotes Ovarian Cancer by Modulating Glycolysis*

In ovarian cancer, the lncRNA ceruloplasmin (*NRCP*) orchestrates glycolytic gene expression by mediating the interaction between two proteins. *NRCP* was shown to act as an intermediary in the interaction between STAT1 and RNA polymerase II (RNA pol II), enabling the transcription of glycolysis-related genes, including *PGI*, *ALDOA*, and *ALDOC*. This *NRCP*-STAT1-RNA pol II axis enhances glycolysis, as demonstrated by increased glucose uptake, lactate production, and glycolytic flux upon *NRCP* overexpression, while its silencing generated the opposite effects, together with reduced proliferation, increased apoptosis, and impaired tumor progression (Rupaimoole et al. 2015).

Collectively, the presented studies establish *lncRNAs* as important regulators of glucose metabolism, offering a deeper insight into how these molecules drive metabolic reprogramming across various biological contexts. The interplay between *lncRNAs* and glycolytic pathways reveals their mechanistic complexity and spatial-temporal specificity. Even though here we focused on well-established mechanistic demonstrations of *lncRNAs* interfering with glucose metabolism, additional links, associations, and correlations between these have been extensively reported in the literature and summarized in multiple reviews (Fan et al. 2017; Zhang et al. 2020a; Zong et al. 2023).

4.3 LncRNAs in Oxygen Sensing and Hypoxia

Oxygen (O_2) is an essential nutrient for cellular metabolism, and its availability varies significantly in human tissues in both physiological and pathological conditions. Variations in oxygen levels naturally occur during embryogenesis and hematopoiesis, but cellular oxygen homeostasis is further challenged by diseases like cancer, inflammation, and cardiovascular disorders. To survive and function under varying O_2 tensions, cells rely on sophisticated oxygen sensing.

The main modulators of cellular adaptation to oxygen deprivation are hypoxia-inducible factors (HIFs), which coordinate modifications in gene expression to preserve homeostasis. These transcription factors are heterodimers that consist of a constitutively stable β subunit and an oxygen-sensitive α subunit. Prolyl hydroxylase domain protein 2 (PHD2) hydroxylates HIFα under normoxic conditions, allowing the von Hippel–Lindau protein (pVHL) to recognize and tag it for degradation via proteasome. In hypoxic conditions, the absence of oxygen prevents hydroxylation, thereby stabilizing HIFα, which then dimerizes with HIFβ to activate transcriptional programs that promote adaptation to oxygen deprivation (Wang et al. 1995; Schofield and Ratcliffe 2004; Majmundar et al. 2010).

A striking example of how hypoxia-driven pathways are hijacked in pathological conditions is seen in cancer metabolism. The Warburg Effect emphasizes how, even when oxygen is present, cancer cells prefer glycolysis to oxidative phosphorylation. By limiting mitochondrial reactive oxygen species (ROS) and giving cancer cells biosynthetic precursors for nucleotide and lipid synthesis, this metabolic shift ultimately promotes tumor growth and survival (Denko 2008; Thompson et al. 2023). Under hypoxia, HIF-1α activation reinforces this effect by upregulating glucose transporter 1 (Glut1), increasing glycolytic flux, inhibiting pyruvate conversion to acetyl-CoA through PDK1, and promoting lactate production via LDHA (Wood et al. 1998; Ryan 1998; Kim et al. 2006; Papandreou et al. 2006).

4.3.1 LincRNA-p21 Modulates the Warburg Effect in Cancer Cells

A fascinating mechanism by which *lncRNAs* contribute to cancer cells adapting to low-oxygen environments was uncovered by Yang et al. (2014). In this study, they identified a hypoxia-responsive lncRNA, *lincRNA-p21*, as a key modulator of cancer cell metabolism under hypoxic conditions. *lincRNA-p21* was shown to interact directly with HIF-1α and pVHL, blocking pVHL-mediated HIF-1α ubiquitination and degradation. This interaction led to HIF-1α stabilization, reinforcing glycolysis through a positive feedback loop. *lincRNA-p21* knockdown in cancer cells caused a reduction in glucose uptake, lactate production, LDHA activity, and Glut1 expression, suggesting a role in maintaining metabolic adaptations under hypoxia. Mechanistically, *lincRNA-p21* was shown to be transcriptionally upregulated by

HIF-1α, as confirmed by chromatin immunoprecipitation (ChIP) assays showing direct HIF-1α binding to hypoxia response elements (HREs) in its promoter. The loss of HIF-1α, but not HIF-2α or p53, led to the abolition of hypoxia-induced *lincRNA-p21* expression, thereby reinforcing its dependence on HIF-1α signaling. In vivo, cancer cells that had been exposed to hypoxia and were then used for xenografts exhibited reduced tumorigenicity upon *lincRNA-p21* knockdown, further supporting its role in metabolic reprogramming and tumor progression (Yang et al. 2014).

4.3.2 *HISLA* is Transferred from Tumor-Associated Macrophages and Regulates Aerobic Glycolysis in Breast Cancer Cells

A similar mechanism has been elegantly described in breast cancer, where a hypoxia-induced lncRNA not only regulates tumor metabolism but also influences tumor-microenvironment interactions. Chen et al. (2019) identified a myeloid-specific lncRNA, *HISLA* (HIF-1α-stabilizing long noncoding RNA), which is transferred from tumor-associated macrophages (TAM) to tumor cells via extracellular vesicles (Fig. 4.2). After its internalization within tumor cells, *HISLA* prevents PHD2 binding to HIF-1α and thereby prevents its hydroxylation and degradation. Nevertheless, *HISLA* was shown to play roles beyond metabolic regulation, also conferring chemotherapy resistance to cancer cells. Chen et al. further demonstrated that tumor cells that received *HISLA*-containing extracellular vesicles showed increased resistance to apoptosis in response to chemotherapy. Intriguingly, this regulatory axis forms a feed-forward loop, in which lactate, a byproduct of glycolysis, upregulates *HISLA* expression in macrophages, thereby sustaining hypoxia adaptation and metabolic reprogramming (Chen et al. 2019).

4.3.3 *HILPS* Orchestrates Oxygen Sensing in Normal and Cancer Cells

While *lincRNA-p21 and HISLA* are examples of how hypoxia-induced *lncRNAs* can modulate cancer metabolism, recent findings suggest these molecules could also play a broader role in general oxygen sensing and cellular adaptation to hypoxia. A study by Chen et al. (2023) introduced a novel lncRNA, *HILPS* (Hypoxia-Induced lncRNA for Polo-like Kinase 1 Stabilization), as a key regulator of global oxygen sensing (Fig. 4.3). *HILPS* was shown to influence HIF-1α stabilization across different cell types, including embryonic stem cells.

HILPS is constitutively expressed at moderate levels in diverse normal and cancerous human cells, and its transcription is strongly upregulated under hypoxia by

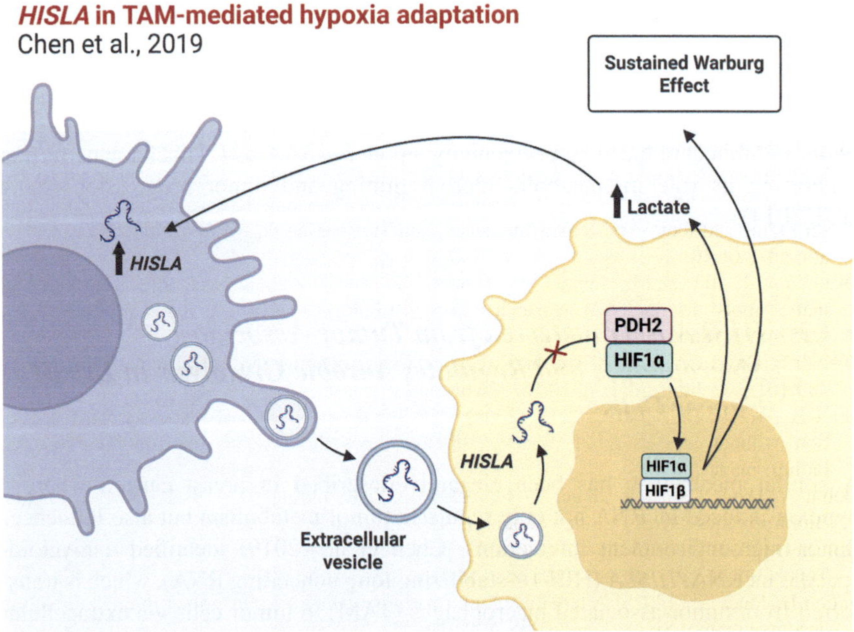

Fig. 4.2 LncRNA *HISLA* in TAM-mediated hypoxia adaptation. *HISLA* is transmitted from TAMs to tumor cells via extracellular vesicles. When incorporated by the tumor cells, *HISLA* prevents PHD2 binding to HIF-1α and thereby prevents HIF-1α degradation, allowing it to proceed to transcriptional activation and therefore sustaining the Warburg Effect. (Adapted from Chen et al. (2019). Created in BioRender. Guil, S. (2025) https://BioRender.com/534xqdh)

HIF-1α. Chen et al. demonstrated that *HILPS* binds to Polo-like kinase 1 (PLK1), antagonizing CDH1 (anaphase-promoting complex or cyclosome activator protein CDH1, APC/CCDH1), an E3 ubiquitin ligase complex that promotes PLK1 degradation. Stabilizing PLK1 prevents the degradation of its substrate, HIF-1α, thereby enhancing the cellular response to hypoxia. In the absence of *HILPS*, cells exhibit defective hypoxic adaptation, affecting both physiological (pluripotency in human ESCs) and pathological (cancer proliferation and engraftment) processes. These findings establish the *HILPS*-PLK1-HIF-1α pathway as a fundamental oxygen-sensing mechanism that extends beyond cancer-specific metabolic adaptations (Chen et al. 2023).

HIF-1α stabilization emerges as a unifying principle in lncRNA-mediated hypoxia responses. Whether through direct interaction with degradation pathways, extracellular transfer from immune cells, or modulation of upstream regulators like PLK1, these *lncRNAs* play critical roles in fine-tuning oxygen adaptation.

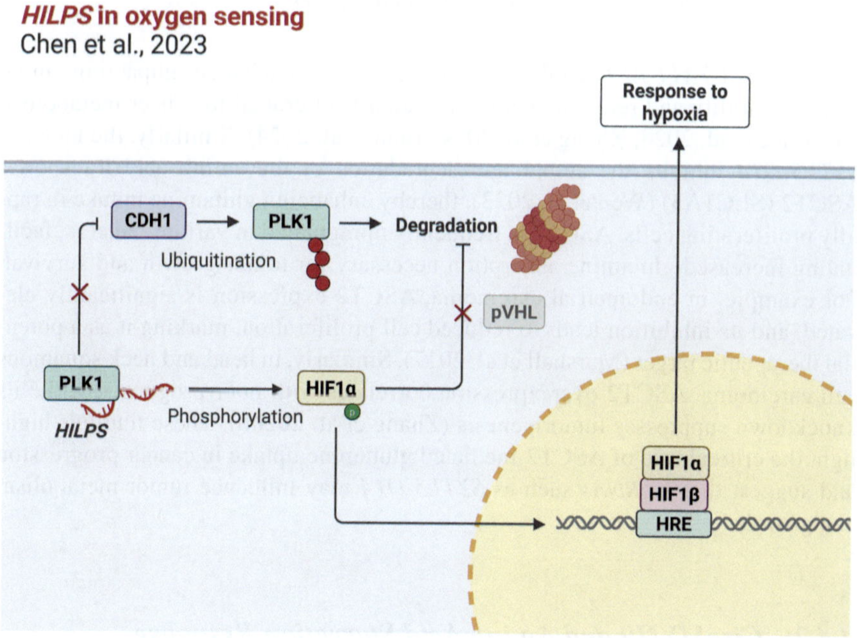

Fig. 4.3 LncRNA *HILPS* in oxygen sensing. *HILPS* is strongly upregulated under hypoxia and binds to PLK1, sequestering it from proteasomal degradation. PLK1 in turn, phosphorylates HIF-1α, enhancing its stability and thereby promoting the cellular response to hypoxia. (Adapted from Chen et al. (2023). Created in BioRender. Guil, S. (2025) https://BioRender.com/t301pey)

4.4 LncRNAs in Amino Acid Sensing

Amino acids serve as essential building blocks for proteins and function as signaling molecules that regulate cellular metabolism. In addition to their structural roles, they contribute 10%–15% of ATP generation as a direct energy source. Cells rely on amino acid-sensing mechanisms to maintain metabolic balance, and recent studies have highlighted the involvement of *lncRNAs* in these pathways. By interacting with amino acid transporters, metabolic enzymes, and signaling pathways, *lncRNAs* facilitate cellular adaptation under nutrient-rich or nutrient-deprived conditions. Their role in amino acid sensing is particularly significant in tumorigenesis, where cancer cells must efficiently utilize amino acids to sustain growth and survival.

4.4.1 LncRNAs and Amino Acid Transport

LncRNA *SLC7A11-AS1* regulates cystine/glutamate exchange, impacting amino acid availability and redox homeostasis—an aspect crucial to cancer metabolism (Banerjee et al. 2024; Zhang et al. 2024; Yuan et al. 2024). Similarly, the lncRNA *SYTL5-OT4* inhibits the autophagic degradation of the amino acid transporter ASCT2 (SLC1A5) (Wen et al. 2023), thereby enhancing glutamine uptake in rapidly proliferating cells. ASCT2 is frequently upregulated in various cancers, facilitating increased glutamine absorption necessary for tumor growth and survival. For example, in endometrial carcinoma, ASCT2 expression is significantly elevated, and its inhibition leads to reduced cell proliferation, marking it as a potential therapeutic target (Marshall et al. 2017). Similarly, in head and neck squamous cell carcinoma, ASCT2 overexpression correlates with poor prognosis, while its knockdown suppresses tumorigenesis (Zhang et al. 2020b). These findings highlight the critical role of ASCT2-mediated glutamine uptake in cancer progression and suggest that *lncRNAs* such as *SYTL5-OT4* may influence tumor metabolism and growth.

4.4.2 CircMYH9 and Amino Acid Starvation Response

A notable example of lncRNA involvement in amino acid metabolism is the circular RNA *circMYH9*, which is highly upregulated in colorectal cancer tissues following amino acid starvation (Fig. 4.4). Its transcriptional regulation is mediated by the reactive oxygen species (ROS)-HIF1α pathway. *CircMYH9* destabilizes p53 pre-mRNA by preventing hnRNPA2B1 from binding to m^6A sites on the 3′ UTR of p53. This leads to p53 downregulation, which in turn enhances the expression of enzymes involved in the serine and glycine (SG) biosynthetic pathway, maintaining SG metabolism, redox homeostasis, and promoting tumor growth (Liu et al. 2021).

4.4.3 LncRNAs in Glutamine Metabolism and HCC Progression

Glutamine metabolism plays a crucial role in hepatocellular carcinoma (HCC). A study identified an *HMGB1*-driven RNA–RNA crosstalk network that enhances glutamine metabolism via two mechanisms: (1) activating the mTORC2-AKT-C-MYC feedback loop to upregulate glutamine synthetase (GS) and (2) stimulating mTORC1 signaling to relieve SIRT4-mediated suppression of glutamate

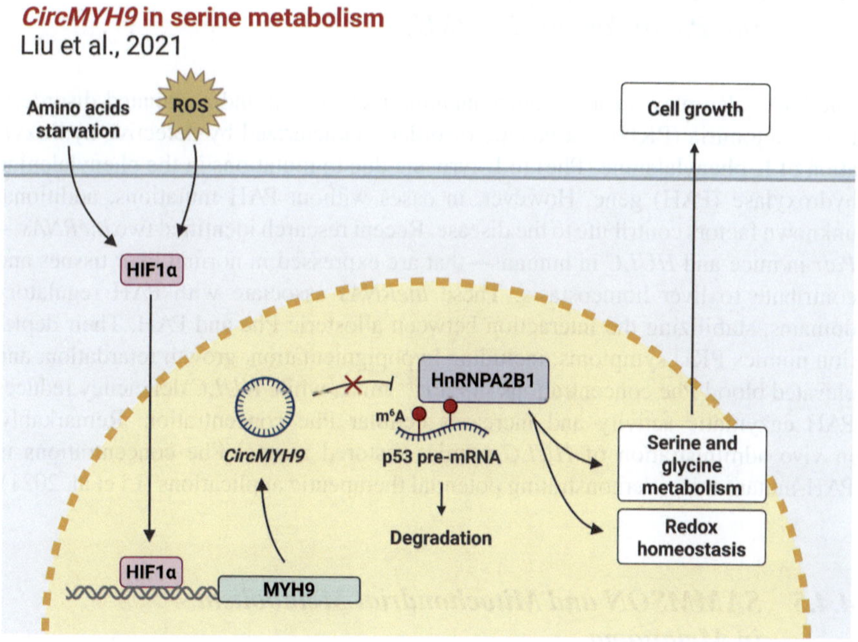

Fig. 4.4 *CircMYH9* in serine metabolism. In colorectal cancer, amino acid starvation increases ROS levels, stabilizing HIF-1α, which binds to the *circMYH9* promoter and induces its expression. *CircMYH9* disrupts the interaction between hnRNPA2B1 and m^6A-modified sites on the 3′ UTR of p53 pre-mRNA, leading to p53 destabilization. The resulting p53 downregulation promotes the expression of enzymes in the SG biosynthetic pathway, supporting SG metabolism, redox balance, and tumor progression. (Adapted from Liu et al. (2021). Created in BioRender. Guil, S. (2025) https://BioRender.com/ykx249e)

dehydrogenase (GDH). Additionally, this network influences immunotherapy response by promoting PD-L1 production through mTORC1-P70S6K signaling and increasing PD-L1+ exosome activity, which may contribute to immune evasion. The lncRNA *RICTOR* was identified as a potential HMGB1 competing endogenous RNA (ceRNA) in early-stage HBV+ HCC. *HMGB1* mRNA enhances *RICTOR* expression by competing for *miR-200* family of microRNAs, particularly *miR-429*. Functional studies confirmed that both *HMGB1* and *RICTOR* 3′UTRs drive malignant traits in HCC cells. Targeting *HMGB1* and *RICTOR* improved responses to anti-PD-L1 immunotherapy, underscoring *HMGB1*'s role in metabolic and immune regulation and highlighting RNA-based therapeutic strategies (Wei et al. 2021).

4.4.4 LncRNAs in Phenylalanine Metabolism and Phenylketonuria (PKU)

LncRNAs also play a role in phenylalanine metabolism and associated disorders. Phenylketonuria (PKU) is a genetic disorder characterized by defective hydroxylation of L-phenylalanine (Phe) to L-tyrosine due to mutations in the phenylalanine hydroxylase (PAH) gene. However, in cases without PAH mutations, additional unknown factors contribute to the disease. Recent research identified two *lncRNAs*—*Pair* in mice and *HULC* in humans—that are expressed in normal liver tissues and contribute to liver homeostasis. These *lncRNAs* associate with PAH regulatory domains, stabilizing the interaction between allosteric Phe and PAH. Their depletion mimics PKU symptoms, including hypopigmentation, growth retardation, and elevated blood Phe concentrations in *Pair*$^{-/-}$ mice, while *HULC* deficiency reduces PAH enzymatic activity and increases cellular Phe concentration. Remarkably, in vivo administration of *HULC* mimics restored normal Phe concentrations in PAH-mutant mice, demonstrating potential therapeutic applications (Li et al. 2021).

4.4.5 SAMMSON and Mitochondrial Metabolism in Melanoma

LncRNA *SAMMSON* plays a key role in cell proliferation by regulating translational complexes. Originally identified in melanoma, where it is highly expressed, *SAMMSON* directly binds to the mitochondrial metabolism protein p32, promoting its mitochondrial localization and function. *SAMMSON* silencing leads to mitochondrial cristae fragmentation and overall mitochondrial dysfunction, making it a promising therapeutic target. Notably, *SAMMSON* inhibition enhances the effectiveness of chemotherapy, such as BRAF inhibitors, in suppressing melanoma growth and survival (Leucci et al. 2016; Han et al. 2021). Further research has elucidated its role in coordinating cytoplasmic and mitochondrial translation to support cancer cell growth. *SAMMSON* modulates the RNA-binding protein CARF, which sequesters XRN2 to limit ribosomal RNA (rRNA) maturation. By disrupting this interaction, *SAMMSON* enhances rRNA maturation and protein synthesis in both the cytoplasm and mitochondria. Additionally, it facilitates an abnormal cytoplasmic RNA-protein complex involving CARF and p32, a key mitochondrial factor for rRNA processing. This dual mechanism enables cancer cells to sustain translational balance and proliferation (Vendramin et al. 2018).

Altogether, *lncRNAs* have emerged as crucial regulators of amino acid metabolism, influencing key processes such as amino acid transport, biosynthesis, and signaling pathways that maintain cellular homeostasis. By modulating the expression and activity of transporters, enzymes, and metabolic sensors, *lncRNAs* fine-tune amino acid availability to meet cellular demands under varying physiological and pathological conditions. As our understanding of lncRNA-mediated metabolic regulation deepens, these molecules hold promise as novel therapeutic targets for metabolic diseases and cancer, offering new opportunities for precision medicine.

4.5 LncRNAs in Lipid Metabolism

Lipid metabolism often refers to the synthesis and breakdown of lipids. The degradation of lipids generates intermediate metabolites that are essential for other metabolic processes, such as glycolysis and the TCA cycle. For instance, glycerol is converted to a glycolysis intermediate, whereas fatty acids undergo beta-oxidation to generate acetyl-CoA. In addition to providing the necessary energy for the TCA cycle, beta-oxidation generates NADH and FADH2, which serve as electron donors for oxidative phosphorylation. Given its essential role in cellular energy balance, lipid metabolism is subject to rigorous regulation by a network of molecular signals, including *lncRNAs* (Yoon et al. 2021).

4.5.1 Blnc1 Drives Thermogenic Adipocyte Differentiation

One of the earliest studies highlighting the role of lncRNA in metabolic regulation was the discovery of brown fat (BAT) lncRNA 1 (*Blnc1*), a critical molecule for brown adipocyte differentiation. Two main types of adipose tissue can be found in humans: brown adipose tissue, which specializes in thermogenesis and energy expenditure to combat cold exposure and metabolic disorders, and white adipose tissue, which mainly stores energy in the form of triglycerides. *Blnc1* forms a ribonucleoprotein complex with early B cell factor 2 (EBF2), a transcription factor that controls brown fat differentiation, and this complex activates the thermogenic gene program. While this study primarily focused on adipocyte differentiation rather than lipid metabolism, it underscored the broader role of adipose tissue in systemic energy homeostasis, linking *lncRNAs* to fat storage and utilization (Zhao et al. 2014).

4.5.2 Lnc-HC Modulates Cholesterol Metabolism

A more direct role for *lncRNAs* in lipid metabolism was established with the discovery of *lnc-HC*, a liver-specific lncRNA that is upregulated in metabolic disorders. *Lnc-HC* was found to influence cholesterol metabolism, a key component of lipid metabolism. Cholesterol contributes to cell membrane formation, hormone production, and fat digestion, while its transport and regulation in the body affect overall lipid balance and health. In their study, Lan et al. (2016) showed that *lnc-HC* binds to hnRNPA2B1, an RNA-binding protein, and together they target the mRNAs of Cyp7a1, a key enzyme in the synthesis of bile acid, and Abca1, a crucial transporter in cholesterol efflux. This interaction results in the inhibition of cholesterol metabolism. The described mechanism provided one of the first insights into how *lncRNAs* can directly influence lipid metabolism (Lan et al. 2016).

4.5.3 *Snhg9* Mediates Intestinal Lipid Absorption

Beyond adipose tissue, *lncRNAs* regulate lipid metabolism in other organs. In a 2023 study, Want et al. described the role of gut microbiota in lipid metabolism, identifying the lncRNA *Snhg9* as a mediator of intestinal lipid absorption. *Snhg9* expression was shown to be suppressed in intestinal epithelial cells in response to microbial signals, leading to metabolic reprogramming that affected lipid uptake and distribution. Mechanistically, they demonstrated that *Snhg9* directly binds to CCAR2, dissociating it from SIRT1, an NAD + -dependent deacetylase. Free SIRT1 then represses PPARγ—a key nuclear receptor involved in regulating lipid metabolism by transcribing lipid metabolic genes—and this interaction leads to reduced lipid metabolism. To further characterize *Snhg9* roles in vivo, they generated transgenic mice in which *Snhg9* was overexpressed specifically in the intestinal epithelial cells using the Villin promoter. Overexpression of *Snhg9* led to increased SIRT1 activity and suppressed the expression of lipid metabolism-related genes, such as *Pparg* and fatty acid transporters, resulting in reduced lipid accumulation in intestinal cells and increased lipid excretion in feces, similar to the observed for germ-free mice (Wang et al. 2023).

4.5.4 *LIPTER* Regulates Cardiac Lipid Homeostasis

In another metabolic context, *LIPTER*, a lipid droplet (LD)-associated lncRNA, was identified as a key regulator of cardiac lipid homeostasis (Fig. 4.5). Once thought to be just lipid storage sites, LDs are now recognized as dynamic organelles essential for lipid trafficking and metabolic balance. In cardiomyocytes, transient LD accumulation can be protective, but prolonged lipid overload leads to lipotoxicity, contributing to cardiac dysfunction, especially in obesity and type 2 diabetes. *LIPTER* is highly expressed in healthy cardiomyocytes but downregulated in type 2 diabetes. Mechanistically, *LIPTER* was shown to facilitate the movement of LDs by interacting with phosphatidic acid (PA) and phosphatidylinositol 4-phosphate (PI4P) on the LD membrane, while also binding to MYH10, anchoring the LDs to the cytoskeleton for intracellular trafficking. This process is essential for maintaining lipid turnover and preventing excessive lipid accumulation. *LIPTER* silencing led to LD buildup and cardiac dysfunction, while its overexpression preserved cardiac function in metabolic disease models. Interestingly, NKX2.5, a transcription factor essential for heart development, was identified as the primary transcriptional regulator of *LIPTER*, with hyperglycemia repressing the NKX2.5–*LIPTER* axis in human

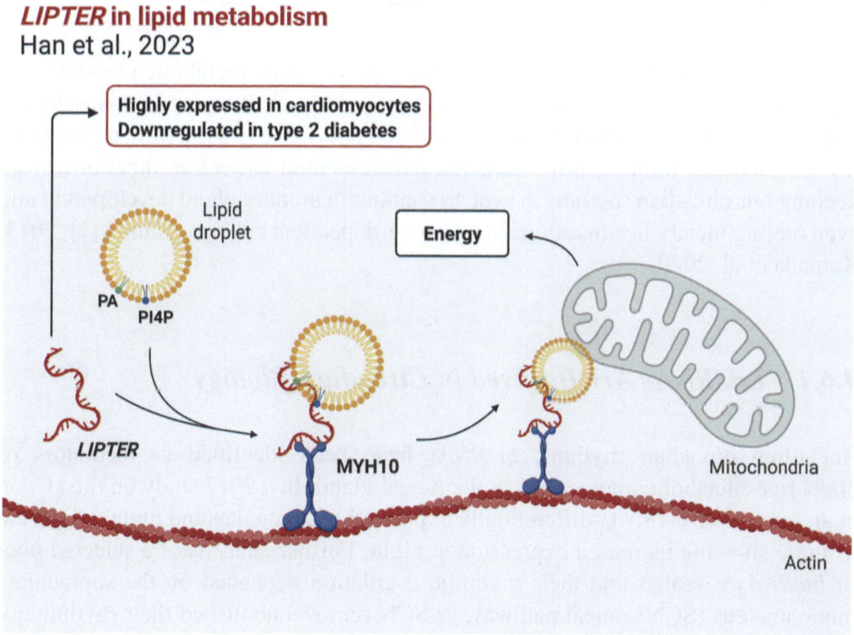

Fig. 4.5 LncRNA *LIPTER* in lipid metabolism. *LIPTER* is highly expressed in healthy cardiomyocytes but downregulated in cardiomyocytes in type 2 diabetes. In healthy contexts, *LIPTER* facilitates the transport of lipid droplets (LD) by interacting with phosphatidic acid (PA) and phosphatidylinositol 4-phosphate (PI4P) on the LD membrane, while also binding to MYH10, anchoring the LDs to the cytoskeleton for intracellular trafficking. In the context of type 2 diabetes, *LIPTER* downregulation leads to LD buildup and cardiac dysfunction. (Adapted from Han et al. (2023). Created in BioRender. Guil, S. (2025) https://BioRender.com/8boixqu)

cardiomyocytes. Similar to *Snhg9* in intestinal lipid metabolism, *LIPTER* functions as a metabolic sensor, adapting lipid utilization to physiological needs and reinforcing the idea that *lncRNAs* actively shape metabolic homeostasis across tissues (Han et al. 2023).

Collectively, these studies highlight the diverse and tissue-specific roles of *lncRNAs* in lipid metabolism, ranging from the regulation of adipocyte differentiation and hepatic cholesterol homeostasis to modulating lipid absorption in the intestine and cardiac lipid trafficking. The importance of these molecules as dynamic regulators of lipid balance is underscored by their capacity to integrate environmental and metabolic cues, such as microbial signals or glucose levels. As seen with *Snhg9* and *LIPTER*, *lncRNAs* do not merely fine-tune lipid metabolism but actively reshape it in response to physiological changes, positioning them as promising targets for metabolic disease intervention.

4.6 LncRNAs in Hormonal Metabolic Regulation

LncRNAs are known to play important roles in endocrine-metabolic crosstalk, but capturing their full functional diversity is particularly challenging in this field due to the highly specialized cell populations within endocrine tissues. Still, recent studies are shedding light on how *lncRNAs* drive essential endocrine functions, from keeping our circadian rhythms in sync to shaping mammary gland development and even fueling metabolic alterations in hormone-dependent cancers (Knoll et al. 2015; Kamada et al. 2020).

4.6.1 LncRNAs Are Involved in Circadian Biology

Regarding circadian rhythms, *lncRNAs* have been identified as mediators of endocrine-metabolic interactions in the pineal gland. In a 2012 study on rats, Coon et al. found 112 *lncRNAs* differentially expressed between day and night, with 59% of these showing increased expression at night. Further analysis of a selected pool of *lncRNAs* revealed that their rhythmic oscillation depended on the suprachiasmatic nucleus (SCN)-pineal pathway, as SCN removal abolished their rhythmicity and constant darkness did not. The control of the expression of the *lncRNAs* that were upregulated at night was shown to be promoted by the norepinephrine-cAMP signaling. These findings point to *lncRNAs* playing a critical role in melatonin synthesis and endocrine signaling, effectively translating environmental cues into metabolic changes (Coon et al. 2012).

4.6.2 Neat1 Orchestrates Circadian mRNA Retention

Neat1, previously introduced as the architectural lncRNA of paraspeckles, has also been shown to be involved in the circadian regulation of gene expression through its role in nuclear RNA retention (Fig. 4.6). In pituitary cells, *Neat1* and all major paraspeckle components exhibited circadian rhythmicity, with fluctuations in their expression levels and nuclear paraspeckle abundance over the course of the day. This circadian assembly of paraspeckles, together with the rhythmic retention of IRAlu-inserted mRNAs in the nucleus, suggests that they contribute to the circadian modulation of gene expression (Torres et al. 2016).

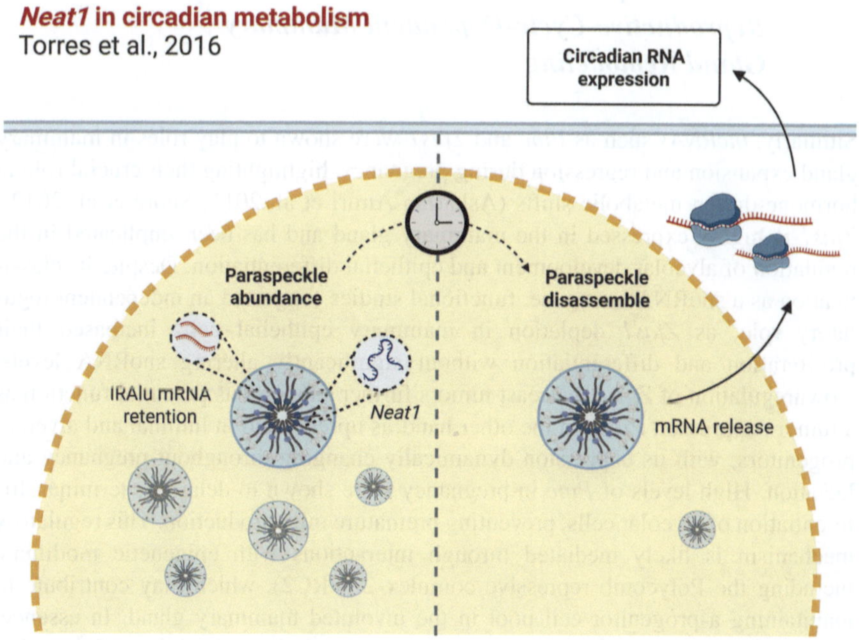

Fig. 4.6 *Neat1* in circadian metabolism. In pituitary cells, *Neat1* and other paraspeckle components exhibit circadian expression. The rhythmic assembly of paraspeckles regulates circadian gene expression by promoting IRAlu-mediated nuclear mRNA retention. (Adapted from Torres et al., 2016. Created in BioRender. Guil, S. (2025) https://BioRender.com/0u19z3m)

4.6.3 *Neat1* Binds Directly to mRNA to Regulate Circadian Expression

In a genome-wide analysis of RNA interactions in pituitary cells, researchers identified thousands of *Neat1*-associated transcripts. Notably, over 30% of paraspeckle-retained RNAs were found to directly bind *Neat1*, a finding that underscored *Neat1*'s contribution to circadian gene regulation through the selective retention and release of transcripts in a rhythmic fashion. In a rat pituitary cell line, the deletion of *Neat1* disrupted the circadian expression of hundreds of genes, leading to the loss, alteration, or emergence of new rhythmic expression patterns. These findings indicate that *Neat1*, in coordination with the circadian system, orchestrates gene expression dynamics, integrating environmental cues with cellular metabolic and endocrine responses (Jacq et al. 2021).

4.6.4 Pinc and Zfas1 Orchestrate Reproductive-Cycle-Dependent Mammary Gland Remodeling

Similarly, *lncRNAs* such as *Pinc* and *Zfas1* were shown to play roles in mammary gland expansion and regression during pregnancy, highlighting their crucial role in hormone-driven metabolic shifts (Askarian-Amiri et al. 2011; Shore et al. 2012). *Zfas1* is highly expressed in the mammary gland and has been implicated in the regulation of alveolar development and epithelial differentiation. Despite its classification as a snoRNA host gene, functional studies suggested an independent regulatory role, as *Zfas1* depletion in mammary epithelial cells increased their proliferation and differentiation without significantly altering snoRNA levels. Downregulation of *Zfas1* in breast tumors further supports its potential function as a tumor suppressor. *Pinc,* on the other hand, is upregulated in luminal and alveolar progenitors, with its expression dynamically changing throughout pregnancy and lactation. High levels of *Pinc* in pregnancy were shown to delay the terminal differentiation of alveolar cells, preventing premature milk production. This regulatory mechanism is likely mediated through interactions with epigenetic modifiers, including the Polycomb repressive complex 2 (PRC2), which may contribute to maintaining a progenitor cell pool in the involuted mammary gland. In essence, these *lncRNAs* guide critical transitions in mammary gland physiology, fine-tuning endocrine and metabolic responses across reproductive cycles.

4.6.5 LncRNAs Regulate Metabolism in Endocrine Cancers

Endocrine cancers help to further illustrate the role of *lncRNAs* in metabolic regulation, particularly in hormone-responsive malignancies. In breast cancer, for instance, estrogen triggers the expression of certain *lncRNAs*, which then drive transcriptional activation. *HOTAIR*, for example, mediates estrogen receptor-driven gene expression changes that promote cancer invasiveness (Hah et al. 2011). Additionally, *H19* was shown to act as a ceRNA, sponging *let-7,* leading to increased HIF-1α expression and boosting glycolytic metabolism and therapy resistance (Basak et al. 2018). Similarly, prostate cancer-associated *lncRNAs* such as *PCGEM1* and *PRNCR1* facilitate androgen receptor-mediated transcription, contributing to disease progression and metabolic alterations (Yang et al. 2013). Another example is *PCAT1*, overexpressed in metastatic prostate cancer, which works by repressing tumor suppressors and promoting proliferative metabolism (Prensner et al. 2011).

Finally, *lncRNAs* have also been demonstrated to influence glucocorticoid receptor signaling and metabolic stress responses. For instance, *Gas5* was shown to repress glucocorticoid receptor activity, impacting cellular energy balance during growth arrest and starvation (Kino et al. 2010). All these findings paint a landscape of *lncRNAs* intricately weaving endocrine signals into metabolic pathways, impacting both normal physiology and disease. As we delve deeper, unraveling the precise mechanisms of lncRNA-mediated endocrine-metabolic crosstalk will be crucial for developing targeted therapies for hormone-dependent malignancies.

4.7 Conclusion

Despite considerable advances in the field, the landscape of *lncRNAs* in metabolic regulation remains extensive and mechanistically complex. A defining characteristic of *lncRNAs* is their highly context-dependent modes of action, which are intrinsically linked to the cellular environment, tissue specificity, and environmental conditions. For instance, *lncRNAs* like *HULC* and *GLCC1* regulate glycolysis in cancer cells through direct interactions with metabolic enzymes, while *lnc-IIC* and *LIPTER* modulate lipid metabolism in hepatocytes and cardiomyocytes, respectively, by binding to RNA-binding proteins or lipid droplets. Similarly, *lncRNAs* such as *lincRNA-p21* and *HISLA* stabilize HIF-1α under hypoxia, but their mechanisms—whether through direct protein interactions or extracellular vesicle transfer—are distinct and context-dependent. This specificity extends to hormonal regulation; *Gas5* and *H19*, for example, fine-tune glucocorticoid and estrogen signaling in ways that are tightly linked to cellular energy status and disease progression.

The functional versatility of *lncRNAs*—serving as scaffolds, guides, decoys, or signaling molecules—underscores their adaptability and highlights the complex nature of their regulatory networks. This context-dependence not only makes the field broad but also reveals the intricate ways in which *lncRNAs* integrate metabolic, environmental, and developmental cues. Future research will continue to unravel the nuanced mechanisms by which *lncRNAs* exert their effects, offering new insights into metabolic regulation and opening avenues for the development of targeted therapies for metabolic diseases.

Acknowledgements Refined with the assistance of the artificial intelligence (AI) tool Grammarly, v1.2.183.1724, 2025.

Funding This study was supported by the Spanish Ministry of Science and Innovation, co-funded by European Regional Development Fund, ERDF, "A way to achieve Europe," [PID2022-142829OB I00, AEI/10.13039/501100011033 to S.G.]; the Departament de Recerca i Universitats/Generalitat de Catalunya [2021 SGR 01309 to S.G.]. We thank the CERCA program (Generalitat de Catalunya) and the Josep Carreras Foundation for institutional support and the Guil lab for discussion and help.Compliance with Ethical Standards

Conflict of Interest Authors declare no conflict of interest.

Ethical Approval This chapter is a review of previously published research; as such, no animal or human studies were performed.

References

Amândio AR, Necsulea A, Joye E et al (2016) Hotair is dispensible for mouse development. PLoS Genet 12:e1006232. https://doi.org/10.1371/journal.pgen.1006232

Askarian-Amiri ME, Crawford J, French JD et al (2011) SNORD-host RNA Zfas1 is a regulator of mammary development and a potential marker for breast cancer. RNA 17:878–891. https://doi.org/10.1261/rna.2528811

Banerjee S, Sarkar R, Mukherjee A et al (2024) Rotavirus-induced lncRNA SLC7A11-AS1 promotes ferroptosis by targeting cystine/glutamate antiporter xCT (SLC7A11) to facilitate virus infection. Virus Res 339:199261. https://doi.org/10.1016/j.virusres.2023.199261

Basak P, Chatterjee S, Bhat V et al (2018) Long non-coding RNA H19 acts as an estrogen receptor modulator that is required for endocrine therapy resistance in ER + breast cancer cells. Cell Physiol Biochem 51:1518–1532. https://doi.org/10.1159/000495643

Brannan CI, Dees EC, Ingram RS, Tilghman SM (1990) The product of the H19 gene may function as an RNA. Mol Cell Biol 10:28–36. https://doi.org/10.1128/mcb.10.1.28

Bridges MC, Daulagala AC, Kourtidis A (2021) LNCcation: lncRNA localization and function. J Cell Biol 220:e202009045

Brockdorff N, Ashworth A, Kay GF et al (1992) The product of the mouse Xist gene is a 15 kb inactive X-specific transcript containing no conserved ORF and located in the nucleus. Cell 71:515–526. https://doi.org/10.1016/0092-8674(92)90519-I

Brown CJ, Hendrich BD, Rupert JL et al (1992) The human XIST gene: analysis of a 17 kb inactive X-specific RNA that contains conserved repeats and is highly localized within the nucleus. Cell 71:527–542. https://doi.org/10.1016/0092-8674(92)90520-M

Chandel NS (2021) Glycolysis. Cold Spring Harb Perspect Biol 13:a040535. https://doi.org/10.1101/CSHPERSPECT.A040535

Chen F, Chen J, Yang L et al (2019) Extracellular vesicle-packaged HIF-1α-stabilizing lncRNA from tumour-associated macrophages regulates aerobic glycolysis of breast cancer cells. Nat Cell Biol 21:498–510. https://doi.org/10.1038/s41556-019-0299-0

Chen Z, Chen C, Xiao L et al (2023) HILPS, a long noncoding RNA essential for global oxygen sensing in humans. Sci Adv 9:eadi1867. https://doi.org/10.1126/sciadv.adi1867

Coon SL, Munson PJ, Cherukuri PF et al (2012) Circadian changes in long noncoding RNAs in the pineal gland. Proc Natl Acad Sci USA 109:13319–13324. https://doi.org/10.1073/pnas.1207748109

Denko NC (2008) Hypoxia, HIF1 and glucose metabolism in the solid tumour. Nat Rev Cancer 8:705–713

Fan C, Tang Y, Wang J et al (2017) Role of long non-coding RNAs in glucose metabolism in cancer. Mol Cancer 16:130

Hah N, Danko CG, Core L et al (2011) A rapid, extensive, and transient transcriptional response to estrogen signaling in breast cancer cells. Cell 145:622–634. https://doi.org/10.1016/j.cell.2011.03.042

Han S, Yan Y, Ren Y et al (2021) LncRNA SAMMSON mediates adaptive resistance to RAF inhibition in BRAF-mutant melanoma cells. Cancer Res 81:2918–2929. https://doi.org/10.1158/0008-5472.CAN-20-3145

Han L, Huang D, Wu S et al (2023) Lipid droplet-associated lncRNA LIPTER preserves cardiac lipid metabolism. Nat Cell Biol 25:1033–1046. https://doi.org/10.1038/s41556-023-01162-4

Jacq A, Becquet D, Bello-Goutierrez MM et al (2021) Genome-wide screening of circadian and non-circadian impact of Neat1 genetic deletion. Comput Struct Biotechnol J 19:2121–2132. https://doi.org/10.1016/j.csbj.2021.04.022

Kamada S, Takeiwa T, Ikeda K et al (2020) Long non-coding RNAs involved in metabolic alterations in breast and prostate cancers. Front Oncol 10:593200

Kim JW, Tchernyshyov I, Semenza GL, Dang CV (2006) HIF-1-mediated expression of pyruvate dehydrogenase kinase: a metabolic switch required for cellular adaptation to hypoxia. Cell Metab 3:177–185. https://doi.org/10.1016/j.cmet.2006.02.002

Kino T, Hurt DE, Ichijo T et al (2010) Noncoding RNA Gas5 is a growth arrest- and starvation-associated repressor of the glucocorticoid receptor. Sci Signal 3:ra8. https://doi.org/10.1126/scisignal.2000568

Knoll M, Lodish HF, Sun L (2015) Long non-coding RNAs as regulators of the endocrine system. Nat Rev Endocrinol 11:151–160

Kutter C, Watt S, Stefflova K et al (2012) Rapid turnover of long noncoding RNAs and the evolution of gene expression. PLoS Genet 8:e1002841. https://doi.org/10.1371/journal.pgen.1002841

Lan X, Yan J, Ren J et al (2016) A novel long noncoding RNA Lnc-HC binds hnRNPA2B1 to regulate expressions of Cyp7a1 and Abca1 in hepatocytic cholesterol metabolism. Hepatology 64:58–72. https://doi.org/10.1002/hep.28391

Leucci E, Vendramin R, Spinazzi M et al (2016) Melanoma addiction to the long non-coding RNA SAMMSON. Nature 531:518–522. https://doi.org/10.1038/nature17161

Li Y, Tan Z, Zhang Y et al (2021) A noncoding RNA modulator potentiates phenylalanine metabolism in mice. Science 373:662–673. https://doi.org/10.1126/science.aba4991

Liu X, Liu Y, Liu Z et al (2021) CircMYH9 drives colorectal cancer growth by regulating serine metabolism and redox homeostasis in a p53-dependent manner. Mol Cancer 20:114. https://doi.org/10.1186/s12943-021-01412-9

Ma L, Bajic VB, Zhang Z (2013) On the classification of long non-coding RNAs. RNA Biol 10:924–933

Majmundar AJ, Wong WJ, Simon MC (2010) Hypoxia-inducible factors and the response to hypoxic stress. Mol Cell 40:294–309

Marchese FP, Raimondi I, Huarte M (2017) The multidimensional mechanisms of long noncoding RNA function. Genome Biol 18:206

Marshall AD, van Geldermalsen M, Otte NJ et al (2017) ASCT2 regulates glutamine uptake and cell growth in endometrial carcinoma. Oncogene 6:e367. https://doi.org/10.1038/oncsis.2017.70

Pang KC, Frith MC, Mattick JS (2006) Rapid evolution of noncoding RNAs: lack of conservation does not mean lack of function. Trends Genet 22:1–5

Papandreou I, Cairns RA, Fontana L et al (2006) HIF-1 mediates adaptation to hypoxia by actively downregulating mitochondrial oxygen consumption. Cell Metab 3:187–197. https://doi.org/10.1016/j.cmet.2006.01.012

Prensner JR, Iyer MK, Balbin OA et al (2011) Transcriptome sequencing across a prostate cancer cohort identifies PCAT-1, an unannotated lincRNA implicated in disease progression. Nat Biotechnol 29:742–749. https://doi.org/10.1038/nbt.1914

Rupaimoole R, Lee J, Haemmerle M et al (2015) Long noncoding RNA ceruloplasmin promotes cancer growth by altering glycolysis. Cell Rep 13:2395–2402. https://doi.org/10.1016/j.celrep.2015.11.047

Ryan HE (1998) HIF-1alpha is required for solid tumor formation and embryonic vascularization. EMBO J 17:3005–3015. https://doi.org/10.1093/emboj/17.11.3005

Schofield CJ, Ratcliffe PJ (2004) Oxygen sensing by HIF hydroxylases. Nat Rev Mol Cell Biol 5:343–354

Shore AN, Kabotyanski EB, Roarty K et al (2012) Pregnancy-induced noncoding RNA (PINC) associates with polycomb repressive complex 2 and regulates mammary epithelial differentiation. PLoS Genet 8:e1002840. https://doi.org/10.1371/journal.pgen.1002840

Tang J, Yan T, Bao Y et al (2019) LncRNA GLCC1 promotes colorectal carcinogenesis and glucose metabolism by stabilizing c-Myc. Nat Commun 10:3499. https://doi.org/10.1038/s41467-019-11447-8

Thompson CB, Vousden KH, Johnson RS et al (2023) A century of the Warburg effect. Nat Metab 5:1840–1843

Torres M, Becquet D, Blanchard MP et al (2016) Circadian RNA expression elicited by 3′-UTR IRAlu-paraspeckle associated elements. elife 5:e14837. https://doi.org/10.7554/eLife.14837

Vendramin R, Verheyden Y, Ishikawa H et al (2018) SAMMSON fosters cancer cell fitness by concertedly enhancing mitochondrial and cytosolic translation. Nat Struct Mol Biol 25:1035–1046. https://doi.org/10.1038/s41594-018-0143-4

Wang KC, Chang HY (2011) Molecular mechanisms of long noncoding RNAs. Mol Cell 43:904–914

Wang GL, Jiang BH, Rue EA, Semenza GL (1995) Hypoxia-inducible factor 1 is a basic-helix-loop-helix-PAS heterodimer regulated by cellular O2 tension. Proc Natl Acad Sci USA 92:5510–5514. https://doi.org/10.1073/pnas.92.12.5510

Wang C, Li Y, Yan S et al (2020) Interactome analysis reveals that lncRNA HULC promotes aerobic glycolysis through LDHA and PKM2. Nat Commun 11:3162. https://doi.org/10.1038/s41467-020-16966-3

Wang Y, Wang M, Chen J et al (2023) The gut microbiota reprograms intestinal lipid metabolism through long noncoding RNA Snhg9. Science 381:851–857. https://doi.org/10.1126/SCIENCE.ADE0522

Warburg O, Wind F, Negelein E (1927) The metabolism of tumors in the body. J Gen Physiol 8:519. https://doi.org/10.1085/jgp.8.6.519

Wei S, Chen H, Dzakah EE et al (2019) Systematic evaluation of C. elegans lincRNAs with CRISPR knockout mutants. Genome Biol 20:7. https://doi.org/10.1186/s13059-018-1619-6

Wei Y, Tang X, Ren Y et al (2021) An RNA–RNA crosstalk network involving HMGB1 and RICTOR facilitates hepatocellular carcinoma tumorigenesis by promoting glutamine metabolism and impedes immunotherapy by PD-L1+ exosomes activity. Signal Transduct Target Ther 6:421. https://doi.org/10.1038/s41392-021-00801-2

Wen Q, Huang M, Xie J et al (2023) lncRNA SYTL5-OT4 promotes vessel co-option by inhibiting the autophagic degradation of ASCT2. Drug Resist Updat 69:100975. https://doi.org/10.1016/j.drup.2023.100975

Wood SM, Wiesener MS, Yeates KM et al (1998) Selection and analysis of a mutant cell line defective in the hypoxia-inducible factor-1 α-subunit (HIF-1α). J Biol Chem 273:8360–8368. https://doi.org/10.1074/jbc.273.14.8360

Xing Z, Lin A, Li C et al (2014) LncRNA directs cooperative epigenetic regulation downstream of chemokine signals. Cell 159:1110–1125. https://doi.org/10.1016/j.cell.2014.10.013

Yang L, Lin C, Jin C et al (2013) LncRNA-dependent mechanisms of androgen-receptor-regulated gene activation programs. Nature 500:598–602. https://doi.org/10.1038/nature12451

Yang F, Zhang H, Mei Y, Wu M (2014) Reciprocal regulation of HIF-1α and LincRNA-p21 modulates the Warburg effect. Mol Cell 53:88–100. https://doi.org/10.1016/j.molcel.2013.11.004

Yoon H, Shaw JL, Haigis MC, Greka A (2021) Lipid metabolism in sickness and in health: emerging regulators of lipotoxicity. Mol Cell 81:3708–3730

Yuan X, Wang Y, Jiao S et al (2024) Identification of SLC7A11-AS1/SLC7A11 pair as a ferroptosis-related therapeutic target for hepatocellular carcinoma. J Cell Mol Med 28:e18496. https://doi.org/10.1111/jcmm.18496

Zhang K, Shi ZM, Chang YN et al (2014) The ways of action of long non-coding RNAs in cytoplasm and nucleus. Gene 547:1–9

Zhang TN, Wang W, Yang N et al (2020a) Regulation of glucose and lipid metabolism by long non-coding RNAs: facts and research progress. Front Endocrinol (Lausanne) 11:457

Zhang Z, Liu R, Shuai Y et al (2020b) ASCT2 (SLC1A5)-dependent glutamine uptake is involved in the progression of head and neck squamous cell carcinoma. Br J Cancer 122:82–93. https://doi.org/10.1038/s41416-019-0637-9

Zhang W, Li Q, Zhang Y et al (2024) Multiple myeloma with high expression of SLC7A11 is sensitive to erastin-induced ferroptosis. Apoptosis 29:412–423. https://doi.org/10.1007/s10495-023-01909-2

Zhao XY, Li S, Wang GX et al (2014) A long noncoding RNA transcriptional regulatory circuit drives thermogenic adipocyte differentiation. Mol Cell 55:372–382. https://doi.org/10.1016/j.molcel.2014.06.004

Zhao Y, Chai X, Peng J et al (2025) Proline exacerbates hepatic gluconeogenesis via paraspeckle-dependent mRNA retention. Nat Metab 7:367–382. https://doi.org/10.1038/s42255-024-01206-5

Zheng X, Han H, Liu G et al (2017) Lnc RNA wires up hippo and hedgehog signaling to reprogramme glucose metabolism. EMBO J 36:3325–3335. https://doi.org/10.15252/embj.201797609

Zong Y, Wang X, Cui B et al (2023) Decoding the regulatory roles of non-coding RNAs in cellular metabolism and disease. Mol Ther 31:1562–1576

Chapter 5
Long Noncoding RNAs in Aging and Cellular Senescence

Carlota Tavares-Marcos and Bruno Bernardes de Jesus

Abstract Less than 2% of the human genome is translated into proteins, supporting the existence of a set of non-protein-coding transcripts with uncovered biological functions. ncRNAs have been stratified and identified according to their size, function, and/or structure. Among them, long noncoding RNAs (lncRNAs) are an extensive class of ncRNAs being characterized for having 200 or more nucleotides, often showing tissue-specific expression and regulatory roles in cellular processes, such as cell state identity or conversion. Here, we provide an overview of the potential role of lncRNAs in aging and cellular senescence. lncRNAs have been linked to seminal pathways regulating different aging hallmarks, supporting a direct function for lncRNAs during aging. We highlight how lncRNAs are involved in key aging hallmarks, including epigenetic regulation, mitochondrial dysfunction, nutrient sensing, or telomere dysfunction. By compiling evidence on these regulatory networks, we aim to provide an integrated view of how lncRNAs contribute to aging biology and to discuss their emerging potential as novel therapeutic targets, paving the way for the development of lncRNA anti-aging therapies.

Keywords lncRNAs · Aging · Senescence · Stem cells · Telomeres

Abbreviations

AMPK	AMP-activated protein kinase
CR	Caloric restriction
DNA	Deoxyribonucleic acid
DNMT	DNA methyltransferases

C. Tavares-Marcos · B. Bernardes de Jesus (✉)
Department of Medical Sciences and Institute of Biomedicine—iBiMED, University of Aveiro, Aveiro, Portugal
e-mail: brunob.jesus@ua.pt

L. M. Vaschetto (ed.), *Long Non-Coding RNAs, Human Health, and Diseases*, Epigenetics and Human Health 16, https://doi.org/10.1007/978-3-032-19668-2_5

ECs	Endothelial cells
GR	Glucocorticoid receptor
HATs	Histone acetyltransferases
HDACs	Histone deacetylases
HDMs	Histone demethylases
HMTs	Histone methyltransferases
IGF-1	Insulin-like growth factor-1
IL	Interleukins
lincRNA	Long intergenic noncoding RNA
lncRNA-LET	Long noncoding RNA-low expression in tumor
lncRNAs	Long noncoding RNAs
LPS	Lipopolysaccharide
miRNA/miR	Micro-ribonucleic acid
mRNA	Messenger ribonucleic acid
mtDNA	Mitochondrial DNA
mTOR	Mammalian target of rapamycin
NSN	Nutrient-sensing network
Pol I, II, or III	RNA polymerase I, II, or III
PRC	Polycomb repressive complex
RNA	Ribonucleic acid
ROS	Reactive oxygen species
SAHF	Senescence-associated heterochromatin foci
SASP	Senescence-associated secretory phenotype
SIRT	Sirtuins
TERC	Telomerase RNA component
TERRA	Telomeric repeat-containing RNA
TET	Ten-eleven translocation
THRIL	TNF- and hnRNPL-related immunoregulatory lincRNA
TNF-α	Tumor necrosis factor-α
VEGF	Vascular endothelial growth factor
VSMCs	Vascular smooth muscle cells

5.1 The Hallmarks of Aging from the Perspective of lncRNAs

Aging is a natural and universal process correlated with the increased incidence of multiple diseases (Hou et al. 2019; Berben et al. 2021; Zhou et al. 2022). Aging can be defined as a progressive decline of the physiological functions required for survival and fertility (López-Otín et al. 2023). The hallmarks of aging comprise a number of processes intimately associated with the aging process and include stem cell exhaustion, altered intercellular communication, genomic instability, telomere attrition, epigenetic alterations, loss of proteostasis, deregulated nutrient sensing, mitochondrial dysfunction, and cellular senescence (López-Otín et al. 2023). Despite the universality of the process, aging can be delayed or even reversed. In fact, modulation of some of the biological or physiological processes described as hallmarks of aging can lead to improvements in the aging process, in terms of both healthspan and lifespan (e.g., telomerase therapies; Bernardes de Jesus and Blasco 2011, 2013; Jaskelioff et al. 2011; Bernardes de Jesus et al. 2012; de Jesus and Blasco 2012; Bär et al. 2014).

Among the characteristics of the aging process, the appearance of senescent cells is one distinctive characteristic that closely relates to the course of the process. Senescence is characterized by an irreversible cell cycle arrest, caused by either critical telomere shortening or damage to the cell (including DNA damage, oncogene activation, oxidative stress, or chemotherapy; de Jesus et al. 2012; Hernandez-Segura et al. 2018). Although senescence plays a crucial role as a brake on cells or tissues that, for some reason, show damage to telomeres or DNA (leading, e.g., to the inhibition of the division of potentially precancerous cells), the accumulation of senescent cells as tissues age leads to the accumulation of nonoptimal and dysfunctional properties, which are a fingerprint of the aging process (Collado et al. 2007).

Genetic information is mostly translated into function through the sequential reading of DNA tracks, into RNA, and, subsequently, into proteins. Most of the human genome includes, however, a myriad of different fragments of information which don't seem to follow the Central Dogma of Molecular Biology (Mattick et al. 2023). Due to a limitation on DNA-sequencing techniques in the past, priority has been given to the study of protein-coding genes and their associated functions. Although proteins are the building blocks that define cellular function and their inherent processes, it has now been realized that their transcription, assembly, or regulation involves delicate regulatory processes. In fact, less than 2% of the genome potentially encodes proteins, while an astonishing 98% seems to be transcribed into noncoding (nc) RNAs: short and long noncoding RNAs (lncRNAs) (Lander et al. 2001; Venter et al. 2001). ncRNAs have been stratified according to their size, function, and/or structure, such as microRNAs, circular RNAs (circRNAs), lncRNAs, PIWI-interacting RNAs (piRNAs), and snoRNAs (Kung et al. 2013). They share similarities with mRNAs, being Pol II–transcribed (although they can be transcribed by RNA polymerase I or III), spliced, 5-capped, and polyadenylated at the 3′ (Clark and Mattick 2011). Similar to protein-coding genes, conservation is usually higher at the promoter regions, which are enriched for A/T nucleotides and diminished in CpG patterns and contain distinct transcription factor–binding sequences and a unique pattern of chromatin marks, i.e., higher density of methylation at transcriptional start sites, regardless of expression levels (Tonmoy et al. 2022).

lncRNA expression seems to better correlate with cell identity than mRNA or protein expression levels, guiding to the hypothesis that they have a function supporting specific traits between cells, tissues, organs, and/or species. lncRNAs have been shown to exert diverse functional roles, including epigenetic regulation, chromatin remodeling, miRNA sponging, or splice regulation affecting various aspects of cellular homeostasis, including proliferation, survival, migration, metabolism, or genomic stability (Ponting et al. 2009; Clark and Mattick 2011; Prensner and

Chinnaiyan 2011; Lee 2012; Rinn and Chang 2012; Kung et al. 2013; Huarte 2015; Degirmenci and Lei 2016; Schmitt and Chang 2016; Yao et al. 2019; Santos et al. 2020, 2022; Fang et al. 2020; Correia et al. 2021, 2022; Han 2023). In fact, coding RNAs can also exert translation-independent functions, being the distinction between mRNA and lncRNA more conditioned than once thought (Li and Liu 2019). Although lncRNAs lack an open reading frame of significant length (less than 100 amino acids) and do not present coding potential, some have shown to be translated into functional or nonfunctional small peptides (Patraquim et al. 2022).

lncRNAs are usually characterized by decreased levels of gene expression, comparing to mRNA expression, in general, or cognate mRNAs, in particular. This may not necessarily be the result of low RNA copy number in all cells but rather be the result from restricted expression in only a subpopulation of cells (Cuevas-Diaz Duran et al. 2019). Indeed, a considerable fraction (29%) of all expressed lncRNAs is detected in only one of the cell lines studied, whereas 10% were detected in all cell lines. Conversely, a considerable fraction (53%) of all expressed protein-coding genes were constitutive, whereas only 7% were cell line–specific (Djebali et al. 2012). Most lncRNAs are mainly nuclear, being this associated with their function either regulating transcription, organizing subnuclear structure, or mediating chromosomal modifications and interactions. Cytoplasmatic lncRNAs are described for modulating mRNA stability, transcription, and splicing (Sana et al. 2012; Watson et al. 2019; Tavares e Silva et al. 2024). lncRNA control of protein expression or translation may be mediated through different processes: guidance of chromatin factors either activating or repressing protein expression through epigenetic modulation, such as miRNA or protein sponges, decoys of transcription factors or regulators, and Pol II inhibitors, through compartmentalization of RNAs within the cell nucleus or mRNA processing/splicing (Sana et al. 2012; Watson et al. 2019; Tavares e Silva et al. 2024).

Being aging a multifactorial process related to all aspects of cellular homeostasis, it is certainly impacted by the expression of lncRNAs. In fact, lncRNAs have been described as key regulators of several aging-related processes. They can control cellular senescence and the senescence-associated secretory phenotype (SASP) (Panda et al. 2017; Puvvula 2019)—a major feature of senescence cells characterized by the production and secretion of interleukins, cytokines, growth factors, angiogenic factors, and matrix metalloproteases (Kuilman and Peeper 2009; Coppé et al. 2010; Montes et al. 2021). They can also influence stem cell function, tissue regeneration (Sousa-Franco et al. 2018; Tavares e Silva et al. 2024), and other molecular hallmarks of aging and disease (Grammatikakis et al. 2014; Sherazi et al. 2023). On the contrary, the impact of aging on biological processes may severely affect the expression of lncRNAs, adding a level of complexity in understanding the interconnection between both processes, and being occasionally difficult to distinguish between the causes and consequences of lncRNA modulation.

5.2 lncRNAs in SASP Regulation

Although senescent cells lost the property to resume cell division, they are metabolically active, exhibiting distinct metabolic profiles, with an increased activity of β-galactosidase and an altered gene expression pattern (Goldstein 1990; Cristofalo et al. 1998; Debacq-Chainiaux et al. 2009). One of the most prominent hallmarks of senescence is the SASP, a secretion of pro-inflammatory cytokines, chemokines, growth factors, and matrix-modifying enzymes that helps expanding the senescence characteristics, promoting tissue dysfunction, tumor progression, or age-related pathologies (Acosta et al. 2008, 2013; Kuilman and Peeper 2009; Coppé et al. 2010).

Recently, several studies have found that the expression of certain lncRNAs is altered in senescent cells, namely, *lncRNA*-low expression in tumor (*LET*), *lincRNA-Cox2, NEAT1, lnc-IL7R, Lethe, ANRIL*, and TNF- and hnRNPL-related immunoregulatory lincRNA (*THRIL*). These lncRNAs can directly or indirectly regulate the expression and secretion of SASP components, either at the transcriptional, post-transcriptional, and translational level, or even by interacting with RNA-binding proteins (Abdelmohsen et al. 2013; Grammatikakis et al. 2014).

lncRNA-LET positively regulates SASP by suppressing nuclear factor 90 (NF90), a protein that inhibits the translation of SASP factors such as MCP1, CXCL1, and IL-6; however, this lncRNA is downregulated in senescent cells. Silencing of *lncRNA-LET* leads to NF90 accumulation and reduced expression of SASP (Tominaga-Yamanaka et al. 2012; Yang et al. 2013). *lincRNA-Cox2* modulates SASP by repressing or enhancing the transcription of pro-inflammatory genes. *lincRNA-Cox2* interacts with hnRNP-A/B and A2/B1 to repress or enhance transcription of certain pro-inflammatory genes, such as IL-6, being its role in SASP context-dependent (Carpenter et al. 2013; Hu et al. 2016; He et al. 2018). The lncRNA *NEAT1* was shown to be upregulated in senescent vascular smooth muscle cells (VSMCs), promoting SASP by increasing the expression of pro-inflammatory cytokines such as IL-8. SASP enhancement seems to be mediated through NEAT1 role in forming nuclear paraspeckles and subsequent sequestering of transcription inhibitors of IL-8 (Naganuma and Hirose 2013; Imamura et al. 2014; Rossi et al. 2023).

lnc-IL7R is upregulated in response to immunomodulatory stimulus such as lipopolysaccharide (LPS) and negatively regulates SASP by reducing the expression of pro-inflammatory cytokines and other molecules like IL-6, IL-8, and VCAM-1 (Cui et al. 2014). lncRNA *Lethe* is regulated by pro-inflammatory signals (such as TNF-α and IL-1β) via NF-κB activation. *Lethe* negatively regulates the transcription of downstream SASP cytokines by inhibiting NF-κB activity through the interaction with the NF-κB subunit p65 (Rapicavoli et al. 2013). The lncRNA *ANRIL* was shown to be significantly upregulated in senescent endothelial cells (ECs). *ANRIL* was shown to regulate miR-146a expression through promoter alterations, enhancing miR-146a delivery to macrophages, indicating that *ANRIL* mediates critical interactions between ECs and macrophages, inhibiting vascular endothelial growth factor (VEGF) secretion and disrupting endothelial neovascularization. By contributing to SASP, which influences macrophage-derived VEGF release, *ANRIL* downregulation may work as a potential target for therapeutic strategies aimed at improving neovascularization in age-related ischemic diseases (Lin et al. 2025). The lncRNA *THRIL* promotes SASP through the formation of a complex including hnRNPL that binds to the TNF promoter and drives its expression and the transcriptional activation of the cytokines TNF-α, IL-8, CXCL10, CCL1, and CSF1 (Li et al. 2014).

5.3 lncRNAs in Epigenetic Alterations

Epigenetic alterations, including DNA methylation and histone modifications (e.g., heterochromatin remodeling), are one prominent molecular hallmark of aging and a feature of cellular senescence (Pal and Tyler 2016). lncRNAs are important regulators of such alterations, whether by interacting with epigenetic regulators or by controlling chromatin topology (Herman et al. 2022; Mangiavacchi et al. 2023). In the nucleus, lncRNAs can guide epigenetic regulators or arrange chromatin folding and compartmentalization to direct enhancer-promoter communication, regulating protein-coding gene epigenetic status and thus their expression (Wang and Chang 2011; Long et al. 2017; Deforzh et al. 2022). lncRNA misregulation can have a significant impact on physiological processes and pathological conditions, like aging (He et al. 2018; Statello et al. 2020; Oo et al. 2022).

DNA methylation is characterized by the addition of a methyl group on a cytosine (C) fifth carbon (5-methylcytosine, 5mC) by DNA methyltransferases (DNMT) in the context of CpG dinucleotides in mammals (Deaton and Bird 2011), and its patterns change in aging and senescence, with DNA losing methylation (hypomethylation) from CpGs located in repeat regions of the genome and concomitant increases in DNA methylation (hypermethylation) in CpGs localized in the promoter regions of specific genes, including tumor suppressors (Baylin and Ohm 2006; Baylin and Jones 2011). DNMT1 is responsible for maintaining DNA methylation status during DNA replication, while DNMT3a/3b are responsible for de novo DNA methylation (Okano et al. 1999; Weber et al. 2007). DNMT action can be opposed through 5mC oxidation mediated by ten-eleven translocation (TET), erasing DNA methylation (Verma et al. 2018). lncRNAs can regulate DNA methylation, controlling the expression of target genes, through different manners. They can recruit (directly or indirectly) or repel DNMTs/TETs to specific genes (Frank et al. 2019; Huang et al. 2022). Besides, they can regulate DNMT activity by controlling the DNMT cofactor (SAM/SAH) levels (Zhou et al. 2015, 2019). Lastly, there are lncRNAs capable of regulating DNMT expression per se at transcriptional or translational levels or stabilizing DNMT mRNA (Jin et al. 2020; Geng et al. 2021). Some of those lncRNAs are *H19, Kcnq1ot, Airn, ecCEBP, pRNA, PAPAS, PTENpg1-AS*, and *TARID* (Fu et al. 2008; Schmitz et al. 2010; Latos et al. 2012; Di Ruscio et al. 2013; Travers et al. 2013; Johnsson et al. 2013; Bierhoff et al. 2014; Arab et al. 2014).

Chromatin consists of a flexible and dynamic structure of DNA and histones and can exist as heterochromatin (transcriptionally inactive) or euchromatin (transcriptionally active) (Swagatika and Tomar 2016; Wang et al. 2022). The remodeling of chromatin can happen on specific chromosomes or chromosome domains. In aged cells, there is a significant chromatin remodeling, influenced by histone modifications and variants, DNA methylation and ncRNAs, associated with the formation of senescence-associated heterochromatin foci (SAHF) (Narita et al. 2003). Heterochromatin is disrupted and lost with aging, leading to chromosomal instability (Larson et al. 2012). lncRNAs can regulate gene expression through their interactions with chromatin-modifying enzymes, especially with histone or DNA-modifying enzymes, including polycomb repressive complex (PRC), MLL/TrxG complex, histone demethylase (HDM) LSD1, DNMT1, and DNA demethylation regulator GADD45a (Tsai et al. 2010; Wang et al. 2011; Di Ruscio et al. 2013; Brockdorff 2013; Arab et al. 2014). This regulation can occur through the direct binding of lncRNAs to chromatin-modifying enzymes or through the incorporation of lncRNAs into the chromatin-modifying complex (Ilik et al. 2013; Maenner et al. 2013; Arab et al. 2014). Additionally, the histone components of chromatin are subject to different posttranslational alterations (acetylation, methylation, ubiquitylation, SUMOylation, and phosphorylation), leading to the creation of binding sites for chromatin proteins that can alter the degree of chromatin compaction (Tan et al. 2011; Rothbart and Strahl 2014). The addition and removal of these modifications are tightly regulated by histone methyltransferases (HMTs), HDMs, histone acetyltransferases (HATs), and histone deacetylases (HDACs); however, in aging and senescence, the activity and function of such enzymes change, causing alterations in the epigenome. The most common epigenetic alterations in aging and senescence are H4K16 acetylation, H4K20 trimethylation, H3K4 trimethylation, H3K9 methylation, and H3K27 trimethylation, whose levels increased or decreased is usually context-dependent (Wang et al. 2022). There are some lncRNAs implicated in these modifications by modifying, guiding, or recruiting histone and DNMT, for example, *Kcnq1ot1*, *NeST*, *ANRIL*, *ANRASSF1*, and *PINT* (Pandey et al. 2008; Yap et al. 2010; Gomez et al. 2013; Beckedorff et al. 2013; Marín-Béjar et al. 2013).

5.4 lncRNAs in Mitochondrial Dysfunction

Mitochondria are fundamental in energy regulation by producing energy through oxidative phosphorylation and in metabolic homeostasis, calcium balance, and dietary substrate interconversion (Kluge et al. 2013). These organelles have their own protein synthesis apparatus, replicate themselves, and are the primary intracellular source of reactive oxygen species (ROS) (Kausar et al. 2018). Although ROS have important roles in cellular differentiation and homeostasis (Sena and Chandel 2012), the highly reactive species can lead to oxidative stress and contribute to age-associated processes and diseases (Warraich et al. 2020), including metabolic syndrome, neurodegenerative and cardiac diseases, and cancer, especially when combined with other factors like epigenetic alterations (Zhu et al. 2022). Thus, mitochondrial and cellular homeostasis maintenance requires tight quality control, not only to guarantee a balance between the generation of new and removal of damaged mitochondria, but also to keep ROS levels low and steady (Palikaras et al. 2015; Srivastava 2017). There is evidence that mitochondrial dysfunction arises with age triggered by accumulation of mitochondrial DNA (mtDNA) mutations, elevated ROS production, protein oxidation, abnormal energy metabolism, and reduced mitochondrial biogenesis, leading to cellular senescence (Natarajan et al. 2020; Amorim et al. 2022).

lncRNAs can regulate different mitochondrial processes and mediate communication between mitochondria and nuclei, whether they are expressed in mitochondria, nuclei, or cytoplasm (Villegas et al. 2007). The lncRNAs encoded in mitochondria (mt-lncRNAs) include *SncmtRNA*, *ASncmtRNA-1* and *ASncmtRNA-2*, *MDL1*, *LncCytb*, *LncND6*, and *LncND5*, and some of them can be translocated to the cytosol or remain within mitochondria and are associated with cell proliferation and replicative state (Villegas et al. 2007; Burzio et al. 2009; Rackham et al. 2011; Liu and Shan 2021; Li et al. 2022). Regarding the nuclear-encoded lncRNAs, they can be transported to the mitochondria or to mitochondrial outer membrane, regulating mitochondrial copy number, status, morphology, and function; some examples are *SAMMSON*, *SRA*, *MALAT1*, and *RMRP*. There are also cytoplasmatic lncRNAs with impact on mitochondria, including *Cerox1*, *Tug1*, and *UCA1*, that can directly target or indirectly influence mitochondrial-related genes, transcripts, or proteins (Gusic and Prokisch 2020). The functions and mechanism of action of these lncRNAs are described in Table 5.1.

5.5 lncRNAs in Deregulated Nutrient Sensing

The nutrient-sensing network (NSN) consists of several essential pathways, centered on insulin and insulin-like growth factor-1 (IGF-1), mammalian target of rapamycin (mTOR), AMP-activated protein kinase (AMPK), and sirtuins (SIRT), that act as molecular links between nutrient availability, energy homeostasis, and aging (López-Otín et al. 2023). Such network is highly conserved during evolution and a crucial regulator of cellular activity, including glucose, nucleotide, and lipid metabolism, autophagy, mitochondrial, mRNA, and ribosome biogenesis, proteasomal activity, and protein synthesis (Wullschleger et al. 2006; Kim et al. 2011; Hardie et al. 2012; Zhao et al. 2015). Its activity relies on nutrition and stress status, activating anabolism in the presence of nutrients and low stress levels, whereas in case of nutrient shortage and stress, cellular defense pathways are induced (Wullschleger et al. 2006; Hardie et al. 2012). NSN deregulation can trigger or accelerate cellular senescence and is recognized as an aging hallmark, contributing to age-related conditions, such as cancer, cardiovascular diseases, and type 2 diabetes (Saxton and Sabatini 2017).

Table 5.1 Summary of lncRNAs that are implicated in senescence and aging processes, and their respective functions and mechanism of action

lncRNA	Species	Function	Mechanism	References
SASP				
lncRNA-LET	Human	Promotes SASP	Suppresses NF90, increasing the translation of SASP factors	Tominaga-Yamanaka et al. (2012) and Yang et al. (2013)
lincRNA-Cox2	Mouse	Promotes/suppresses SASP	Interacts with hnRNP-A/B and A2/B1 to repress or enhance transcription of certain pro-inflammatory genes	Carpenter et al. (2013), Hu et al. (2016), and He et al. (2018)
NEAT1	Human	Promotes SASP	Promotes the formation of nuclear paraspeckles, increasing the expression of pro-inflammatory cytokines	Naganuma and Hirose (2013), Imamura et al. (2014), and Rossi et al. (2023)
Lnc-IL7R	Human	Suppresses SASP	Regulates the expression of pro-inflammatory mediators in response to LPS	Cui et al. (2014)
Lethe	Mouse	Suppresses SASP	Interacts with the NF-κB subunit RelA, inhibiting the DNA-binding activity of NF-κB	Rapicavoli et al. (2013)
THRIL	Human	Promotes SASP	Interacts with RBP hnRNP L, promoting TNF-α transcription	Li et al. (2014)
ANRIL	Human	Promotes SASP	Regulates the expression of miR146a at the transcriptional level, facilitating the transfer of miR146a to macrophages as a component of the SASP	Lin et al. (2025)
Epigenetic alterations				
DNA methylation				
H19	Human/mouse	Imprinting control	Recruits methylation regulators to Igf2/H19 locus	Fu et al. (2008)
Kcnq1ot	Mouse	Chromatin regulation	Guides DNMT1 to Kcnq1 locus	Travers et al. (2013)
Airn	Mouse	Gene silencing	Recruits G9a to silence Igf2r	Latos et al. (2012)
ecCEBP	Human	Methylation regulation	Interacts with DNMT1 at CEBPα	Di Ruscio et al. (2013)
pRNA	Mouse	rDNA silencing	Guides DNMT3b to rDNA promoter	Schmitz et al. (2010)
PAPAS	Human	rDNA repression	Recruits Suv4-20 h2 and HDACs	Bierhoff et al. (2014)

(continued)

Table 5.1 (continued)

lncRNA	Species	Function	Mechanism	References
PTENpg1-AS	Human	Tumor suppressor regulation	Modulates PTEN via DNMT3a	Johnsson et al. (2013)
TARID	Human	DNA demethylation	Recruits GADD45A to TCF21 promoter	Arab et al. (2014)
Histone modifications				
Kcnq1ot1	Human/mouse	Chromatin silencing	Recruits PRC2 and G9a	Pandey et al. (2008)
NeST	Mouse	Histone methylation	Promotes H3K4me3 at IFN-γ	Gomez et al. (2013)
ANRIL	Human	INK4 regulation	Interacts with PRC1/2	Yap et al. (2010)
ANRASSF1	Human	Tumor suppression	Regulates RASSF1A chromatin state	Beckedorff et al. (2013)
PINT	Human	P53 response	Recruits PRC2 to repress genes	Marín-Béjar et al. (2013)
Mitochondrial dysfunction				
SncmtRNA	Human	Cell proliferation	Involved in mitochondrial biogenesis	Villegas et al. (2007) and Burzio et al. (2009)
ASncmtRNA-1	Human	Cell cycle arrest	Regulates apoptosis	Burzio et al. (2009)
ASncmtRNA-2	Human	Tumor suppression	Induces apoptosis	Burzio et al. (2009)
MDL1	Human	mtDNA transcription	Regulates H-strand promoters	Liu and Shan (2021) and Li et al. (2022)
LncCytb	Human	mRNA regulation	Stabilizes Cytb mRNA	Rackham et al. (2011)
LncND6	Human	Complex I regulation	Regulates ND6 mRNA stability	Rackham et al. (2011)
LncND5	Human	Complex I regulation	Modulates ND5 expression	Rackham et al. (2011)
SAMMSON	Human	Mitochondrial function in melanoma	Regulates p32 metabolism	Vendramin et al. (2018)
SRA	Human	Mitochondrial gene expression	Modulates nuclear receptors	Colley et al. (2008)
MALAT1	Human	Mitochondrial dynamics	Regulates oxidative stress response	Zhao et al. (2019)
RMRP	Human	RNA processing	Component of RNase MRP	Noh et al. (2016)
Cerox1	Mouse	Complex I activity	miRNA sponge modulating complex I	Sirey et al. (2019)
TUG1	Mouse	Mitochondrial biogenesis	Enhances PGC-1α-mediated gene expression	Jianyin et al. (2016)

(continued)

Table 5.1 (continued)

lncRNA	Species	Function	Mechanism	References
UCA1	Human	Mitochondrial regulation	Controls membrane potential	Li et al. (2017)
Deregulated nutrient sensing				
H19	Human	Glucose metabolism	Inhibits let-7 to regulate IGF signaling	Geng et al. (2018)
NBR2	Human	AMPK activation	Activated during stress	Li et al. (2024) and Yang et al. (2024)
MALAT1	Human	mTOR activation	Enhances mTORC1 signaling	Zhu et al. (2019) and Peng et al. (2020)
TUG1	Mouse	Metabolic regulation	Regulates SIRT1 and mitochondria	Wei et al. (2022)
GAS5	Human	mTOR suppression	Decoy for GR	Kino et al. (2010) and Xagorari et al. (2021)
Telomerase modulation, telomere attrition, and cellular senescence				
GUARDIN	Human	Telomere maintenance	Stabilizes TRF2	Rossi and Gorospe (2020)
TERRA	Human	Telomere transcription	Modulates telomerase access	Bettin et al. (2024)
TERC	Human	Telomerase RNA component	Provides RNA template	Blasco et al. (1997) and Wong et al. (2008)

In more recent studies, lncRNAs have been described as important regulators of specific pathways within NSN. The insulin/IGF-1 and mTOR pathways are key drivers of aging, with their inhibition being associated with extended lifespan in several organisms. The lncRNA *H19* is highly expressed during development, and this expression decreases in aged tissues; besides, it has been shown to negatively affect the IGF-2 axis and modulate insulin sensitivity. This lncRNA sponges let-7 microRNAs, which target insulin receptor substrates, indirectly sustaining insulin/IGF signaling. In aging, the decreased levels of *H19* may contribute to metabolic dysfunction and insulin resistance (Geng et al. 2018). In the case of the lncRNA *NBR2*, it is induced under energy stress and acts as a positive regulator of AMPK signaling by interacting with AMPKα, increasing its kinase activity and promoting cellular adaptation to nutrient scarcity and inhibiting mTORC1. *NRB2* downregulation can lead to impaired energy stress responses and to age-related metabolic diseases, including type 2 diabetes and cancer (Li et al. 2024; Yang et al. 2024). Additionally, *MALAT1* has been shown to be implicated in both mTOR and autophagy pathways. The AKT/mTOR pathway is activated by *MALAT1*, leading to autophagy inhibition in gastric cancer cells, which suggests that this lncRNA has a role in nutrient sensing and cellular homeostasis. However, its function in aging appears to be context-dependent, with autophagy suppression occurring only under certain conditions, while in others, like oxidative stress, *MALAT1* promotes cell viability (Zhu et al. 2019; Peng et al. 2020).

SIRT is a NAD^+-dependent deacetylase with a crucial role in nutrient sensing, whose activity decreases with aging as a consequence of altered NAD^+ metabolism. Emerging evidence suggests that SIRT activity and expression can be controlled by certain lncRNAs, as is the case of *TUG1*. It was demonstrated that *TUG1* can regulate the expression of SIRT1, promoting mitochondrial biogenesis and oxidative metabolism. Thus, this lncRNA contributes to improve metabolic efficiency and to reduce oxidative damage during aging, especially in caloric restriction (CR) and exercise situations. However, the decrease in *TUG1* levels associated with age can impair SIRT-mediated cellular defenses (Wei et al. 2022).

Additionally, CR has been described as a modulator of lifespan expansion, due to its impact on nutrient-sensing pathways, with lncRNAs emerging as mediators of CR beneficial effects. For example, the lncRNA *GAS5* functions as a decoy for the glucocorticoid receptor (GR), inhibiting its transcriptional activity and impacting cell survival and metabolic activity. Under starvation, *GAS5* accumulates, decreasing GR-mediated metabolic gene expression, leading to a shift from anabolic to catabolic metabolism, that mimics CR by preventing activation of stress-related transcription by glucocorticoids. On the other hand, the expression of *GAS5* declines with age and GR activity is disinhibited, favoring pathways that increase inflammation, oxidative damage, and insulin resistance (Kino et al. 2010; Xagorari et al. 2021). The roles of lncRNAs in nutrient sensing and metabolic regulation are detailed in Chap. 4 by Guil's research group.

5.6 lncRNAs in Telomerase Modulation and Telomere Attrition

Telomeres are repetitive DNA sequences at the chromosomal ends. In humans, telomeres have the conserved sequence TTAGGG, repeated between 12 k and 15 k times (Shore 1998; Blasco 2002). If the sequence seems to be phylogenetically conserved, the number of repetitions seems to vary depending on the organism or, within the same organism, between different cell types. In mammalian cells, telomeres are shielded by the shelterin/telosome complex comprising six core proteins: TRF1 and TRF2, which bind double-stranded telomeric DNA to regulate length and promote T-loop formation; TIN2, which connects TRF1 and TRF2; RAP1, which prevents inappropriate recombination; TPP1, which recruits and activates telomerase; and POT1, which binds the single-stranded overhang to prevent DNA damage signaling (de Lange 2005; Palm and De Lange 2008; Martinez and Blasco 2010). This complex orchestrates the recruitment of several factors from diverse cellular pathways to ensure their protection and maintenance. The telomeric DNA adopts a safeguarding T-loop structure, wherein the single-stranded 3′ overhang penetrates the double-stranded telomeric region (de Lange 2004; Wang et al. 2004), protecting single-strand DNA from degradation, and preventing the activation of a DNA damage response (Hemann et al. 2001; Hao et al. 2005; Armanios 2009). The attrition over successive cell divisions can induce chromosomal instability, significantly contributing to genomic rearrangements and potentially fostering tumorigenesis or senescence (Hemann et al. 2001; Hao et al. 2005; Armanios 2009; Vera et al. 2012; Tomas-Loba et al. 2013). Here, the telomeric repeat-containing RNA (*TERRA*) lncRNA plays key roles in supporting telomere maintenance (Rivosecchi et al. 2024). *TERRA* levels are another factor essential to maintain the integrity of telomeric DNA (Luke and Lingner 2009; Lopez de Silanes et al. 2010). In addition to cell division, accelerated telomere shortening has also been identified in cells exposed to external insults such as oxidative damage (Barnes et al. 2019). It is assumed that the peculiar sequential structure of telomeres (TTAGGG), rich in guanine, is a preferential target for some type of damage, serving as an active dock for accumulation and recognition of DNA damage (Barnes et al. 2019). Therapies directed to decrease the number of telomere-damaged cells (for instance, through telomere extension) are therefore viewed as potentially effective in slowing the aging progression. In this regard, telomerase constitutive expression by using mouse transgenesis in adult tissues has pinpointed a role of telomerase in tissue fitness and prevention of aging (Tomas-Loba et al. 2008). Telomerase expression, in a cancer protective background, has demonstrated to increase overall healthspan and lifespan (Bernardes de Jesus and Blasco 2011, 2013; Bernardes de Jesus et al. 2011, 2012; de Jesus and Blasco 2012; Vera et al. 2013). Additionally, when this therapy was adapted to a particular disease (such as myocardial infarction) it also demonstrated protective capacities, demonstrating that the expression of telomerase, and subsequent telomere extension, leads to healthy cardiac tissues and an increase in post-infarction survival (Bär et al. 2014). Telomere length is regulated by a ribonucleoprotein complex that contains the protein TERT and the lncRNA *TERC* (telomerase RNA component), as well as by *TERRA* lncRNA. In fact, ncRNAs are emerging as major regulators of telomere length homeostasis (Rossi and Gorospe 2020).

TERC is the lncRNA component that, together with TERT, forms the telomerase extension ribonucleoprotein complex (Jafri et al. 2016). lncRNA *TERC* is an integral part of the telomere extension complex and is necessary for the canonical function of telomerase in telomere extension. The main function of *TERC* is to position the extension complex on the telomere, serving as a docking station and template, and to support telomere extension through its catalytic function. Supporting its crucial role in telomere extension, alterations in its functioning have a severe impact on the maintenance of telomere homeostasis and, consequently, on various aspects related to it, such as survival, DNA damage, or senescence (Blasco et al. 1997; Wong et al. 2008; Sarek et al. 2015; Jafri et al. 2016; Scarabino et al. 2017). Recently, telomere-independent roles of TERC in senescence have been described through its involvement in mitochondrial functions (Cheng et al. 2018; Zheng et al. 2019). TERC seems to be imported into mitochondria and processed to a shorter form named TERC-53. TERC-53 acts downstream of mitochondria as an indicator of mitochondrial functions.

The lncRNA *TERRA* is transcribed from telomeres in several organisms including humans, yeasts, and nematodes (Azzalin et al. 2007; Azzalin and Lingner 2008; Schoeftner and Blasco 2009; Lopez de Silanes et al. 2010). *TERRA* expression has been reported in normal mammalian cells, including fibroblasts, embryonic stem cells, and germ cells (Azzalin et al. 2007; Azzalin and Lingner 2008; Schoeftner and Blasco 2009; Lopez de Silanes et al. 2010). *TERRA* transcripts are RNA Pol II–transcribed from sequences arising from the subtelomeric region toward the telomeres, favoring a 3′-extremity G-rich. Contrary to *TERC*, *TERRA* seems to negatively affect telomere length through a direct binding and competitive behavior through the telomere sequences. Recent reports demonstrated that *TERRA* transcripts relocate away from chromosome ends during telomere lengthening, with a consequent increase in "TERRA-free" telomerase molecules at telomeres. Upon transcription, *TERRA* relocates to (long) chromosome ends, showing that telomeric *TERRA* transcripts inhibit telomere elongation by telomerase acting in trans. *TERRA* transcription in mammals is tightly regulated by multiple mechanisms. *TERRA* expression is regulated by epigenetic alterations in the CpG-rich promoters at the subtelomeric regions, in particular through the DNMT1- and DNMT3b-dependent CpG methylation. *TERRA* transcription seems also to depend on the chromatin assessment, where a decreased level of repressive marks associates with higher expression of *TERRA* transcripts (Schoeftner and Blasco 2008, 2009; Marion et al. 2009). From a variety of functions, it has been described that *TERRA* may model chromatin, promote telomere capping, and assist the recruitment of *TERRA*-binding proteins with a role in replication and DNA damage response pathways with an influence in telomere homeostasis, thereby impacting senescence (Bettin et al. 2024).

lncRNA *GUARDIN* plays an important role at telomeres in stabilizing the shelterin complex, regulating telomere extension through telomerase recruitment, responding to DNA damage, and modulating chromatin. *GUARDIN* is a lncRNA identified for its role in genomic stability (Hu et al. 2018). It is transcribed from a *locus* next to the TERT gene, which encodes the catalytic subunit of telomerase. *GUARDIN* role in telomeres can depend on distinct pathways: (1) through interaction and maintenance of some shelterin components, such as TRF2, a core shelterin protein, enhancing its recruitment to telomeres; (2) through regulation of telomere length, by the direct modulation of TERT expression and/or activity; (3) through the recruitment of chromatin-modifying complexes to telomeres, leading to a protective milieu that supports telomere function and represses recombination events; or (4) through the displacement of the DNA repair machinery (such as BRCA1 protein) away from telomeres (Rossi and Gorospe 2020). Despite the consolidating evidence that *GUARDIN* may play at telomeres, the precise molecular interactions between *GUARDIN* and the shelterin components or telomerase remain to be further explored. On the other hand, the regulation of *GUARDIN* expression is also not fully understood.

5.7 Conclusions

In the present book chapter, we highlight the involvement of lncRNAs in biological and physiological processes related to aging. Although lncRNAs have extensive functions in each of the described hallmarks of aging, we decided to focus on a selection of processes. It could have been equally valid to look at the role of lncRNAs in the aging of tissues, namely, how their altered expression is associated with the normal aging of organs, or to focus on the role of lncRNAs in age-related diseases, which are ultimately related to increased senescence. The role of lncRNAs in these areas has been widely discussed in the literature, for example, the association between lncRNA and tissue homeostasis (Amaral et al. 2013; Salviano-Silva et al. 2018), lncRNAs and stemness (Sousa-Franco et al. 2018; Mirzadeh Azad et al. 2021; Emarati et al. 2025), lncRNAs and heart (Sallam et al. 2018; Fang et al. 2020; Hu et al. 2023; Kim et al. 2023; Zhao et al. 2025; Xie et al. 2025), and lncRNAs in cancer (Huarte and Rinn 2010; Prensner and Chinnaiyan 2011; Huarte 2015; Jiang et al. 2019; Qian et al. 2020; Ahmad et al. 2023).

Taking into account what has been discussed here, we may support lncRNAs' role in health and disease, particularly in the processes that control aging (it should be noted here that these processes are transversal to several organs). Understanding their function remains a challenge, given their low levels of expression and their often-local mode of action, acting frequently at the site of transcription. Some of the techniques employed for lncRNA understanding still alter the system and do not unequivocally prove their role. However, we are expecting rapid progress in the field of analyzing and characterizing lncRNAs, with the prospect of new therapies based on these transcripts in the near future.

Acknowledgments We thank members of IBIMED for insightful discussions and advice.

Funding This work was supported by Fundação para a Ciência e Tecnologia, project co-funded by the European Regional Development Fund (FEDER) through the POR Lisboa 2020—Programa Operacional Regional de Lisboa, PORTUGAL 2020, and by Fundação para a Ciência e a Tecnologia (FCT), through the project 2022.01199.PTDC (DOI:10.54499/2022.01199.PTDC) and Institute of Biomedicine—iBiMED (UIDP/04501/2020, UIDB/04501/2020, and UID 4501). C. Tavares-Marcos was funded by an individual predoctoral fellowship (BI/REIT/10332/2024).Compliance with Ethical Standards

Conflict of Interest Statement The authors declare that the research was conducted in the absence of any commercial or financial relationships that could be construed as a potential conflict of interest.

Ethical Approval This chapter is a review of previously published accounts, as such no animal or human studies were performed.

Author Contributions C.T-M and B. BdJ conceptualized, revised the paper, and wrote the manuscript. All authors contributed to the article and approved the submitted version.

References

Abdelmohsen K, Panda A, Kang MJ et al (2013) Senescence-associated lncRNAs: senescence-associated long noncoding RNAs. Aging Cell 12:890–900. https://doi.org/10.1111/ACEL.12115

Acosta JC, O'Loghlen A, Banito A et al (2008) Chemokine signaling via the CXCR2 receptor reinforces senescence. Cell 133:1006–1018. https://doi.org/10.1016/j.cell.2008.03.038

Acosta JC, Banito A, Wuestefeld T et al (2013) A complex secretory program orchestrated by the inflammasome controls paracrine senescence. Nat Cell Biol 15:978–990. https://doi.org/10.1038/NCB2784

Ahmad M, Weiswald L-B, Poulain L et al (2023) Involvement of lncRNAs in cancer cells migration, invasion and metastasis: cytoskeleton and ECM crosstalk. J Exp Clin Cancer Res 42:173. https://doi.org/10.1186/s13046-023-02741-x

Amaral PP, Dinger ME, Mattick JS (2013) Non-coding RNAs in homeostasis, disease and stress responses: an evolutionary perspective. Brief Funct Genomics 12:254–278. https://doi.org/10.1093/bfgp/elt016

Amorim JA, Coppotelli G, Rolo AP et al (2022) Mitochondrial and metabolic dysfunction in ageing and age-related diseases. Nat Rev Endocrinol 18(4):243–258. https://doi.org/10.1038/s41574-021-00626-7

Arab K, Park YJ, Lindroth AM et al (2014) Long noncoding RNA TARID directs demethylation and activation of the tumor suppressor TCF21 via GADD45A. Mol Cell 55:604–614. https://doi.org/10.1016/j.molcel.2014.06.031

Armanios M (2009) Syndromes of telomere shortening. Annu Rev Genomics Hum Genet 10:45–61. https://doi.org/10.1146/annurev-genom-082908-150046

Azzalin CM, Lingner J (2008) Telomeres: the silence is broken. Cell Cycle 7:1161–1165

Azzalin CM, Reichenbach P, Khoriauli L et al (2007) Telomeric repeat containing RNA and RNA surveillance factors at mammalian chromosome ends. Science 318:798–801. https://doi.org/10.1126/science.1147182

Bär C, De Jesus BB, Serrano R et al (2014) Telomerase expression confers cardioprotection in the adult mouse heart after acute myocardial infarction. Nat Commun 5:1–14. https://doi.org/10.1038/ncomms6863

Barnes RP, Fouquerel E, Opresko PL (2019) The impact of oxidative DNA damage and stress on telomere homeostasis. Mech Ageing Dev 177:37–45. https://doi.org/10.1016/j.mad.2018.03.013

Baylin SB, Jones PA (2011) A decade of exploring the cancer epigenome — biological and translational implications. Nat Rev Cancer 11(10):726–734. https://doi.org/10.1038/nrc3130

Baylin SB, Ohm JE (2006) Epigenetic gene silencing in cancer – a mechanism for early oncogenic pathway addiction? Nat Rev Cancer 6:107–116. https://doi.org/10.1038/nrc1799

Beckedorff FC, Ayupe AC, Crocci-Souza R et al (2013) The Intronic Long noncoding RNA ANRASSF1 recruits PRC2 to the RASSF1A promoter, reducing the expression of RASSF1A and increasing cell proliferation. PLoS Genet 9:e1003705. https://doi.org/10.1371/JOURNAL.PGEN.1003705

Berben L, Floris G, Wildiers H, Hatse S (2021) Cancer and aging: two tightly interconnected biological processes. Cancers (Basel) 13:1–20. https://doi.org/10.3390/CANCERS13061400

Bernardes de Jesus B, Blasco MA (2011) Aging by telomere loss can be reversed. Cell Stem Cell 8:3–4. https://doi.org/10.1016/j.stem.2010.12.013

Bernardes de Jesus B, Blasco MA (2013) Telomerase at the intersection of cancer and aging. Trends Genet 29:513–520. https://doi.org/10.1016/j.tig.2013.06.007

Bernardes de Jesus B, Schneeberger K, Vera E et al (2011) The telomerase activator TA-65 elongates short telomeres and increases health span of adult/old mice without increasing cancer incidence. Aging Cell 10:604–621. https://doi.org/10.1111/j.1474-9726.2011.00700.x

Bernardes de Jesus B, Vera E, Schneeberger K et al (2012) Telomerase gene therapy in adult and old mice delays aging and increases longevity without increasing cancer. EMBO Mol Med 4:691–704. https://doi.org/10.1002/emmm.201200245

Bettin N, Querido E, Gialdini I et al (2024) TERRA transcripts localize at long telomeres to regulate telomerase access to chromosome ends. Sci Adv 10:eadk4387. https://doi.org/10.1126/sciadv.adk4387

Bierhoff H, Dammert MA, Brocks D et al (2014) Quiescence-induced LncRNAs trigger H4K20 trimethylation and transcriptional silencing. Mol Cell 54:675–682. https://doi.org/10.1016/j.molcel.2014.03.032

Blasco MA (2002) Telomerase beyond telomeres. Nat Rev Cancer 2:627–633. https://doi.org/10.1038/nrc862

Blasco MA, Lee H-W, Hande MP et al (1997) Telomere shortening and tumor formation by mouse cells lacking telomerase RNA. Cell 91:25–34. https://doi.org/10.1016/S0092-8674(01)80006-4

Brockdorff N (2013) Noncoding RNA and Polycomb recruitment. RNA 19:429–442. https://doi.org/10.1261/RNA.037598.112

Burzio VA, Villota C, Villegas J et al (2009) Expression of a family of noncoding mitochondrial RNAs distinguishes normal from cancer cells. Proc Natl Acad Sci USA 106:9430–9434. https://doi.org/10.1073/PNAS.0903086106/SUPPL_FILE/0903086106SI.PDF

Carpenter S, Aiello D, Atianand MK et al (2013) a Long noncoding RNA mediates both activation and repression of immune response genes. Science 341:789–792. https://doi.org/10.1126/science.1240925

Cheng Y, Liu P, Zheng Q et al (2018) Mitochondrial trafficking and processing of telomerase RNA TERC. Cell Rep 24:2589–2595. https://doi.org/10.1016/j.celrep.2018.08.003

Clark MB, Mattick JS (2011) Long noncoding RNAs in cell biology. Semin Cell Dev Biol 22:366–376. https://doi.org/10.1016/J.SEMCDB.2011.01.001

Collado M, Blasco MA, Serrano M (2007) Cellular senescence in cancer and aging. Cell 130:223–233. https://doi.org/10.1016/J.CELL.2007.07.003

Colley SM, Iyer KR, Leedman PJ (2008) The RNA coregulator SRA, its binding proteins and nuclear receptor signaling activity. IUBMB Life 60:159–164. https://doi.org/10.1002/IUB.22

Coppé JP, Desprez PY, Krtolica A, Campisi J (2010) The senescence-associated secretory phenotype: the dark side of tumor suppression. Ann Rev Pathol Mech Dis 5:99–118. https://doi.org/10.1146/ANNUREV-PATHOL-121808-102144

Correia M, Bernardes de Jesus B, Nóbrega-Pereira S (2021) Novel insights linking lncRNAs and metabolism with implications for cardiac regeneration. Front Physiol 12:586927. https://doi.org/10.3389/FPHYS.2021.586927/BIBTEX

Correia M, Santos F, da Silva Ferreira R et al (2022) Metabolic determinants in cardiomyocyte function and heart regenerative strategies. Meta 12:500. https://doi.org/10.3390/metabo12060500

Cristofalo VJ, Volker C, Francis MK, Tresini M (1998) Age-dependent modifications of gene expression in human fibroblasts. Crit Rev Eukaryot Gene Expr 8:43–80. https://doi.org/10.1615/CRITREVEUKARGENEEXPR.V8.I1.30

Cuevas-Diaz Duran R, Wei H, Kim DH, Wu JQ (2019) Invited review: long non-coding RNAs: important regulators in the development, function and disorders of the central nervous system. Neuropathol Appl Neurobiol 45:538–556. https://doi.org/10.1111/nan.12541

Cui H, Xie N, Tan Z et al (2014) The human long noncoding RNA lnc-IL7R regulates the inflammatory response. Eur J Immunol 44:2085–2095. https://doi.org/10.1002/eji.201344126

de Jesus BB, Blasco MA (2012) Potential of telomerase activation in extending health span and longevity. Curr Opin Cell Biol 24(6):739–743. https://doi.org/10.1016/j.ceb.2012.09.004

de Jesus BB, Blasco MA, Bernardes de Jesus B, Blasco MA (2012) Assessing cell and organ senescence biomarkers. Circ Res 111:97–109. https://doi.org/10.1161/CIRCRESAHA.111.247866

de Lange T (2004) T-loops and the origin of telomeres. Nat Rev Mol Cell Biol 5:323–329. https://doi.org/10.1038/nrm1359

de Lange T (2005) Shelterin: the protein complex that shapes and safeguards human telomeres. Genes Dev 19:2100–2110. https://doi.org/10.1101/gad.1346005

Deaton AM, Bird A (2011) CpG islands and the regulation of transcription. Genes Dev 25:1010–1022. https://doi.org/10.1101/GAD.2037511

Debacq-Chainiaux F, Erusalimsky JD, Campisi J, Toussaint O (2009) Protocols to detect senescence-associated beta-galactosidase (SA-βgal) activity, a biomarker of senescent cells in culture and in vivo. Nat Protoc 4:1798–1806. https://doi.org/10.1038/NPROT.2009.191
Deforzh E, Uhlmann EJ, Das E et al (2022) Promoter and enhancer RNAs regulate chromatin reorganization and activation of miR-10b/HOXD locus, and neoplastic transformation in glioma. Mol Cell 82:1894–1908.e5. https://doi.org/10.1016/J.MOLCEL.2022.03.018
Degirmenci U, Lei S (2016) Role of lncRNAs in cellular aging. Front Endocrinol (Lausanne) 7:151. https://doi.org/10.3389/fendo.2016.00151
Di Ruscio A, Ebralidze AK, Benoukraf T et al (2013) DNMT1-interacting RNAs block gene-specific DNA methylation. Nature 503:371–376. https://doi.org/10.1038/NATURE12598
Djebali S, Davis CA, Merkel A et al (2012) Landscape of transcription in human cells. Nature 489:101–108. https://doi.org/10.1038/nature11233
Emarati M, Azizi H, Ghasemi N, Skutella T (2025) Long non-coding RNAs in stem cell regulation and regenerative medicine: stemness, differentiation, and therapeutic innovation. In: Long non-coding RNAs – function, mechanisms, and applications [Working title]. IntechOpen
Fang Y, Xu Y, Wang R et al (2020) Recent advances on the roles of LncRNAs in cardiovascular disease. J Cell Mol Med 24:12246–12257. https://doi.org/10.1111/jcmm.15880
Frank S, Ahuja G, Bartsch D et al (2019) yylncT defines a class of divergently transcribed lncRNAs and safeguards the T-mediated mesodermal commitment of human PSCs. Cell Stem Cell 24:318–327.e8. https://doi.org/10.1016/j.stem.2018.11.005
Fu VX, Dobosy JR, Desotelle JA et al (2008) Aging and cancer-related loss of insulin-like growth factor 2 imprinting in the mouse and human prostate. Cancer Res 68:6797–6802. https://doi.org/10.1158/0008-5472.CAN-08-1714
Geng T, Liu Y, Xu Y et al (2018) H19 lncRNA promotes skeletal muscle insulin sensitivity in part by targeting AMPK. Diabetes 67:2183–2198. https://doi.org/10.2337/DB18-0370/-/DC1
Geng X, Zhao J, Huang J et al (2021) Lnc-MAP3K13-7:1 inhibits ovarian GC proliferation in PCOS via DNMT1 downregulation-mediated CDKN1A promoter Hypomethylation. Mol Ther 29:1279–1293. https://doi.org/10.1016/j.ymthe.2020.11.018
Goldstein S (1990) replicative senescence: the human fibroblast comes of age. Science 249:1129–1133. https://doi.org/10.1126/SCIENCE.2204114
Gomez JA, Wapinski OL, Yang YW et al (2013) The NeST long ncRNA controls microbial susceptibility and epigenetic activation of the interferon-γ locus. Cell 152:743–754. https://doi.org/10.1016/j.cell.2013.01.015
Grammatikakis I, Panda AC, Abdelmohsen K, Gorospe M (2014) Long noncoding RNAs (lncRNAs) and the molecular hallmarks of aging. Aging 6:992–1009. https://doi.org/10.18632/aging.100710
Gusic M, Prokisch H (2020) ncRNAs: new players in mitochondrial health and disease? Front Genet 11:510070. https://doi.org/10.3389/FGENE.2020.00095/XML/NLM
Han JDJ (2023) LncRNAs: the missing link to senescence nuclear architecture. Trends Biochem Sci 48:618–628. https://doi.org/10.1016/J.TIBS.2023.03.007
Hao LY, Armanios M, Strong MA et al (2005) Short telomeres, even in the presence of telomerase, limit tissue renewal capacity. Cell 123:1121–1131. https://doi.org/10.1016/j.cell.2005.11.020
Hardie DG, Ross FA, Hawley SA (2012) AMPK: a nutrient and energy sensor that maintains energy homeostasis. Nat Rev Mol Cell Biol 13(4):251–262. https://doi.org/10.1038/nrm3311
He J, Tu C, Liu Y (2018) Role of lncRNAs in aging and age-related diseases. Aging Med 1:158–175. https://doi.org/10.1002/agm2.12030
Hemann MT, Strong MA, Hao LY, Greider CW (2001) The shortest telomere, not average telomere length, is critical for cell viability and chromosome stability. Cell 107:67–77
Herman AB, Tsitsipatis D, Gorospe M (2022) Integrated lncRNA function upon genomic and epigenomic regulation. Mol Cell 82:2252–2266. https://doi.org/10.1016/J.MOLCEL.2022.05.027
Hernandez-Segura A, Nehme J, Demaria M (2018) Hallmarks of cellular senescence. Trends Cell Biol 28:436–453. https://doi.org/10.1016/J.TCB.2018.02.001

Hou Y, Dan X, Babbar M et al (2019) Ageing as a risk factor for neurodegenerative disease. Nat Rev Neurol 15(10):565–581. https://doi.org/10.1038/s41582-019-0244-7

Hu G, Gong A-Y, Wang Y et al (2016) LincRNA-Cox2 promotes late inflammatory gene transcription in macrophages through modulating SWI/SNF-mediated chromatin remodeling. J Immunol 196:2799–2808. https://doi.org/10.4049/jimmunol.1502146

Hu WL, Jin L, Xu A et al (2018) GUARDIN is a p53-responsive long non-coding RNA that is essential for genomic stability. Nat Cell Biol 20:492–502. https://doi.org/10.1038/s41556-018-0066-7

Hu B, Chen W, Zhong Y, Tuo Q (2023) The role of lncRNA-mediated pyroptosis in cardiovascular diseases. Front Cardiovasc Med 10:1217985. https://doi.org/10.3389/fcvm.2023.1217985

Huang W, Li H, Yu Q et al (2022) LncRNA-mediated DNA methylation: an emerging mechanism in cancer and beyond. J Exp Clin Cancer Res 41(1):1–12. https://doi.org/10.1186/S13046-022-02319-Z

Huarte M (2015) The emerging role of lncRNAs in cancer. Nat Med 21:1253–1261. https://doi.org/10.1038/nm.3981

Huarte M, Rinn JL (2010) Large non-coding RNAs: missing links in cancer? Hum Mol Genet 19:R152–R161. https://doi.org/10.1093/hmg/ddq353

Ilik IA, Quinn JJ, Georgiev P et al (2013) Tandem stem-loops in roX RNAs act together to mediate X Chromosome dosage compensation in Drosophila. Mol Cell 51:156–173. https://doi.org/10.1016/j.molcel.2013.07.001

Imamura K, Imamachi N, Akizuki G et al (2014) Long noncoding RNA NEAT1-dependent SFPQ relocation from promoter region to Paraspeckle mediates IL8 expression upon immune stimuli. Mol Cell 53:393–406. https://doi.org/10.1016/j.molcel.2014.01.009

Jafri MA, Ansari SA, Alqahtani MH, Shay JW (2016) Roles of telomeres and telomerase in cancer, and advances in telomerase-targeted therapies. Genome Med 8:69. https://doi.org/10.1186/s13073-016-0324-x

Jaskelioff M, Muller FL, Paik JH et al (2011) Telomerase reactivation reverses tissue degeneration in aged telomerase-deficient mice. Nature 469:102–106. https://doi.org/10.1038/nature09603

Jiang M-C, Ni J-J, Cui W-Y et al (2019) Emerging roles of lncRNA in cancer and therapeutic opportunities. Am J Cancer Res 9:1354–1366

Jianyin L, Badal SS, Zengchun Y et al (2016) Long noncoding RNA Tug1 regulates mitochondrial bioenergetics in diabetic nephropathy. J Clin Invest 126:4205–4218. https://doi.org/10.1172/JCI87927

Jin C, Zhao J, Zhang ZP et al (2020) Long non-coding RNA GAS5, by up-regulating PRC2 and targeting the promoter methylation of miR-424, suppresses multiple malignant phenotypes of glioma. J Neuro-Oncol 148:529–543. https://doi.org/10.1007/S11060-020-03544-2

Johnsson P, Ackley A, Vidarsdottir L et al (2013) A pseudogene long-noncoding-RNA network regulates PTEN transcription and translation in human cells. Nat Struct Mol Biol 20:440–446. https://doi.org/10.1038/NSMB.2516

Kausar S, Wang F, Cui H (2018) The role of mitochondria in reactive oxygen species generation and its implications for neurodegenerative diseases. Cells 7:274. https://doi.org/10.3390/CELLS7120274

Kim J, Kundu M, Viollet B, Guan KL (2011) AMPK and mTOR regulate autophagy through direct phosphorylation of Ulk1. Nat Cell Biol 13:132–141. https://doi.org/10.1038/NCB2152

Kim N, Chung W-Y, Cho J-Y (2023) The role and medical prospects of long non-coding RNAs in cardiovascular disease. Heart Fail Rev 28:1437–1453. https://doi.org/10.1007/s10741-023-10342-1

Kino T, Hurt DE, Ichijo T et al (2010) Noncoding RNA Gas5 is a growth arrest- and starvation-associated repressor of the glucocorticoid receptor. Sci Signal 3:ra8. https://doi.org/10.1126/SCISIGNAL.2000568/SUPPL_FILE/3_RA8_SM.PDF

Kluge MA, Fetterman JL, Vita JA (2013) Mitochondria and endothelial function. Circ Res 112:1171–1188. https://doi.org/10.1161/CIRCRESAHA.111.300233

Kuilman T, Peeper DS (2009) Senescence-messaging secretome: SMS-ing cellular stress. Nat Rev Cancer 9:81–94. https://doi.org/10.1038/nrc2560

Kung JTY, Colognori D, Lee JT (2013) Long noncoding RNAs: past, present, and future. Genetics 193:651–669. https://doi.org/10.1534/genetics.112.146704

Lander ES, Linton LM, Birren B et al (2001) Initial sequencing and analysis of the human genome. Nature 409:860–921. https://doi.org/10.1038/35057062

Larson K, Yan SJ, Tsurumi A et al (2012) Heterochromatin formation promotes longevity and represses ribosomal RNA synthesis. PLoS Genet 8:e1002473. https://doi.org/10.1371/JOURNAL.PGEN.1002473

Latos PA, Pauler FM, Koerner MV et al (2012) Airn transcriptional overlap, but not its lncRNA products, induces imprinted Igf2r silencing. Science 338:1469–1472. https://doi.org/10.1126/SCIENCE.1228110

Lee JT (2012) Epigenetic regulation by long noncoding RNAs. Science 338:1435–1439. https://doi.org/10.1126/science.1231776

Li J, Liu C (2019) Coding or noncoding, the converging concepts of RNAs. Front Genet 10:496. https://doi.org/10.3389/fgene.2019.00496

Li Z, Chao T-C, Chang K-Y et al (2014) The long noncoding RNA THRIL regulates TNFα expression through its interaction with hnRNPL. Proc Natl Acad Sci 111:1002–1007. https://doi.org/10.1073/pnas.1313768111

Li HJ, Sun XM, Li ZK et al (2017) LncRNA UCA1 promotes mitochondrial function of bladder cancer via the MiR-195/ARL2 signaling pathway. Cell Physiol Biochem 43:2548–2561. https://doi.org/10.1159/000484507

Li J, Bai R, Yang W et al (2022) The mitochondrial-derived lncRNA MDL1 mediates a mitochondria-to-nucleus retrograde regulation by inhibiting the nuclear translocation of p53. MedComm—Oncol 1:e15. https://doi.org/10.1002/MOG2.15

Li J, Xiao F, Wang S et al (2024) LncRNAs are involved in regulating ageing and age-related disease through the adenosine monophosphate-activated protein kinase signalling pathway. Genes Dis 11:101042. https://doi.org/10.1016/J.GENDIS.2023.06.014

Lin K, Luo X, Du C et al (2025) ANRIL modulates endothelial senescence and angiogenesis through SASP-driven miR146a regulation in age-related vascular dysfunction. Mech Ageing Dev 225:112058. https://doi.org/10.1016/j.mad.2025.112058

Liu X, Shan G (2021) Mitochondria encoded non-coding RNAs in cell physiology. Front Cell Dev Biol 9:713729. https://doi.org/10.3389/FCELL.2021.713729/BIBTEX

Long Y, Wang X, Youmans DT, Cech TR (2017) How do lncRNAs regulate transcription? Sci Adv 3:eaao2110. https://doi.org/10.1126/SCIADV.AAO2110/

Lopez de Silanes I, Stagno d'Alcontres M, Blasco MA (2010) TERRA transcripts are bound by a complex array of RNA-binding proteins. Nat Commun 1:33. https://doi.org/10.1038/ncomms1032

López-Otín C, Blasco MA, Partridge L et al (2023) Hallmarks of aging: an expanding universe. Cell 186:243–278. https://doi.org/10.1016/j.cell.2022.11.001

Luke B, Lingner J (2009) TERRA: telomeric repeat-containing RNA. EMBO J 28:2503–2510. https://doi.org/10.1038/emboj.2009.166

Maenner S, Müller M, Fröhlich J et al (2013) ATP-dependent roX RNA remodeling by the Helicase maleless enables specific association of MSL proteins. Mol Cell 51:174–184. https://doi.org/10.1016/j.molcel.2013.06.011

Mangiavacchi A, Morelli G, Orlando V (2023) Behind the scenes: how RNA orchestrates the epigenetic regulation of gene expression. Front Cell Dev Biol 11:1123975. https://doi.org/10.3389/FCELL.2023.1123975

Marín-Béjar O, Marchese FP, Athie A et al (2013) Pint lincRNA connects the p53 pathway with epigenetic silencing by the polycomb repressive complex 2. Genome Biol 14(9):R104. https://doi.org/10.1186/GB-2013-14-9-R104

Marion RM, Strati K, Li H et al (2009) Telomeres acquire embryonic stem cell characteristics in induced pluripotent stem cells. Cell Stem Cell 4:141–154. https://doi.org/10.1016/j.stem.2008.12.010

Martinez P, Blasco MA (2010) Role of shelterin in cancer and aging. Aging Cell 9:653–666. https://doi.org/10.1111/j.1474-9726.2010.00596.x

Mattick JS, Amaral PP, Carninci P et al (2023) Long non-coding RNAs: definitions, functions, challenges and recommendations. Nat Rev Mol Cell Biol 24:430–447. https://doi.org/10.1038/s41580-022-00566-8

Mirzadeh Azad F, Polignano IL, Proserpio V, Oliviero S (2021) Long noncoding RNAs in human stemness and differentiation. Trends Cell Biol 31:542–555. https://doi.org/10.1016/j.tcb.2021.02.002

Montes M, Lubas M, Arendrup FS et al (2021) The long non-coding RNA MIR31HG regulates the senescence associated secretory phenotype. Nat Commun 12:2459. https://doi.org/10.1038/s41467-021-22746-4

Naganuma T, Hirose T (2013) Paraspeckle formation during the biogenesis of long non-coding RNAs. RNA Biol 10:456–461. https://doi.org/10.4161/rna.23547

Narita M, Nũnez S, Heard E et al (2003) Rb-mediated heterochromatin formation and silencing of E2F target genes during cellular senescence. Cell 113:703–716. https://doi.org/10.1016/S0092-8674(03)00401-X

Natarajan V, Chawla R, Mah T et al (2020) Mitochondrial dysfunction in age-related metabolic disorders. Proteomics 20:1800404. https://doi.org/10.1002/PMIC.201800404;SUBPAGE:STRING:FULL

Noh JH, Kim KM, Abdelmohsen K et al (2016) HuR and GRSF1 modulate the nuclear export and mitochondrial localization of the lncRNA RMRP. Genes Dev 30:1224–1239. https://doi.org/10.1101/GAD.276022.115/-/DC1

Okano M, Bell DW, Haber DA, Li E (1999) DNA methyltransferases Dnmt3a and Dnmt3b are essential for de novo methylation and mammalian development. Cell 99:247–257. https://doi.org/10.1016/S0092-8674(00)81656-6

Oo JA, Brandes RP, Leisegang MS (2022) Long non-coding RNAs: novel regulators of cellular physiology and function. Pflugers Arch 474:191–204. https://doi.org/10.1007/S00424-021-02641-Z/FIGURES/2

Pal S, Tyler JK (2016) Epigenetics and aging. Sci Adv 2:e1600584. https://doi.org/10.1126/SCIADV.1600584

Palikaras K, Lionaki E, Tavernarakis N (2015) Coordination of mitophagy and mitochondrial biogenesis during ageing in C. elegans. Nature 521:525–528. https://doi.org/10.1038/NATURE14300

Palm W, De Lange T (2008) How shelterin protects mammalian telomeres. Annu Rev Genet 42:301–334

Panda AC, Abdelmohsen K, Gorospe M (2017) SASP regulation by noncoding RNA. Mech Ageing Dev 168:37–43. https://doi.org/10.1016/j.mad.2017.05.004

Pandey RR, Mondal T, Mohammad F et al (2008) Kcnq1ot1 antisense noncoding RNA mediates lineage-specific transcriptional silencing through chromatin-level regulation. Mol Cell 32:232–246. https://doi.org/10.1016/j.molcel.2008.08.022

Patraquim P, Magny EG, Pueyo JI et al (2022) Translation and natural selection of micropeptides from long non-canonical RNAs. Nat Commun 13:6515. https://doi.org/10.1038/s41467-022-34094-y

Peng N, He J, Li J et al (2020) Long noncoding RNA MALAT1 inhibits the apoptosis and autophagy of hepatocellular carcinoma cell by targeting the microRNA-146a/PI3K/Akt/mTOR axis. Cancer Cell Int 20:1–11. https://doi.org/10.1186/S12935-020-01231-W/FIGURES/7

Ponting CP, Oliver PL, Reik W (2009) Evolution and functions of long noncoding RNAs. Cell 136:629–641. https://doi.org/10.1016/j.cell.2009.02.006

Prensner JR, Chinnaiyan AM (2011) The emergence of lncRNAs in cancer biology. Cancer Discov 1:391–407. https://doi.org/10.1158/2159-8290.CD-11-0209

Puvvula PK (2019) LncRNAs regulatory networks in cellular senescence. Int J Mol Sci 20:2615. https://doi.org/10.3390/ijms20112615

Qian Y, Shi L, Luo Z (2020) Long non-coding RNAs in cancer: implications for diagnosis, prognosis, and therapy. Front Med (Lausanne) 7:612393. https://doi.org/10.3389/fmed.2020.612393

Rackham O, Shearwood AMJ, Mercer TR et al (2011) Long noncoding RNAs are generated from the mitochondrial genome and regulated by nuclear-encoded proteins. RNA 17:2085–2093. https://doi.org/10.1261/RNA.029405.111

Rapicavoli NA, Qu K, Zhang J et al (2013) A mammalian pseudogene lncRNA at the interface of inflammation and anti-inflammatory therapeutics. elife 2:e00762. https://doi.org/10.7554/eLife.00762

Rinn JL, Chang HY (2012) Genome regulation by long noncoding RNAs. Annu Rev Biochem 81:145–166. https://doi.org/10.1146/annurev-biochem-051410-092902

Rivosecchi J, Jurikova K, Cusanelli E (2024) Telomere-specific regulation of TERRA and its impact on telomere stability. Semin Cell Dev Biol 157:3–23. https://doi.org/10.1016/j.semcdb.2023.11.001

Rossi M, Gorospe M (2020) Noncoding RNAs controlling telomere homeostasis in senescence and aging. Trends Mol Med 26:422–433. https://doi.org/10.1016/j.molmed.2020.01.010

Rossi C, Venturin M, Gubala J et al (2023) PURPL and NEAT1 Long non-coding RNAs are modulated in vascular smooth muscle cell replicative senescence. Biomedicine 11:3228. https://doi.org/10.3390/biomedicines11123228

Rothbart SB, Strahl BD (2014) Interpreting the language of histone and DNA modifications. Biochim Biophys Acta Gene Regul Mech 1839:627–643. https://doi.org/10.1016/J.BBAGRM.2014.03.001

Sallam T, Sandhu J, Tontonoz P (2018) Long noncoding RNA discovery in cardiovascular disease. Circ Res 122:155–166. https://doi.org/10.1161/CIRCRESAHA.117.311802

Salviano-Silva A, Lobo-Alves S, Almeida R et al (2018) Besides pathology: long non-coding RNA in cell and tissue homeostasis. Noncoding RNA 4:3. https://doi.org/10.3390/ncrna4010003

Sana J, Faltejskova P, Svoboda M, Slaby O (2012) Novel classes of non-coding RNAs and cancer. J Transl Med 10:1–21. https://doi.org/10.1186/1479-5876-10-103/TABLES/3

Santos F, Correia M, Nóbrega-Pereira S, Bernardes de Jesus B (2020) Age-related pathways in cardiac regeneration: a role for lncRNAs? Front Physiol 11:583191. https://doi.org/10.3389/fphys.2020.583191

Santos F, Capela AM, Mateus F et al (2022) Non-coding antisense transcripts: fine regulation of gene expression in cancer. Comput Struct Biotechnol J 20:5652–5660. https://doi.org/10.1016/j.csbj.2022.10.009

Sarek G, Marzec P, Margalef P, Boulton SJ (2015) Molecular basis of telomere dysfunction in human genetic diseases. Nat Struct Mol Biol 22:867–874. https://doi.org/10.1038/nsmb.3093

Saxton RA, Sabatini DM (2017) mTOR signaling in growth, metabolism, and disease. Cell 168:960–976. https://doi.org/10.1016/J.CELL.2017.02.004

Scarabino D, Broggio E, Gambina G et al (2017) Common variants of human TERT and TERC genes and susceptibility to sporadic Alzheimer's disease. Exp Gerontol 88:19–24. https://doi.org/10.1016/j.exger.2016.12.017

Schmitt AM, Chang HY (2016) Long noncoding RNAs in cancer pathways. Cancer Cell 29:452–463. https://doi.org/10.1016/j.ccell.2016.03.010

Schmitz KM, Mayer C, Postepska A, Grummt I (2010) Interaction of noncoding RNA with the rDNA promoter mediates recruitment of DNMT3b and silencing of rRNA genes. Genes Dev 24:2264–2269. https://doi.org/10.1101/GAD.590910

Schoeftner S, Blasco MA (2008) Developmentally regulated transcription of mammalian telomeres by DNA-dependent RNA polymerase II. Nat Cell Biol 10:228–236. https://doi.org/10.1038/ncb1685

Schoeftner S, Blasco MA (2009) A "higher order" of telomere regulation: telomere heterochromatin and telomeric RNAs. EMBO J 28:2323–2336. https://doi.org/10.1038/emboj.2009.197

Sena LA, Chandel NS (2012) Physiological roles of mitochondrial reactive oxygen species. Mol Cell 48:158–167. https://doi.org/10.1016/J.MOLCEL.2012.09.025

Sherazi SM, Abbasi A, Jamil A et al (2023) Molecular hallmarks of long non-coding RNAs in aging and its significant effect on aging-associated diseases. Neural Regen Res 18:959. https://doi.org/10.4103/1673-5374.355751

Shore D (1998) Telomeres–unsticky ends. Science 281:1818–1819

Sirey TM, Roberts K, Haerty W et al (2019) The long non-coding rna cerox1 is a post transcriptional regulator of mitochondrial complex i catalytic activity. elife 8:e45051. https://doi.org/10.7554/ELIFE.45051

Sousa-Franco A, Rebelo K, da Rocha ST, Bernardes de Jesus B (2018) LncRNAs regulating stemness in aging. Aging Cell 18:e12870. https://doi.org/10.1111/acel.12870

Srivastava S (2017) The mitochondrial basis of aging and age-related disorders. Genes (Basel) 8:398. https://doi.org/10.3390/GENES8120398

Statello L, Guo CJ, Chen LL, Huarte M (2020) Gene regulation by long non-coding RNAs and its biological functions. Nat Rev Mol Cell Biol 22(2):96–118. https://doi.org/10.1038/s41580-020-00315-9

Swagatika S, Tomar RS (2016) Modulation of epigenetics by environmental toxic molecules. Adv Mol Toxicol 10:361–389. https://doi.org/10.1016/B978-0-12-804700-2.00008-8

Tan M, Luo H, Lee S et al (2011) Identification of 67 histone marks and histone lysine crotonylation as a new type of histone modification. Cell 146:1016–1028. https://doi.org/10.1016/J.CELL.2011.08.008/ATTACHMENT/BAEA3374-62D0-4670-ACD2-545019B1A0B9/MMC7.PDF

Tavares e Silva J, Pessoa J, Nóbrega-Pereira S, Bernardes de Jesus B (2024) The impact of long noncoding RNAs in tissue regeneration and senescence. Cells 13:119. https://doi.org/10.3390/cells13020119

Tomas-Loba A, Flores I, Fernandez-Marcos PJ et al (2008) Telomerase reverse transcriptase delays aging in cancer-resistant mice. Cell 135:609–622. https://doi.org/10.1016/j.cell.2008.09.034

Tomas-Loba A, Bernardes de Jesus B, Mato JM, Blasco MA (2013) A metabolic signature predicts biological age in mice. Aging Cell 12:93–101. https://doi.org/10.1111/acel.12025

Tominaga-Yamanaka K, Abdelmohsen K, Martindale JL et al (2012) NF90 coordinately represses the senescence-associated secretory phenotype. Aging 4:695–708. https://doi.org/10.18632/aging.100497

Tonmoy MIQ, Fariha A, Hami I et al (2022) Computational epigenetic landscape analysis reveals association of CACNA1G-AS1, F11-AS1, NNT-AS1, and MSC-AS1 lncRNAs in prostate cancer progression through aberrant methylation. Sci Rep 12:10260. https://doi.org/10.1038/s41598-022-13381-0

Travers ME, Mackay DJG, Nitert MD et al (2013) Insights into the molecular mechanism for type 2 diabetes susceptibility at the KCNQ1 locus from temporal changes in imprinting status in human islets. Diabetes 62:987–992. https://doi.org/10.2337/DB12-0819

Tsai MC, Manor O, Wan Y et al (2010) Long noncoding RNA as modular scaffold of histone modification complexes. Science 329:689–693. https://doi.org/10.1126/SCIENCE.1192002

Vendramin R, Verheyden Y, Ishikawa H et al (2018) SAMMSON fosters cancer cell fitness by concertedly enhancing mitochondrial and cytosolic translation. Nat Struct Mol Biol 25:1035–1046. https://doi.org/10.1038/S41594-018-0143-4

Venter JC, Adams MD, Myers EW et al (2001) The sequence of the human genome. Science 291:1304–1351. https://doi.org/10.1126/science.1058040

Vera E, Bernardes de Jesus B, Foronda M et al (2012) The rate of increase of short telomeres predicts longevity in mammals. Cell Rep 2:732–737. https://doi.org/10.1016/j.celrep.2012.08.023

Vera E, Bernardes de Jesus B, Foronda M et al (2013) Telomerase reverse transcriptase synergizes with calorie restriction to increase health span and extend mouse longevity. PLoS One 8:e53760. https://doi.org/10.1371/journal.pone.0053760

Verma N, Pan H, Doré LC et al (2018) TET proteins safeguard bivalent promoters from de novo methylation in human embryonic stem cells. Nat Genet 50:83–95. https://doi.org/10.1038/S41588-017-0002-Y

Villegas J, Burzio V, Villota C et al (2007) Expression of a novel non-coding mitochondrial RNA in human proliferating cells. Nucleic Acids Res 35:7336–7347. https://doi.org/10.1093/NAR/GKM863

Wang KC, Chang HY (2011) Molecular mechanisms of long noncoding RNAs. Mol Cell 43:904–914. https://doi.org/10.1016/J.MOLCEL.2011.08.018

Wang RC, Smogorzewska A, de Lange T (2004) Homologous recombination generates T-loop-sized deletions at human telomeres. Cell 119:355–368. https://doi.org/10.1016/j.cell.2004.10.011

Wang KC, Yang YW, Liu B et al (2011) A long noncoding RNA maintains active chromatin to coordinate homeotic gene expression. Nature 472:120–126. https://doi.org/10.1038/NATURE09819

Wang K, Liu H, Hu Q et al (2022) Epigenetic regulation of aging: implications for interventions of aging and diseases. Signal Transduct Target Ther 7(1):1–22. https://doi.org/10.1038/s41392-022-01211-8

Warraich U e A, Hussain F, Kayani HUR (2020) Aging – oxidative stress, antioxidants and computational modeling. Heliyon 6:e04107. https://doi.org/10.1016/J.HELIYON.2020.E04107

Watson CN, Belli A, Di Pietro V (2019) Small non-coding RNAs: new class of biomarkers and potential therapeutic targets in neurodegenerative disease. Front Genet 10:448151. https://doi.org/10.3389/FGENE.2019.00364/BIBTEX

Weber M, Hellmann I, Stadler MB et al (2007) Distribution, silencing potential and evolutionary impact of promoter DNA methylation in the human genome. Nat Genet 39:457–466. https://doi.org/10.1038/NG1990

Wei W, Wang X, Wei Y et al (2022) lncRNA TUG1 protects intestinal epithelial cells from damage induced by high glucose and high fat via AMPK/SIRT1. Mol Med Rep 25:139. https://doi.org/10.3892/MMR.2022.12655

Wong LSM, Oeseburg H, de Boer RA et al (2008) Telomere biology in cardiovascular disease: the TERC−/− mouse as a model for heart failure and ageing. Cardiovasc Res 81:244–252. https://doi.org/10.1093/cvr/cvn337

Wullschleger S, Loewith R, Hall MN (2006) TOR signaling in growth and metabolism. Cell 124:471–484. https://doi.org/10.1016/j.cell.2006.01.016

Xagorari M, Marmarinos A, Kossiva L et al (2021) Overexpression of the GR riborepressor LncRNA GAS5 results in poor treatment response and early relapse in childhood B-ALL. Cancers (Basel) 13:6064. https://doi.org/10.3390/CANCERS13236064/S1

Xie X, Huang M, Ma S et al (2025) The role of long non-coding RNAs in cardiovascular diseases: a comprehensive review. Noncoding RNA Res 11:158–187. https://doi.org/10.1016/j.ncrna.2024.12.009

Yang F, Huo X, Yuan S et al (2013) Repression of the long noncoding RNA-LET by histone deacetylase 3 contributes to hypoxia-mediated metastasis. Mol Cell 49:1083–1096. https://doi.org/10.1016/j.molcel.2013.01.010

Yang J, Gao L, Wang Z et al (2024) Effects of the lncRNA NBR2 on the proliferation and autophagy of breast cancer cells under starvation conditions. Sci Rep 14:22624. https://doi.org/10.1038/S41598-024-72181-W

Yao RW, Wang Y, Chen LL (2019) Cellular functions of long noncoding RNAs. Nat Cell Biol 21:542–551. https://doi.org/10.1038/s41556-019-0311-8

Yap KL, Li S, Muñoz-Cabello AM et al (2010) Molecular interplay of the noncoding RNA ANRIL and methylated histone H3 lysine 27 by Polycomb CBX7 in transcriptional silencing of INK4a. Mol Cell 38:662–674. https://doi.org/10.1016/j.molcel.2010.03.021

Zhao J, Zhai B, Gygi SP, Goldberg AL (2015) MTOR inhibition activates overall protein degradation by the ubiquitin proteasome system as well as by autophagy. Proc Natl Acad Sci USA 112:15790–15797. https://doi.org/10.1073/PNAS.1521919112

Zhao Y, Liu S, Zhou L et al (2019) Aberrant shuttling of long noncoding RNAs during the mitochondria-nuclear crosstalk in hepatocellular carcinoma cells. Am J Cancer Res 9:999

Zhao L, Gu M, Sun Z et al (2025) The role of exosomal lncRNAs in cardiovascular disease: emerging insights based on molecular mechanisms and therapeutic target level. Noncoding RNA Res 10:198–205. https://doi.org/10.1016/j.ncrna.2024.10.001

Zheng Q, Liu P, Gao G et al (2019) Mitochondrion-processed TERC regulates senescence without affecting telomerase activities. Protein Cell 10:631–648. https://doi.org/10.1007/s13238-019-0612-5

Zhou J, Yang L, Zhong T et al (2015) H19 lncRNA alters DNA methylation genome wide by regulating S-adenosylhomocysteine hydrolase. Nat Commun 6:10221. https://doi.org/10.1038/NCOMMS10221

Zhou J, Xu J, Zhang L et al (2019) Combined single-cell profiling of lncRNAs and functional screening reveals that H19 is pivotal for embryonic hematopoietic stem cell development. Cell Stem Cell 24:285–298.e5. https://doi.org/10.1016/j.stem.2018.11.023

Zhou M, Zhao G, Zeng Y et al (2022) Aging and cardiovascular disease: current status and challenges. Rev Cardiovasc Med 23:135. https://doi.org/10.31083/J.RCM2304135

Zhu Y, Yang T, Duan J et al (2019) *MALAT1*/miR-15b-5p/*MAPK1* mediates endothelial progenitor cells autophagy and affects coronary atherosclerotic heart disease via mTOR signaling pathway. Aging 11:1089–1109. https://doi.org/10.18632/AGING.101766

Zhu D, Li X, Tian Y (2022) Mitochondrial-to-nuclear communication in aging: an epigenetic perspective. Trends Biochem Sci 47:645–659. https://doi.org/10.1016/J.TIBS.2022.03.008

Chapter 6
Long Non-coding RNAs in Cancer: From Diagnosis to Therapy

Mateus Nóbrega Aoki, Anelis Maria Marin, Jaqueline Carvalho de Oliveira, Maria Augusta Poersch, and Dalila Luciola Zanette

Abstract This chapter reviews the roles of long non-coding RNAs (lncRNAs) in cancer, from diagnosis to therapy. It begins by exploring the potential of lncRNAs as diagnostic markers in various cancers, including prostate, gastric, lung, colorectal, breast, and liver cancer, highlighting specific lncRNAs and multi-lncRNA panels with promising diagnostic accuracy. The chapter then delves into the prognostic value of lncRNAs, discussing their association with clinical outcomes across different malignancies. Finally, it addresses lncRNA-based therapeutic strategies, including lncRNA manipulation, targeting lncRNA interactions, CRISPR/Cas9 technology, nanoparticle delivery systems, and combination therapies, while also acknowledging the therapeutic challenges related to delivery, stability, specificity, mechanisms of action, regulatory hurdles, patient heterogeneity, and ethical considerations. By covering the diagnostic, prognostic, and therapeutic applications of lncRNAs, this chapter provides a comprehensive overview of their significance in cancer research and clinical practice.

Keywords lncRNA · Cancer · Diagnosis · Prognosis · Therapy

Abbreviations

ABC-DLBCL	B-cell-like diffuse large B-cell lymphoma
AFP	Alpha-fetoprotein

M. N. Aoki (✉) · A. M. Marin · D. L. Zanette
Laboratory for Applied Science and Technology in Health, Carlos Chagas Institute, Oswaldo Cruz Foundation (Fiocruz),
Curitiba, PR, Brazil
e-mail: mateus.aoki@fiocruz.br

J. Carvalho de Oliveira · M. A. Poersch
Department of Genetics, Federal University of Paraná (UFPR), Curitiba, PR, Brazil

L. M. Vaschetto (ed.), *Long Non-Coding RNAs, Human Health, and Diseases*, Epigenetics and Human Health 16, https://doi.org/10.1007/978-3-032-19668-2_6

ALP	Alkaline Phosphatase
ALT	Alanine Aminotransferase
AML	Acute Myeloid Leukemia
APL	Acute Promyelocytic Leukemia
ASP	Aspartate Aminotransferase
AUC	Area Under the Curve
BC	Breast Cancer
CA125	Cancer Antigen 125
CA15-3	Cancer Antigen 15-3
CA19–9	Carbohydrate Antigen 19–9
CEA	Carcinoembryonic Antigen
CESC	Cervical Squamous Cell Carcinoma
CRC	Colorectal Cancer
DFS	Disease-Free Survival
DLBCL	Diffuse Large B-Cell Lymphoma
EAU	European Association of Urology
EFS	Event-Free Survival
EMT	Epithelial-Mesenchymal Transition
EOGC	Early-Onset Gastric Cancer
ESCC	Esophageal Squamous Cell Carcinoma
FDA	U.S. Food and Drug Administration
GC	Gastric Cancer
GGT	Gamma-Glutamyl Transferase
HBV	Human Papillomavirus
HCC	Hepatocellular Carcinoma
HCH	Chronic Hepatitis and Cirrhosis
HNSC	Head and Neck Squamous Cell Carcinoma
HR	Hazard Ratio
KIRC	Kidney Renal Clear Cell Carcinoma
KIRP	Kidney Renal Papillary Cell Carcinoma
LIHC	Liver Hepatocellular Carcinoma
lncRNA	Long Non-coding RNA
LNM	Lymph Node Metastasis
MDSC	Myeloid-Derived Suppressor Cells
NCCN	National Comprehensive Cancer Network
NSCLC	Non-Small Cell Lung Cancer
OS	Overall Survival
PBMC	Peripheral Blood Mononuclear Cells
PCa	Prostate Cancer
PCA3	Prostate Cancer Antigen-3
PCAN	Pan-Cancer Associated lncRNAs
PCPG	Parenchymal Hemangioblastomas
PD-L1	Programmed Death-Ligand 1
PSA	Prostate-Specific Antigen

R-CHOP	Combination drug treatment; Rituximab (R), and three chemotherapy drugs: Cyclophosphamide (C), Doxorubicin (H), and Vincristine (O), and Prednisone (P)
ROC	Receiver Operating Characteristic
RT-qPCR	Reverse Transcription Quantitative Polymerase Chain Reaction
siRNAs	Small interfering RNAs
SKCM	Human Skin Cutaneous Melanoma
STAD	Stomach Adenocarcinoma
TAM	Tumor-Associated Macrophages
TNBC	Triple-Negative Breast Cancer
TNM	Tumor-Node-Metastasis

6.1 Introduction

Cancer remains one of the most complex and persistent challenges in contemporary medicine and is a leading cause of mortality worldwide. The relentless pursuit of more precise and individualized approaches to oncology has brought non-coding regions of the genome, once dismissed as transcriptional "noise," into the spotlight of cancer research. Among them, long non-coding RNAs (lncRNAs) have emerged as central players in the fine regulation of gene expression, chromatin dynamics, and signal transduction pathways involved in tumorigenesis. This class of molecules plays an important role in physiological and pathological contexts. As a rational and logical approach, after elucidating lncRNA roles in cancer initiation and progression, a translational goal focuses on the intervention of this molecule in oncological patients, looking for lncRNA-based therapies, both pharmacological and genetic. This area faces a great challenge: initially to depict lncRNA roles in cancer and subsequently to apply novel therapies that consider specificity, toxicity, side effects, and cost-effectiveness. In this study, the potential of lncRNAs as biomarkers is discussed in several cancer types, followed by the prognostic value of selected lncRNAs, providing a comprehensive overview of the involvement of lncRNAs in cancer therapy (Fig. 6.1). By understanding the complexity, versatility, and application of lncRNAs in cancer, basic and translational knowledge is required for innovative therapeutic modalities that enhance treatment efficacy and improve patient outcomes against cancer.

The use of RNA expression analyses in onco-molecular diagnosis has become increasingly crucial in refining the accuracy of cancer detection and guiding personalized treatment strategies. Several RNA-based tests have been developed to address diagnostic challenges in specific types of cancer. For example, Oncotype DX and MammaPrint are widely used for breast cancer assessment, analyzing a select number of protein-coding genes to provide prognostic and predictive insights that assist in determining the most appropriate treatment approach (Xin et al. 2017). Additionally, the Afirma® test is utilized for distinguishing malignant from benign

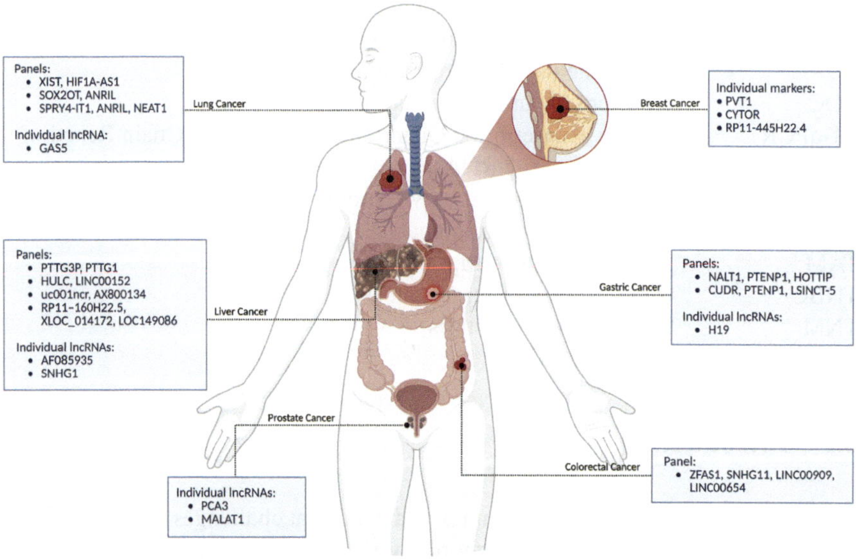

Fig. 6.1 lncRNAs with diagnostic potential in cancer: Panels and individual lncRNAs are described as role in lung, liver, prostate, breast, gastric, and colorectal cancer

thyroid nodules, based on RNA sequencing of 142 genes (Chudova et al. 2010). Currently, the clinical application of lncRNA expression analysis remains limited; however, in prostate cancer, the lncRNA PCA3 has been approved for diagnostic use in specific scenarios, such as when prostate-specific antigen (PSA) levels are elevated, but the initial biopsy results are negative (Albala et al. 2024). Despite its limited sensitivity, this biomarker is particularly useful when there is a persistent suspicion of cancer before repeating the biopsy. Although its clinical application remains limited, numerous scientific studies highlight the analysis of lncRNA expression as highly promising biomarkers, with some showing particular utility in specific tumor types.

6.2 Diagnostic Value of lncRNAs

In this section, we describe lncRNAs with diagnostic value in an oncological context, as summarized in Table 6.1.

Table 6.1 lncRNAs described as diagnostic biomarkers

lncRNA	Cancer type	Expression in malignancy	Material	AUC	Sensitivity and specificity	Reference
PCA3	Prostate cancer	Upregulated	Urine	0.72	62% and 75%, respectively	Goode et al. (2013)
		Upregulated	Urine	0.75	65%, and 73%, respectively	Cui et al. (2016)
MALAT1	Prostate cancer	Upregulated	Urine	0.79	ND	Li et al. (2021)
	Colorectal cancer	Upregulated	Meta-study	0.61	67.5 and 77.1%, respectively	Masrour et al. (2024)
H19	Gastric cancer	Upregulated	Plasma	0.64–0.84	38.1%–82.9%, and 68.75%–87.2%, respectively	Arita (2013), Zhou et al. (2015), Hashad et al. (2016), Yörüker et al. (2018)
LINC00152	Gastric cancer	Upregulated	Plasma	0.66	48.1%, and 85.2%, respectively	Zheng et al. (2015)
	Liver cancer	Uperegulated	Serum	0.78	–	Li et al. (2015)
UCA1	Gastric cancer	Upregulated	Gastric juice	0.72	67.2%, and 80.3%, respectively	Zheng et al. (2015)
NALT1, PTENP1, HOTTIP	Gastric cancer	Upregulated	Blood	0.91–0.93	82.6%–95.1%, and 80.5%–88.9%, respectively	Guo et al. (2025)
TINCR, CCAT2, AOC4P, BANCR, LINC00857	Gastric cancer	Upregulated	Plasma	0.93	82%, and 87%, respectively	Zhang et al. (2017b)
CUDR, PTENP1, LSINCT-5	Gastric cancer	Upregulated	Serum	0.832 (stage I), 0.828 (stages II–IV)	77.8%, 97.0% (stage I), 85.7%–84.8% (stages II–IV)	Dong et al. (2015)
GAS5	Lung cancer	Downregulated	Plasma	0.91	86.7% and 90.9%, respectively	Liang et al. (2016)

(continued)

Table 6.1 (continued)

lncRNA	Cancer type	Expression in malignancy	Material	AUC	Sensitivity and specificity	Reference
SOX2OT, ANRIL, CYFRA 21-1, SCCA	Lung cancer	Upregulated	Serum	0.88	91% and 70%, respectively	Xie et al. (2018)
SPRY4-IT1, ANRIL, NEAT1	Lung cancer	Upregulated	Plasma	0.876	82.8 and 91.3%, respectively	Hu et al. (2016)
ZFAS1, SNHG11, LINC00909, LINC00654	Colorectal cancer	Upregulated	Plasma	0.93	–	Xu et al. (2020)
B3GALT5-AS1	Colorectal cancer	Upregulated	Serum	0.76	89% and 53.4%, respectively	Ding et al. (2020)
HOTAIR	Colorectal cancer	Upregulated	PBMC	0.87	67% and 92.5%, respectively	Svoboda et al. (2014)
CCAT1	Colorectal cancer	Upregulated	Plasma	0.78	–	Abedini et al. (2019)
ATB	Colorectal cancer	Upregulated	Plasma	0.64	82% and 75%, respectively	
PVT1	Breast cancer	Upregulated	Serum	0.91	84.17% and 97.5%, respectively	Sayed et al. (2025)
CYTOR	Breast cancer	Upregulated	Plasma	0.91	–	Moradi et al. (2020)
RP11-445H22.4	Breast cancer	Upegulated	Serum	0.9	92% and 74%, respectively	M. Kamel et al. (2024)
T376626	Breast cancer	Upregulated	Serum	0.84	73.7% and 87.5%, respectively	He et al. (2023)
PTTG3P in combination	Liver cancer	Upregulated	Serum	0.96	90.4% and 98%, respectively	Cao et al. (2025)
uc001ncr and AX800134	Liver cancer	Upregulated	Serum	0.95	95% and 88%, respectively	Wang et al. (2015a)
RP11–160H22.5, XLOC_014172 and LOC149086	Liver cancer	Upregulated	Serum	0.9	82% and 73%, respectively	Tang et al. (2015)
SNHG1	Liver cancer	Upregulated	Plasma	0.92	87.3% and 86%, respectively	Gao et al. (2018)
uc003wbd and AF085935	Liver cancer	Upregulated	Serum	0.86	–	Lu et al. (2015)

6.2.1 *Prostate Cancer*

Prostate cancer (PCa) represents a unique case in the field of lncRNA diagnostics due to the clinical implementation of the PCA3 (Prostate Cancer Antigen 3) test, the first FDA-approved lncRNA assay for cancer diagnosis (Day et al. 2011), named ProgensaTM PCA3 test (NCT01024959; https://www.accessdata.fda.gov/scripts/cdrh/cfdocs/cfpma/pma.cfm?id=P100033 accessed on May 19, 2025). PCA3 is a prostate-specific, non-coding RNA significantly overexpressed in malignant prostate tissues but not in benign prostatic hyperplasia or normal tissue (Goode et al. 2013). The PCA3 test involves measuring urinary RNA following a digital rectal exam (DRE), which releases prostate cells into the urine (Goode et al. 2013).

Numerous clinical trials and meta-analyses have validated PCA3's diagnostic value (Goode et al. 2013; Cui et al. 2016). A meta-analysis of 46 studies including over 12,000 patients reported a pooled sensitivity of 65% and specificity of 73%, with an AUC of 0.75 for PCa diagnosis (Cui et al. 2016). Although it is not superior to PSA in sensitivity, it is more specific for not being elevated by other prostatic diseases, and it is particularly valuable in reducing unnecessary biopsies in men with borderline or elevated PSA levels (Day et al. 2011; Goode et al. 2013; Tan et al. 2019; Qin et al. 2020).

In addition to PCA3, the lncRNA MALAT1 has also been identified as differentially expressed in patients with PCa (Li et al. 2021). When analyzed in exosome RNAs in urine, the expression of MALAT1 had a slightly higher value than PCA3 (AUC = 0.791 and AUC = 0.751, respectively), and the combination of both lncRNAs was superior in predicting PCa than the combination of fPSA/PSA (AUC = 0.828 and AUC = 0.661) (Li et al. 2021). This indicates that the implementation of lncRNA assays in PCa diagnosis could not only improve the detection of the disease but also avoid unnecessary biopsies.

6.2.2 *Gastric Cancer*

The utility of lncRNA analysis in the diagnosis of gastric cancer (GC) is also promising, mainly because diagnosis of this cancer type has been largely dependent on endoscopy, which has low efficiency in detecting early GC. One of these new emerging biomarkers is H19, a maternally imprinted transcript that is among the most studied lncRNAs in GC, is expressed in GC tissues, and whose circulating levels are significantly higher in patients compared to healthy individuals (Arita et al. 2013; Zhou et al. 2015; Hashad et al. 2016; Yörüker et al. 2018). Notably, H19 levels have been shown to drop postoperatively, supporting its potential use for monitoring disease recurrence (Arita et al. 2013; Zhou et al. 2015; Yörüker et al. 2018). It has reported specificity between 38.1% and 82.9% and sensitivity between 68.75% and 87.2%, with an AUC varying between 0.64 and 0.84 (Arita et al. 2013; Zhou et al. 2015; Hashad et al. 2016; Yörüker et al. 2018). H19 levels can detect early-stage GC, with a higher sensitivity and specificity than conventional biomarkers like CEA and CA19–9 (Zhou et al. 2015). The sole detection of H19 has a moderate diagnostic performance, but in combination with CEA, it greatly improves the diagnosis of GC, suggesting it may serve as a new complementary tumor marker for GC (Hashad et al. 2016).

Apart from H19, other lncRNAs that have also been suggested as potential biomarkers for the early detection of gastric cancer (GC) include LINC00152 (sensitivity: 48.1%, specificity: 85.2%, AUC = 0.66), which shows upregulation in the plasma of patients with GC compared to controls, and UCA1 (sensitivity: 67.2%, specificity: 80.3%, AUC = 0.72), which is more upregulated in the gastric juice of patients with GC compared to controls (Zheng et al. 2015).

However, the most effective lncRNA-based diagnostic biomarkers are panels comprising multiple lncRNAs, given that the diagnostic accuracy of a single gene might be limited. A panel comprising three extracellular vesicle-derived long non-coding RNAs—NALT1, PTENP1, and HOTTIP—identified in blood samples, was able to distinguish patients with early-onset gastric cancer (EOGC) in a training cohort and two validation cohorts. It showed a sensitivity between 82.6% and 95.1% and a specificity between 80.5% and 88.9%, resulting in AUC values ranging from 0.91 to 0.93. This performance greatly outperformed traditional gastric cancer biomarkers, reliably identifying EOGC while the disease was still in resectable stages (Guo et al. 2025).

Another panel, proposed by Zhang et al. (2017b), identified five upregulated lncRNAs—TINCR, CCAT2, AOC4P, BANCR, and LINC00857—in the plasma of patients with GC, achieving a sensitivity of 82% and specificity of 87%, with an AUC of 0.93, outperforming CEA, which had an AUC of 0.71.

A third panel, based on RT-qPCR detection of lncRNAs CUDR, PTENP1, and LSINCT-5 in serum samples, also showed strong diagnostic potential. It achieved a sensitivity and specificity of 77.8% and 97.0% (AUC = 0.832) in stage I GC and 85.7% and 84.8% (AUC = 0.828) in stages II–IV, outperforming CA19–9 and CEA, and further highlighting the potential of circulating lncRNAs in cancer diagnostics and management (Dong et al. 2015).

6.2.3 Lung Cancer

Lung cancer—especially non-small cell lung cancer (NSCLC)—is one of the leading causes of cancer-related deaths worldwide, in part due to delayed diagnosis resulting from its often-asymptomatic progression in the early stages (Knight et al. 2017). Several deregulated lncRNAs have been identified in NSCLC, including MALAT1, GAS5, XIST, HIF1A-AS1, SOX2OT, ANRIL, SPRY4-IT1, and NEAT1. Among these, the combination of multiple lncRNAs and the integration of lncRNAs with conventional circulating biomarkers has demonstrated the highest sensitivity and specificity for NSCLC diagnosis (Lu et al. 2018b).

The combined detection of GAS5, which has lower expression in patients with NSCLC, and carcinoembryonic antigen (CEA) in plasma significantly improved diagnostic performance compared to either biomarker alone, showing a sensitivity and specificity of 86.7% and 90.9%, respectively (AUC = 0.91). Remarkably, GAS5 expression levels demonstrated much higher sensitivity than CEA in predicting NSCLC (Liang et al. 2016).

A second panel evaluated the expression of the lncRNAs SOX2OT and ANRIL in serum samples, in combination with the common tumor markers CEA, CYFRA 21-1, and SCCA, also demonstrated the superiority of the combined cancer markers over each marker individually, achieving a sensitivity of 91% and a specificity of 70% (AUC = 0.88), with improved detection accuracy as the tumor progresses. The expression levels of the two lncRNAs were also associated with overall survival (OS), and high SOX2OT levels were correlated with lymph node metastasis (Xie et al. 2018).

Circulating lncRNAs SPRY4-IT1, ANRIL, and NEAT1 were also found to be upregulated in the plasma of NSCLC patients. A panel comprising these three lncRNAs demonstrated strong diagnostic performance, with a sensitivity of 82.8% and a specificity of 92.3% (AUC = 0.876). Their stable expression in blood further supports their potential as early, non-invasive biomarkers for NSCLC detection (Hu et al. 2016). Significantly elevated levels of XIST and HIF1A-AS1 have also been observed in tumor tissues and serum samples from NSCLC, with serum levels being markedly reduced following surgical resection. Diagnostic evaluation using ROC curve analysis demonstrated that XIST and HIF1A-AS1 could reliably distinguish NSCLC patients from healthy individuals, but when combined, these two biomarkers achieve a higher diagnostic accuracy (AUC = 0.93) than either lncRNA alone (Tantai et al. 2015).

6.2.4 Colorectal Cancer

To date, several studies have identified lncRNAs that are differentially expressed in colorectal cancer (CRC) patient samples, with diagnostic accuracy ranging from moderate to high, especially when used in combination with different biomarkers, including in pre-cancerous samples. One such example is a panel comprising four lncRNAs, ZFAS1, SNHG11, LINC00909, and LINC00654, which are overexpressed in the plasma of CRC patients and have shown high accuracy (AUC = 0.94), even in early stages (AUC = 0.93). Plasma levels of these lncRNAs decreased significantly after surgical resection, indicating their potential as diagnostic and disease-monitoring biomarkers (Xu et al. 2020).

Another circulating lncRNA with strong diagnostic potential is B3GALT5-AS1. It is upregulated in the serum of CRC patients compared to healthy individuals and is positively correlated with tumor-node-metastasis (TNM) stage and histological differentiation. Although its diagnostic performance was acceptable (AUC = 0.76), when combined with conventional markers such as CEA and CA19–9, sensitivity increased to 94.9%, despite specificity remaining moderate (Ding et al. 2020).

A few individual lncRNAs have also shown diagnostic value, such as MALAT1 (AUC = 0.62), HOTAIR (AUC = 0.87), MEG3 (AUC = 0.80), CCAT1 (AUC = 0.86/0.64), and ATB (AUC = 0.78) (Svoboda et al. 2014; Abedini et al. 2019; Masrour et al. 2024). While these lncRNAs show promise, they are supported by limited data.

6.2.5 *Breast Cancer*

In breast cancer (BC) the upregulation of the lncRNA PVT1 in patients' serum compared to controls and breast fibroadenoma is a promising diagnostic marker, associated with metastatic BC. PVT1, in combination with miR-146a, demonstrated significant diagnostic accuracy for BC cases (AUC = 0.91), highlighting the potential of ncRNAs in cancer diagnostics (Sayed et al. 2025).

In addition to PVT1, the lncRNA CYTOR, which is markedly overexpressed in the plasma of BC patients, has also been identified as a potential diagnostic biomarker (AUC = 0.91), with greater diagnostic power than conventional cancer markers such as CA15–3 (Moradi et al. 2020). Another example is the lncRNA RP11-445H22.4, which is upregulated in BC serum samples compared to controls and demonstrates high diagnostic performance, with a sensitivity of 92% and specificity of 74% (AUC = 0.90), outperforming other common biomarkers such as AFP, CEA, CA125, and CA15–3 (M. Kamel et al. 2024).

Positively correlated with more aggressive molecular subtypes, the lncRNA T376626 is highly expressed in both serum and tissues. For triple-negative breast cancer (TNBC), it demonstrated a sensitivity of 73.7% and specificity of 87.5% (AUC = 0.84). Combining T376626 with CEA and CA15–3 only slightly increased the model's accuracy (AUC = 0.84 and 0.86, respectively) (He et al. 2023).

Furthermore, other lncRNAs upregulated in plasma or serum showed moderate diagnostic potential, such as H19, with a sensitivity of 56.7% and specificity of 86.7% (AUC = 0.81) (Zhang et al. 2016b), and HOTAIR, with a sensitivity of 80.0% and specificity of 68.3% (AUC = 0.79) (Zhou et al. 2015).

6.2.6 *Liver Cancer*

Recent studies have also highlighted the diagnostic potential of various circulating lncRNAs for hepatocellular carcinoma (HCC), even in patients with underlying hepatopathies. LncRNA PTTG3P and its target mRNA PTTG1, both upregulated in the serum of patients with HCC and chronic hepatitis B (CHB), were detected using RT-qPCR and showed moderate predictive power for HCC diagnosis (AUC = 0.64 and 0.63, respectively). However, when combined with conventional biomarkers such as AFP, ALT, AST, GGT, and ALP, diagnostic accuracy significantly improved, achieving a sensitivity of 90.4% and specificity of 98.0% (AUC = 0.96), underscoring the potential of this multi-marker panel as a noninvasive diagnostic tool (Cao et al. 2025).

Similarly, the serum levels of two upregulated lncRNAs, uc001ncr and AX800134, were evaluated for HBV-related HCC. Their combined expression yielded a sensitivity of 95% and specificity of 88% (AUC = 0.95) in distinguishing HBV-positive HCC patients from controls. Importantly, in AFP-negative HCC patients, these markers maintained high diagnostic accuracy (AUC = 0.94), suggesting their usefulness in early-stage detection when traditional markers may fail (Wang et al. 2015a).

In another study, the lncRNAs RP11–160H22.5, XLOC_014172, and LOC149086, were found to be significantly upregulated in HCC patients compared to cancer-free controls. The combined diagnostic signature demonstrated strong performance (sensitivity = 82%, specificity = 73%, AUC = 0.9). Notably, the combination of XLOC_014172 and LOC149086 was particularly effective for identifying metastatic HCC, with a sensitivity of 91% and specificity of 90% (AUC = 0.93), suggesting utility in both diagnosis and risk stratification (Tang et al. 2015). Plasma levels of the lncRNAs HULC and LINC00152 were also significantly elevated in HCC patients compared to healthy individuals. Individually, these lncRNAs demonstrated moderate diagnostic value (AUC = 0.78 for HULC and 0.85 for LINC00152), which increased when used in combination (AUC = 0.87). Incorporating AFP into the model further improved performance (AUC = 0.89) (Li et al. 2015).

The diagnostic value of SNHG1 was assessed in plasma from HCC patients, HBV-positive individuals with chronic hepatitis and cirrhosis (HCH), and healthy controls. SNHG1 levels correlated with clinical characteristics and successfully differentiated HCC from healthy individuals (sensitivity = 87.3%, specificity = 86%, AUC = 0.92), outperforming AFP. When distinguishing HCC from HCH, SNHG1 alone showed moderate performance (AUC = 0.74), but combining it with AFP enhanced diagnostic accuracy in both comparisons: HCC vs. healthy controls (sensitivity = 96.4%, specificity = 87%, AUC = 0.97) and HCC vs. HCH (sensitivity = 73.4%, specificity = 86.1%, AUC = 0.86), supporting its value for differentiating HCC from other liver pathologies (Gao et al. 2018).

Finally, the lncRNAs uc003wbd and AF085935, both of which are upregulated in the serum of patients with HCC and CHB, have been shown to have potential as non-invasive diagnostic markers. AF085935 exhibited superior diagnostic performance compared to uc003wbd for distinguishing HCC from healthy individuals (AUC = 0.96 vs. 0.86). When differentiating HCC from CHB, AF085935 remained robust (AUC = 0.86), while uc003wbd had lower discriminative power (AUC = 0.70). These findings highlight AF085935 as a promising biomarker for early HCC detection, even in CHB-positive individuals (Lu et al. 2015).

6.2.7 Multiple Cancer Markers

Most studies aiming to identify cancer diagnostic biomarkers have focused on specific tumor types. However, several of the lncRNAs discussed have also been found to be deregulated across multiple cancers. For example, MALAT1 is deregulated in PCa (Li et al. 2021), NSCLC (Lu et al. 2018b), and CRC (Masrour et al. 2024); PVT1 in PCa (Liu et al. 2021) and BC (Moradi et al. 2020); H19 in GC (Arita et al. 2013; Zhou et al. 2015; Hashad et al. 2016; Yörüker et al. 2018) and BC (Zhang et al. 2016b); NEAT1 in PCa (Liu et al. 2021) and NSCLC (Hu et al. 2016; Lu et al. 2018b); XIST in NSCLC (Tantai et al. 2015; Lu et al. 2018b) and CRC (Dana et al. 2025); and HOTAIR in CRC (Svoboda et al. 2014) and BC (Zhang et al. 2015). These findings raise an important question: is it possible to develop a single diagnostic panel capable of detecting multiple cancer types by targeting commonly deregulated lncRNAs and pathways?

While there is still no consensus, promising candidates have emerged. One such example is the Pan-Cancer Associated lncRNAs (PCAN), PCAN-1, PCAN-2, PCAN-3, PCAN-4, and PCAN-6, identified through multiple cancer analyses. This panel demonstrated strong diagnostic performance, with an AUC of 0.95, a sensitivity of 81.7%, and a specificity of 97.0% (Ching et al., 2016), supporting the feasibility of multi-cancer detection through lncRNA profiling. However, further studies are needed to confirm the viability of its clinical implementation. Interestingly, some of these lncRNAs tend to be deregulated in cancers that share an embryological origin, suggesting that analyzing such lncRNAs could not only facilitate the detection of multiple cancer types but also provide clues about the tissue or organ of origin (Ching et al. 2016).

6.3 Prognostic Value of lncRNAs

Particularly compelling is their potential role as clinical biomarkers; lncRNAs demonstrate a remarkable combination of tissue specificity, stability in body fluids, and functional relevance to cancer biology. These properties not only make them attractive for mechanistic studies but also place them at the frontier of biomarker discovery for prognostic stratification. Growing evidence supports the association between lncRNA expression levels and various clinical features of cancer, including tumor size, histological grade, and metastatic spread. Some lncRNAs can be upregulated or downregulated as cancer progresses, and therefore, the variation in their expression levels can be used as a marker of advanced stages of disease. Some subtypes of cancer can present a specifically bad prognosis, and their aggressive behavior may be associated with up- or downregulation of specific lncRNAs (Mirzaei et al. 2022). This chapter explores the prognostic potential of selected lncRNAs—HOTAIR, MALAT, TUG1, ANRIL, UCA1, CASC2, HULC, XIST, H19, MEG3, PCAT, and

PCA3—highlighting the evidence supporting their clinical utility in different tumor contexts. Table 6.2 describes the principal lncRNAs related as prognostic biomarkers.

6.3.1 HOTAIR (HOX Transcript Antisense Intergenic RNA)

HOTAIR is generally classified as an oncogene, interfering in oncogenesis signaling pathways such as EMT (Epithelial-Mesenchymal Transition), cell growth, invasiveness, and migration, positioning this lncRNA as a promising biomarker for tumor staging and outcome prediction (Hakami et al. 2024). A strong positive correlation exists between serum HOTAIR expression and the occurrence of metastasis in breast cancer, and its levels effectively distinguish patients with later stages from those with earlier stages (Abdel-Hamid et al. 2023). Additionally, high circulating HOTAIR levels are associated with larger tumor size, increased number of positive lymph nodes, more distant metastasis, less clinical response to treatment, and worse DFS (Disease-Free Survival) in breast cancer (Lu et al. 2018a). HOTAIR expression has been classified as an independent risk factor for predicting the prognosis of higher-grade gliomas (Zhao et al. 2017) and for predicting a poor prognosis in Diffuse Large B-Cell Lymphoma (DLBCL) (Abdelhafiz et al. 2024).

6.3.2 MALAT1 (Metastasis-Associated Lung Adenocarcinoma Transcript 1)

MALAT1 is considered an oncogene. In human gastric cancer, high MALAT1 levels predict lymph node and distant metastasis, and are associated with TNM stage and poorer overall survival (Bai et al. 2019). Cases of colorectal cancer (CRC) with upregulated MALAT1 expression correlate with advanced TNM stage, lymph node metastasis, and worse prognosis. MALAT1 has been shown to promote CRC progression by sponging miR-508-5p and enhancing RAB14 expression (Zhang et al. 2020b). Elevated MALAT1 expression in esophageal tumors is significantly associated with lymphatic invasion, distant metastasis, and tumor differentiation. Moreover, a relatively shorter survival was described when MALAT1 is upregulated, reported as an independent predictor for survival of esophageal cancer patients (Huang et al. 2016). MALAT1 is upregulated in serum exosomes from NSCLC patients, and exosome MALAT1 levels are positively associated with TNM stage and lymphatic node metastasis (Zhang et al. 2017c). In acute monocytic leukemia, a significantly decreased overall survival is observed in patients with high MALAT1 levels, and it was considered an independent prognostic factor for overall survival (Huang et al. 2017).

Table 6.2 lncRNAs described as prognostic biomarkers

lncRNA	Cancer type	Expression	Material	Clinical association	Kaplan-Meier OS (*p* log-rank)	Cox regression			Reference
						Hazard Ratio (OS)	95% CI	*p* value	
HOTAIR	Breast cancer	Upregulated	Serum	Worse OS	0.005	–	–	–	Abdel-Hamid et al. (2023)
		Upregulated	Serum	Worse OS	0.012	–	–	–	Lu et al. (2018a)
	Diffuse large B-cell lymphoma	Upregulated	Plasma	Worse OS	0.004	1.15	1.04–1.27	0.01	Abdelhafiz et al. (2024)
MALAT1	Esophageal cancer	Upregulated	Tissue	Worse OS	0.033	6.63	2.948–14.947	<0.001	Huang et al. (2016)
	Non-small cell lung cancer	Upregulated	Serum exosome	Worse OS	<0.05	–	–	–	Zhang et al. (2017c)
	AML	Upregulated	Bone marrow	Worse OS	0.001	–	–	–	Huang et al. (2017)
TUG1	Glioma	Downregulated	Tissue	Worse OS	<0.001	–	–	–	Li et al. (2016)
		Downregulated	Cell line/ clinical samples	Worse OS	<0.05	–	–	–	Cao et al. (2021)
	Gastric cancer	Upregulated	Tissue	Worse OS/ metastasis	0.013	1.066	1.023–1.112	0.003	Zhang et al. (2016)
	Bladder cancer	Upregulated	Patients samples (meta-analysis)	Worse OS	–	2.67	1.47–4.87	0.001	Li et al. (2017a)

	Various tumors	Upregulated	Patients samples (meta-analysis)	Worse OS	–	2.06	1.23–3.45	0.006	Liu et al. (2017b)
	Various tumors	Upregulated	Patients samples (meta-analysis)	Worse OS	–	1.41	1.01–1.98	–	Ma et al. (2017)
	Various tumors	Upregulated	Patients samples (meta-analysis)	Worse OS	–	1.37	1.07–1.76	0.01	Zhou et al. (2017a)
	Various tumors	Upregulated	Patients samples (meta-analysis)	Worse OS	–	1.7	1.48–1.95	<0.001	Huang et al. (2021)
ANRIL	Cervical cancer	Upregulated	Tissue	Worse OS	0.013	2.97	1.386–7.364	<0.001	Zhang et al. (2017a)
	AML	Upregulated	Bone marrow	Worse OS	–	1.859–3.300	2.260	<0.001	Tan et al. (2021)
	Glioma	Upregulated	Serum	Worse OS	0.025	–	–	–	Sun et al. (2021)

(continued)

Table 6.2 (continued)

lncRNA	Cancer type	Expression	Material	Clinical association	Kaplan-Meier OS (*p* log-rank)	Cox regression			Reference
						Hazard Ratio (OS)	95% CI	*p* value	
UCA1	Gastric cancer	Upregulated	Tissue	Worse OS	0.017	Multivariated 2.351	1.222–4.521	0.01	Zheng et al. (2015)
	Gastric cancer	Upregulated	Tissue	Worse OS	–	2.693	1.563–4.641	<0.001	He et al. (2019)
	Pancreatic cancer	Upregulated	Data from TCGA and GTEx	Worse OS	0.008	1.759	1.153–2.684	0.009	Wang et al. (2023)
	Glioma	Upregulated	Tissue	Worse OS	<0.001	7.368	3.032–17.90	0.001	Zhao et al. (2017)
	Various tumors	Upregulated	Patients samples (meta-analysis)	Worse OS	–	2.05	1.77–2.38	<0.00001	Liu et al. (2019)
	Various tumors	Upregulated	Patients samples (meta-analysis)	Worse OS	–	1.65	1.44–1.88	<0.00001	He et al. (2017)

CASC2	Various tumors	Upregulated	Patients samples (meta-analysis)	Better OS	–	0.437	0.345–0.554	–	Cai et al. (2018)
	Various tumors	Upregulated	Patients samples (meta-analysis	Better OS	–	0.41	0.32–0.50	–	Liang et al. (2018)
	AML	Downregulated	Bone marrow	Worse OS	0.01	2.574	1.224–5.414	0.013	Peng et al. (2024)
	Epithelial ovarian cancer	Downregulated	Tissue	Worse OS	–	0.417	0.251–0.699	0.001	Xue et al. (2018).
	Lung cancer	Upregulated	Patients samples (meta-analysis)	Better OS	–	0.469	0.271–0.668	<0.0001	Han et al. (2024)
	Various cancers	Upregulated	Patients samples (meta-analysis)	Better OS	–	0.43	0.33–0.55	$P = 0$	Yan et al. (2018)
	Various cancers	Upregulated	Patients samples (meta-analysis)	Better OS	–	0.40	029–0.54	<0.0001	Yu et al. (2018)

(continued)

Table 6.2 (continued)

lncRNA	Cancer type	Expression	Material	Clinical association	Kaplan-Meier OS (*p* log-rank)	Cox regression			Reference
						Hazard Ratio (OS)	95% CI	*p* value	
HULC	Osteosarcoma	Upregulated	Tissue	Worse OS	0.021	2.587	1.613–6.058	0.004	Sun et al. (2015)
	Liver cancer	Upregulated	Patients samples (meta-analysis)	Worse OS	–	1.89	1.33–2.47	$P = 0$	Ding et al. (2019)
XIST	Activated B-cell-like diffuse large B-cell lymphoma	Upregulated	Tumor tissue	Worse OS	0.029	2.133	1.059–4.294	0.034	Li et al. (2024)
	Gastric cancer	Upregulated	Tissue	Worse OS	0.017	–	–	–	Chen et al. (2016)
H19	Bladder cancer	Upregulated	Exosome from serum	Worse OS	0.002	2.701	(1.715–4.458)	<0.001	Wang et al. 2018b)
	Colorectal cancer	Upregulated	Tissue	Worse OS	0.002	1.732	1.036–2.933	0.036	Han et al. (2016).
	Non-small cell lung cancer	Upregulated	Tissue	Worse OS	<0.001	1.076	1.042–1.111	<0.001	Zhang et al. (2016a)
MEG3	Pancreatic adenocarcinoma	Downregulated	Tissue	Worse OS	–	2.805	1.108–7.105	0.011	Han et al. (2020)
PCAT-1	Prostate cancer	Upregulated	Tissue	Worse OS	<0.001	10.623	5.798–19.464	0	Li et al. (2020)
	Breast cancer	Upregulated	Tissue	Worse OS	<0.001	2.364	1.458–3.833	<0.001	Wang et al. (2021)

6.3.3 TUG1 (Taurine Upregulated Gene 1)

TUG1 does not conform to a singular oncogenic or tumor suppressor identity; instead, its role appears to be context-dependent (Li et al. 2016; Ren et al. 2017; Zhang et al. 2016; Cao et al. 2021). Nevertheless, several meta-analyses have consistently shown that elevated TUG1 levels are associated with poor overall survival (OS), advanced clinical stage, and greater metastatic burden across various malignancies (Li et al. 2017a; Liu et al. 2017b; Ma et al. 2017; Zhou et al. 2017a). In a meta-analysis encompassing 13 studies and 1274 patients (Ma et al. 2017), the authors demonstrated a significant correlation between TUG1 overexpression and worse clinical outcomes, including larger tumor size, presence of distant metastases, and advanced staging. Another comprehensive meta-analysis involving 57 studies reinforced these findings, linking high TUG1 expressions to unfavorable prognoses in cancers such as osteosarcoma, gastrointestinal, urinary, gynecological, and hematological tumors, while interestingly noting a protective effect in respiratory cancers (Huang et al. 2021).

6.3.4 ANRIL (Antisense Non-coding RNA in the INK4 Locus)

ANRIL is transcribed from the *CDKN2B-AS1* gene at chromosome 9p21, a region that harbors key tumor suppressor genes such as *CDKN2A* and *CDKN2B*, thus playing a significant role in cell cycle regulation and tumor progression (Sanchez et al. 2023).

ANRIL is mostly characterized as an oncogenic lncRNA. Its upregulation has been associated with poor overall survival (OS) in multiple cancers, including breast cancer (Alkhathami et al. 2022), cervical cancer (Zhang et al. 2017a), acute myeloid leukemia (AML) (Tan et al. 2021), hepatocellular carcinoma (Ji et al. 2019), glioma (Sun et al. 2021), and other solid tumors such as intrahepatic cholangiocarcinoma, renal clear cell carcinoma, and endometrial carcinoma (Angenard et al. 2019).

The multifaceted mechanisms through which ANRIL contributes to malignancy, coupled with consistent associations with clinical outcomes, underscore its utility as a potential biomarker for prognostic stratification across diverse tumor types.

6.3.5 UCA1 (Urothelial Carcinoma-Associated 1)

Overexpression of UCA1 is now recognized as a consistent indicator of poor prognosis in multiple malignancies, including gastric (Zheng et al. 2015; He et al. 2019), hepatocellular (Zhang et al. 2020a, b), pancreatic (Wang et al. 2023), and brain tumors such as gliomas (Zhao et al. 2017).

In hepatocellular carcinoma (HCC), high UCA1 levels are associated with advanced tumor stage (III–IV), larger tumors (≥5 cm), and multifocal disease (Zhao et al. 2017). In gliomas its expression correlates with tumor grade, further supporting its role in malignancy (Zhao et al. 2017). Similar patterns have been observed in gastric cancer, linking upregulation of UCA1 to poor differentiation, invasive growth, and worse OS and DFS outcomes (Liu et al. 2019).

Beyond individual studies, meta-analyses have corroborated these findings. A meta-analysis (He et al. 2019) encompassing 38 studies and over 3400 patients found a strong correlation between UCA1 overexpression and increased lymph node metastasis (LNM) and worse OS. Similarly, another meta-analysis (He et al. 2017) reported consistent associations with metastasis and poor outcomes. Notably, UCA1 has also been integrated into multi-lncRNA prognostic panels, such as in AML, reinforcing its value within broader molecular signatures (Wang 2023). These findings highlight UCA1 as a robust and multifaceted biomarker candidate, with strong potential for integration into clinical prognostic models across diverse cancer types.

6.3.6 CASC2 (Cancer Susceptibility Candidate 2)

CASC2 functions predominantly as a tumor suppressor. Its downregulation has been documented in a wide array of cancers, and its diminished expression is often associated with worse clinical outcomes (Cai et al. 2018; Liang et al. 2018; Ghafouri-Fard et al. 2020).

CASC2 has been implicated in the prognosis of thyroid carcinoma (Xiong et al. 2017), AML (Peng et al. 2024), epithelial ovarian cancer (Xue et al. 2018), and lung cancer (Han et al. 2024). One study from 2018 (Cai et al. 2018) included 663 patients and demonstrated that upregulation of CASC2 was an independent protective factor for overall survival (HR = 0.450; 95% CI: 0.336–0.602), although subgroup analyses by cancer type and other variables did not yield statistically significant differences.

In 2024, a study proposed an integrative model involving several CASC family members in lung cancer, grouping patients by expression profiles (Han et al. 2024). Strikingly, those with low CASC2 and CASC7 expressions had significantly better outcomes (HR = 0.469), whereas upregulation of CASC2/CASC7 in combination with other elevated CASC members predicted poor survival (HR = 1.910). In 2018 (Yan et al. 2018; Yu et al. 2018), another study expanded the evidence base with a meta-analysis of 15 studies and 1158 patients, showing that upregulation of CASC2 is correlated with smaller tumors and lower TNM stages. These associations were consistently linked to improved OS and DFS. Additional meta-analyses (Yan et al. 2018; Yu et al. 2018) reinforced these findings with similar effect sizes.

CASC2's tumor-suppressive role, reproducibly associated with better prognosis, positions it as an important lncRNA to consider in prognostic modeling, particularly in cancers where loss of regulatory control is a hallmark of disease progression.

6.3.7 HULC (Highly Upregulated in Liver Cancer)

HULC was one of the first lncRNAs identified with pronounced overexpression in malignant tissue (Panzitt et al. 2007). Since then, HULC has been implicated in the progression of various solid tumors, including osteosarcoma, pancreatic, and gastric cancer (Yu et al. 2017).

In osteosarcoma, upregulation of HULC is linked to distant metastases, advanced clinical stage, and shortened overall survival (Sun et al. 2015). In pancreatic cancer, HULC levels are associated with larger tumor size, lymph node metastasis, and vascular invasion (Peng et al. 2014), while in gastric cancer, its expression correlates with TNM staging (Zhao et al. 2014).

Curiously, the pattern in HCC is not as straightforward. Some studies have observed upregulation of HULC primarily in early-stage tumors, with an inverse association with vascular invasion (Yu et al. 2017), suggesting that the prognostic value of HULC may vary depending on the tumor's molecular subtype or microenvironment.

Meta-analytical evidence supports its prognostic role: Ding et al. (2019) reviewed 11 studies involving 1134 patients, reporting that HULC overexpression was significantly associated with poor OS (HR = 1.89). Fan et al. 2017 and others similarly noted associations between HULC levels and LNM, distant metastasis, and advanced tumor stages. These findings reinforce HULC's role as a prognostic biomarker in a variety of cancers, although its exact function may be modulated by tissue-specific factors and tumor heterogeneity.

6.3.8 XIST (X-Inactive Specific Transcript)

XIST is an lncRNA involved in X-chromosome inactivation in females with complex and context-dependent patterns of expression and roles in cancer, as noticed in a pan-cancer study showing that the prognostic role of XIST varied across cancer types. Upregulation of XIST indicated longer overall survival rates in CESC and SKCM, and a good prognosis in ESCC, PCPG, and STAD. However, upregulation of XIST was linked to poor outcomes for cancers of HNSC, KIRC, KIRP, LIHC, and PDAC (Han et al. 2022).

In activated B-cell-like diffuse large B-cell lymphoma (ABC-DLBCL), upregulation of XIST was an independent prognostic element, worsening overall survival and progression-free survival and contributing to resistance to R-CHOP treatment (Li et al. 2024).

A systematic review and meta-analysis revealed that elevated XIST expression levels were associated with positive distant metastasis and clinical stages and thus with poor prognosis in digestive system tumors (Chen et al. 2022).

In breast cancer, the expression of XIST was negatively related to lymph node invasion and TNM stage but was unrelated to age, tumor size, and histopathologic grade. However, the authors emphasize that in several previous studies, XIST was shown as upregulated in breast cancer and negatively linked with survival rates (Han et al. 2022).

The conflicting information regarding XIST's role in different cancers may result from tumor heterogeneity and differences in the study designs. This demonstrates the need for further evaluation of XIST expression and its association with prognosis.

6.3.9 H19

Typically downregulated in normal tissues, H19 is frequently overexpressed in a variety of cancers, including bladder, breast, gastric, and colorectal cancer. Upregulation of H19 expression in gastric cancer correlates with invasion depth, advanced TNM stage, and lymph node metastasis, contributing to a poor prognosis as evidenced by shorter overall and disease-free survival rates (Chen et al. 2016). Similarly, higher serum exosomal H19 levels in bladder cancer patients were correlated with poorer survival and clinical TNM stage and were independent prognostic factors affecting OS in bladder cancer (Wang et al. 2018b).

In thyroid cancer (TC), a lower 5-year survival rate has been observed in patients with high levels of H19 and low levels of miRNA let-7a. LncRNA H19 and miRNA let-7a expression, as well as tumor diameter, TNM stage, and lymph node metastasis, are independent prognostic factors for thyroid cancer (Liu et al. 2017c).

Elevated expression of H19 correlates with a worse prognosis in breast cancer, which is positively correlated with SOX4 and negatively correlated with miR-138 (Si et al. 2019). H19 expression is significantly correlated with TNM stage, and patients with higher H19 levels exhibit shorter progression-free survival and a negative correlation with response to trastuzumab (Sun et al. 2019). A strong correlation between increased expression of H19, tumor differentiation, and advanced TNM stage was also found in CRC, in which H19 expression was considered an independent prognostic factor for overall and disease-free survival (Han et al. 2016).

Increased expression of H19 is well documented in NSCLC tumor tissues, and it has been suggested as an independent prognostic indicator in this cancer type and a specific target of the oncogene c-Myc (Zhang et al. 2016a). H19 expression is most significant in stages III and IV of NSCLC and may serve as a partial indicator of invasiveness and metastasis (Zhou et al. 2017b). A significant positive association between FOXF2 and H19 has been shown in NSCLC, associated with a worse outcome (Zhou et al. 2017b).

6.3.10 MEG3 (Maternally Expressed Gene 3)

MEG3 has been identified as a tumor suppressor in a wide variety of cancers. In AML, MEG3 is a potential marker of poor prognosis, as complete remission rate, event-free survival (EFS), and overall survival (OS) were reduced when MEG3 is downregulated (Zhang et al. 2024). Conversely, increased MEG3 expression after treatment has been correlated with increased complete remission rate and prolonged EFS and OS (Zhang et al. 2024). In Acute Promyelocytic Leukemia (APL), the expression of MEG3 was significantly reduced in patients whose PML/RARα fusion gene turned negative (Jiang et al. 2020), indicating that MEG3 could be a biomarker of treatment efficiency.

In breast cancer, MEG3 is significantly downregulated and associated with tumor size, metastasis, and pathological stage (Ahmadi et al. 2024).

In pancreatic cancer clinical samples, MEG3 expression was negatively correlated with clinicopathological features such as tumor size, metastasis, and invasion (Gu et al. 2017a). MEG3 showed marked expression in metastatic cluster cells of Pancreatic Ductal Adenocarcinoma (PDAC) cancer cells (Pan et al. 2021). However, another report showed lower MEG3 expression levels in PDAC patients when compared with adjacent normal tissues. Additionally, it was shown that upregulation of MEG3 and downregulation of miR-374a-5p were more evident in patients with better prognoses (Han et al. 2020), showing a complex interplay and highlighting the context-dependent role of MEG3.

6.3.11 PCAT-1 (Prostate Cancer-Associated lncRNA Transcript 1)

PCAT-1 is upregulated in both high-grade localized and metastatic forms of PCa, in which it is intricately linked with the extent of infiltration, lymph node metastasis, and TNM staging (Dahiya et al. 2024). PCAT-1 upregulation was associated with high Gleason score, positive serum PSA, and advanced TNM stage. The overall survival rate of patients with high PCAT-1 level was significantly lower than those with low PCAT-1 expression (Li et al. 2020). PCAT-1 amplification reduced overall survival in patients with primary breast cancer. PCAT-1 amplified gene was mostly altered in grade 3 breast cancer cells, in a co-occurring state with estrogen receptor (Wang et al. 2021).

Overexpression of PCAT-1 was linked to cancer tissue invasion, lymph node metastasis, advanced tumor stage, and low survival in esophageal squamous carcinoma, colorectal cancer, and hepatocellular cancer, highlighting the strong potential of PCAT-1 as a biomarker of poor prognosis (Arriaga-Canon et al. 2018).

6.3.12 PCA3

PCA3 is an lncRNA that is useful as a biomarker to determine if a patient should undergo a repeat biopsy for prostate cancer. The advantage of PCA3 over prostate-specific antigen (PSA) is that its expression is not affected by other prostate conditions, such as prostatitis and benign variations in prostate size. PCA3 was included in the 2020 guidelines from the European Association of Urology (EAU) and the National Comprehensive Cancer Network (NCCN), especially for its ability to reduce the number of unnecessary repeat biopsies. However, the ideal cutoff value for PCA3 varies among studies, which has impaired its application as a gold standard. PCA3 can be associated with other biomarkers, such as PSA levels and the presence of TMPRSS2-ERG fusion. Combining urinary TMPRSS2-ERG and PCA3 at specified thresholds significantly improves the specificity for detecting aggressive PCa (Agbetuyi-Tayo et al. 2024).

6.4 LncRNA Therapeutic Strategies

As lncRNAs may play a protective or oncogenic role in cancer, its therapeutic application should be personalized and targeted to the specific type of cancer (Fig. 6.2). After showing lncRNA roles in cancer diagnosis and prognosis, this chapter finishes directly aiming at translational oncology in this context, depicting lncRNA as therapeutic agents, considering its application strategies and challenges, and using MALAT and HOTAIR as candidates for therapy players.

- *LncRNA manipulation*: Using methods to manipulate lncRNA availability, targeting down or overexpression according to each case. It can be used with small interfering RNAs (siRNAs) or antisense oligonucleotides to inhibit the expression of oncogenic lncRNAs and using synthetic lncRNA or transfection with expression plasmids to overexpress tumor-suppressive lncRNAs. The first approach aims to reduce lncRNAs that promote cancer progression, and the second to reintroduce the tumor-suppressive effects of specific lncRNAs.
- *Targeting lncRNA interactions*: Another rational approach relies upon disrupting the interactions between lncRNAs and their target, even proteins or other RNA. This can be achieved using chemicals, small molecules, or engineered peptides that interfere with these interactions, such as RNA aptamers and antibodies.
- *CRISPR/Cas9 technology*: One of the most promising genetic therapies, CRISPR/Cas9 targets manipulation of lncRNA genes for precise editing to either knock out oncogenic lncRNAs or activate tumor-suppressive lncRNAs, offering a tool for therapeutic intervention.
- *Nanoparticle delivery systems*: Employing nanoparticles to encapsulate lncRNA-based therapeutics can enhance delivery to tumor cells while protecting the lncRNA from enzymatic degradation, thereby improving the therapeutic effi-

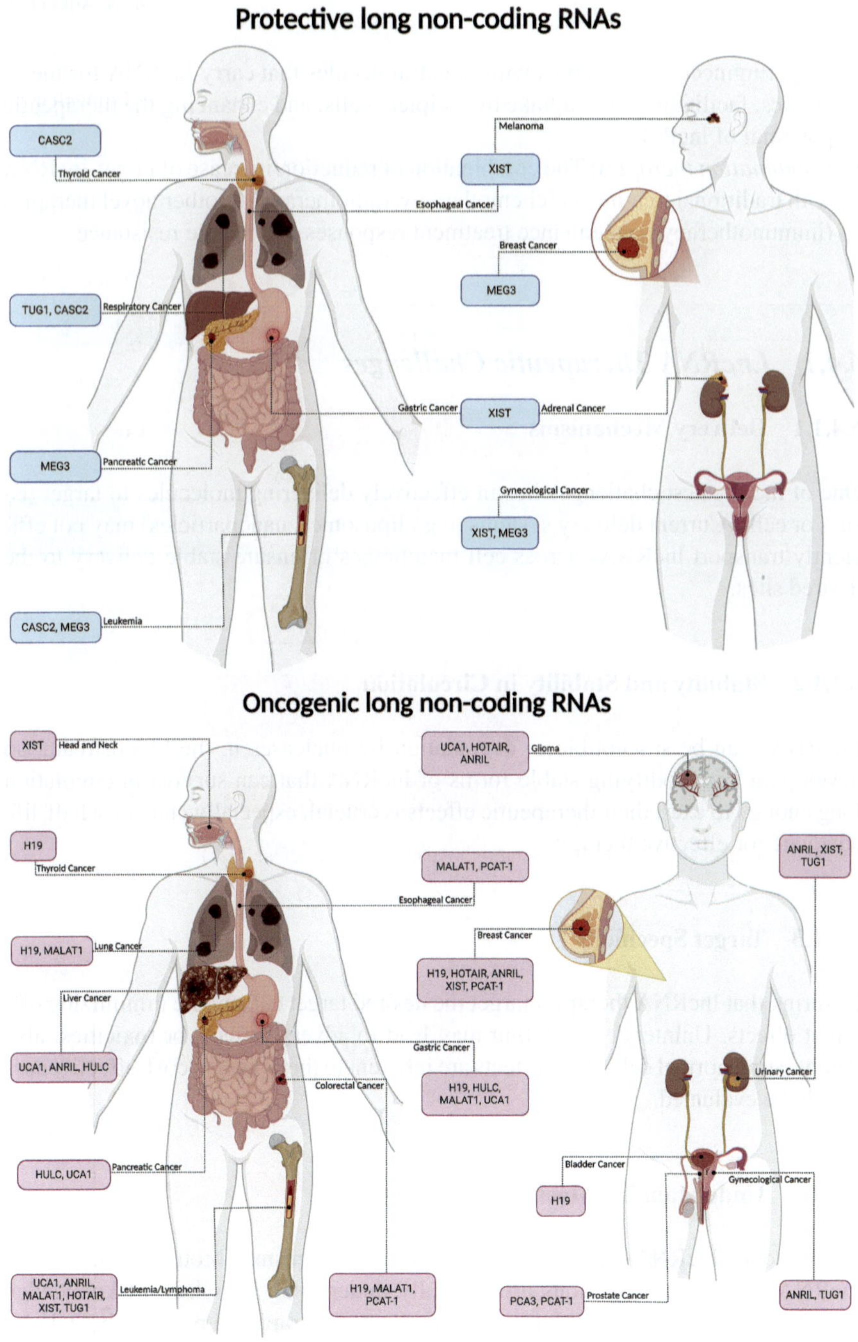

Fig. 6.2 lncRNA with protective effects (panel **a**): CASC2 in thyroid cancer, lung cancer, and leukemia; MEG3 in pancreatic cancer, leukemia, breast cancer, and gynecological cancer; XIST in melanoma, esophageal cancer, gastric cancer, adrenal cancer, and gynecological cancer; TUG1 in lung cancer. LncRNAs with oncogenic effects (panel **b**): highlighting H19 inthyroid, lung, breast, gastric, bladder cancer; MALAT1 in lung, esophageal, gastric cancer, leukemia; ANRIL in leukemia, glioma, liver, breast, urinary, and gynecological cancer

cacy. Engineering exosomes with target molecules that carry lncRNA for therapeutics, facilitating their uptake by recipient cells, and enhancing the therapeutic potential of lncRNAs.

- *Combination therapies*: The combination of reduction/increase of target-lncRNA with traditional treatments (chemotherapy, radiotherapy) or other novel therapies (immunotherapy) can enhance treatment responses and reduce resistance.

6.4.1 LncRNA Therapeutic Challenges

6.4.1.1 Delivery Mechanisms

One of the greatest challenges lies in effectively delivering molecules to target tissues or cells. Current delivery systems (e.g., liposomes, nanoparticles) may not efficiently transport lncRNAs across cell membranes or ensure stable delivery to the desired sites.

6.4.1.2 Stability and Stability in Circulation

LncRNAs can be susceptible to degradation by nucleases in the bloodstream, so developing and modifying stable forms of lncRNA that can survive in circulation long enough to exert their therapeutic effects is crucial, especially chasing a half-life adequate for effective therapy.

6.4.1.3 Target Specificity

Ensuring that lncRNA therapies target the desired target cells while minimizing off-target effects. Unintended targeting may lead to adverse effects or toxicities, also deregulating normal cells. Side effects are inherent to therapies, so cost-effectiveness should be evaluated.

6.4.1.4 Understanding Mechanisms of Action

While some lncRNAs are well-studied, the mechanisms through which many lncRNAs exert their functions are not fully understood. This lack of knowledge must be overcome for the development of targeted therapies, a process that takes a long time for basic and clinical research. More than that, lncRNAs often interact with multiple proteins and RNA molecules, showing a complex network whose knowledge is essential for predicting outcomes in therapeutic contexts.

6.4.1.5 Regulatory Challenges

As with any novel therapeutic approach, lncRNA therapies face regulatory scrutiny. This relies on an important step, evaluating safety and efficacy in preclinical and clinical testing, which can be time-consuming and costly.

6.4.1.6 Patient Heterogeneity

Individual variability in each patient's genetic and cancer biology may affect the efficacy of lncRNA-based therapies. This should require genetic characterization, personalized approaches for acquiring maximum therapeutic efficacy, and lower side effects and security.

6.4.1.7 Ethical and Safety Concerns

As with any genetic manipulation or therapeutic approach, there are ethical considerations and potential safety concerns related to modifying gene expression using lncRNAs. Every preclinical and clinical research study should be approved by an ethical committee, a necessary process that is time-consuming.

6.4.2 MALAT1

The first example of lncRNA correlated with cancer therapy is MALAT1, described in non-small cell lung cancer (Ji et al. 2003). This lncRNA is mostly reported as overexpressed in cancer patients and associated with tumor size, worse progression, and poor prognosis (Zhang et al. 2017d), so its silencing and expression/availability reduction showed interesting results. This lncRNA interacts directly with proteins, mRNA, and sponge microRNAs, thereby affecting all cellular behavior. Mechanistic research showed that MALAT1 expression was negatively correlated with miR-200a-3p and positively correlated with programmed death-ligand 1 (PD-L1) in non-small cell lung cancer, promoting in vitro proliferation, mobility, migration, and invasion (Wei et al. 2019). Another interesting in vitro data demonstrated that non-small cell lung cancer showed significant upregulation of MALAT1 compared to normal lung cells, and its knockdown decreased cell survival, proliferation, migration, and invasion (Roh et al. 2023). MALAT1-deficient cells presented reduced migration and formed fewer tumor nodules in a mouse xenograft, while blocking with antisense oligonucleotides prevented metastasis formation after tumor implantation (Gutschner et al. 2013). This lncRNA also showed an interesting connection with the tumor suppressor protein p53 pathway, with a functional interaction between MALTA1 and MDM4, a negative regulator of the TP53 pathway. NSCLC cell growth, metastasis, and survival were drastically reduced when

MALAT1 and MDM4 were deleted in a miR-185–5p/MDM4 axis (Wang et al. 2020). Another interesting axis relies on MALAT1-miR-124 and STAT3 in NSCLC, where A549 and H460 NSCLC cell lines showed upregulation of MALAT1 and downregulation of miR-124 functionally, and downregulation of MALAT1 decreased colony formation, proliferation, and triggered apoptosis in NSCLC cells. Utilizing miR-124 inhibitors allowed for reversing this effect, indicating that miR-124 was necessary for the anticancer effects of MALAT1 inhibition (Li et al. 2018).

Another interesting context is related to the MALAT1 role in chemotherapy resistance, emerging as an important player in treatment efficiency. MALAT1 upregulation was demonstrated in gemcitabine-resistant NSCLC patients, and when MALAT1 levels were artificially increased in the A549 NSCLC cell line, it showed a significant enhancement to gemcitabine, which was reversed after MALAT1 down-expression, in a process partially supported by miR-27a-5p (Chen et al. 2022). Upregulation of MALAT1 was reported in cisplatin-resistant lung cancer patients by a miR-101-3p pathway (Wang et al. 2018a; Cui et al. 2018). A direct and active reduction of intracellular drug concentration is a major and direct way to tumoral chemotherapy resistance, also linked with MALAT1, and a report also showed upregulation of MRP1 and MDR1 and activation of STST3 for cisplatin resistance in NSCLC (Fang et al. 2018).

MALAT1 is also explored in oncohematology, with overexpression in acute myeloid leukemia and chronic myeloid leukemia, associated with disease therapy. The knockdown of MALAT1 inhibited proliferation, induced apoptosis, and enhanced cytarabine sensitivity of AML cells in a miR-96 and miR-328-3p axis, contributing to the depiction and critical role of MALAT1 in AML therapy (Hu et al. 2019a; Xue et al. 2021). Similarly, MALAT1 knockdown enhanced the imatinib sensitivity of K562 cells, a chronic myeloid leukemia cell line, in a pathway with miR-328's role (Wen et al. 2018).

Despite the knowledge and science behind MALAT1 in cancer, there are no clinical trials for direct therapy evaluations; however, several reports demonstrate its use in in vivo models. MALAT1 knockdown expression in *Tp53*-null triple-negative breast cancer (TNBC) mouse results in a delay in primary tumor growth, decreased proliferation, and increased apoptosis. It also altered the tumor immune microenvironment with a decrease in immunosuppressive tumor-associated macrophages (TAM) and myeloid-derived suppressor cells (MDSC) as well as an increase in cytotoxic CD8+ T cells, inflammatory secretions, and T-cell proliferation (Adewunmi et al. 2023). In an oncohematological context, MALAT1 was demonstrated to inhibit multiple myeloma cells in vivo, using a xenograft model, where tumor volumes and weights were significantly reduced in knockdown MALAT1 animals, showing it as a promising target in the genetic therapy (Gu et al. 2017b; Liu et al. 2017a; Amodio et al. 2018). Another great finding showed that MALAT1 knockdown therapy combined with a PARP1 inhibitor or proteasome inhibitor in multiple myeloma cells showed a synergistic effect in vitro (Hu et al. 2018).

6.4.3 HOTAIR

This lncRNA is emerging as an important player in cancer, especially in initiation and progression, but is also correlated with therapeutics, highlighting HOTAIR and treatment resistance. HOTAIR expression is associated with doxorubicin, taxanes, 5-fluoruracil, and cisplatin resistance in gastric cancer and is related to the activation of the PI3K/AKT signaling pathways via miR-34a and miR-126 (Chen et al. 2015; Shang et al. 2016; Yan et al. 2016; Wang et al. 2018a; Cheng et al. 2018). For breast cancer, the role of chemotherapy was also described and associated with doxorubicin and trastuzumab by PI3K/Akt/mTOR and MEK/MAPK pathways (Li et al. 2019a; Chen et al. 2019).

This lncRNA is also correlated with radiotherapy resistance, where it was described that HOTAIR knockdown upregulates the p21 pathway and increases the radiosensitivity of cervical cancer cells (Jing et al. 2015) and influences autophagy in pancreatic cancer cells (Wu et al. 2018). The interplay between lncRNAs and radiotherapy is detailed in Chap. 7 by Yousefnia et al. In a similar finding, HOTAIR knockdown can slow the proliferation of BC cells, DNA damage accumulation, cell cycle arrest in the G2/M phase, and an increase in radiation-induced cell apoptosis, directly related to the recruitment of miR-218 (Hu et al. 2019b).

The potential application of HOTAIR as a therapeutic has direct involvement in sponging microRNAs. The overexpression of HOTAIR in cervical cancer cell lines and patients is correlated with lower miR-331-3p, one target of this lncRNA, and suggested as a therapeutic target (Buranjiang et al. 2023). In colorectal cancer, lncRNA and microRNAs interaction involves miR-1277-5p, miR-218, and miR-93, which are clinically correlated with resistance to oxaliplatin, fluorouracil, and radiosensitivity, respectively (Li et al. 2017b; Liu et al. 2020; Weng et al. 2022).

Some reports suggested direct involvement of HOTAIR in therapy for breast cancer. Delphinidin, a major anthocyanidin present in pigmented fruits and vegetables, upregulated miR-34a by inhibition of HOTAIR in breast cancer xenograft rats (Han et al. 2019). Another report showed that delphinidin reduces HOTAIR and AKT in the xenograft model of triple-negative breast cancer (Chen et al. 2015). The isoflavone calycosin reduces the expression of HOTAIR, followed by reduction of HuR and IGF2BP1 expression in breast cancer cells, inhibiting cellular proliferation and inducing apoptosis (Li et al. 2023). Other chemical compounds such as the combination of imatinib and lapatinib, AC1Q3QWB (AQB), and metformin were described as having the ability to reduce HOTAIR expression and were associated with reduced breast cancer tumorigenic properties, suggesting a promising effect for oncological therapy (Wang et al. 2015b; Ren et al. 2019; Li et al. 2019b; Golshan et al. 2021).

6.5 Conclusion

LncRNAs have emerged as significant players in cancer biology, demonstrating potential as diagnostic and prognostic biomarkers and as therapeutic targets. As diagnostic tools, specific lncRNAs and multi-lncRNA panels have shown promising accuracy in detecting various cancers, including prostate, gastric, lung, colorectal, breast, and liver cancer, and in some cases, surpassing traditional biomarkers in sensitivity and specificity. Their prognostic value is underscored by associations with clinical outcomes across malignancies, with certain lncRNAs indicating aggressive disease and poorer survival. Moreover, lncRNA-based therapeutic strategies, including manipulation of lncRNA expression, targeting lncRNA interactions, and the use of CRISPR/Cas9 technology, offer innovative approaches to cancer treatment. Despite these advancements, challenges and limitations remain in terms of delivery mechanisms, stability, target specificity, understanding mechanisms of action, regulatory hurdles, patient heterogeneity, and ethical considerations. Future research should focus on addressing these challenges to fully realize the potential of lncRNAs in improving cancer diagnostics, prognostics, and therapeutics, thereby enhancing treatment efficacy and patient outcomes in the battle against cancer.

Compliance with Ethical Standards This chapter does not contain any studies with human participants performed by any of the authors.

Conflict of Interest The authors declare no conflict of interest.

References

Abdelhafiz AS, Nabil R, Ghareeb M et al (2024) Plasma long non-coding RNAs as biomarkers for bone marrow infiltration and stage in diffuse large B-cell lymphoma. Int J Immunopathol Pharmacol 38:3946320241292665. https://doi.org/10.1177/03946320241292665

Abdel-hamid NR, Mohammed EA, Toraih EA et al (2023) Circulating ESR1, long non-coding RNA HOTAIR and microRNA-130a gene expression as biomarkers for breast cancer stage and metastasis. Sci Rep 13:22654. https://doi.org/10.1038/s41598-023-50007-5

Abedini P, Fattahi A, Agah S et al (2019) Expression analysis of circulating plasma long noncoding RNAs in colorectal cancer: the relevance of lncRNAs ATB and CCAT1 as potential clinical hallmarks. J Cell Physiol 234:22028–22033. https://doi.org/10.1002/jcp.28765

Adewunmi O, Shen Y, Zhang XHF, Rosen JM (2023) Targeted inhibition of lncRNA Malat1 alters the tumor immune microenvironment in preclinical syngeneic mouse models of triple-negative breast cancer. Cancer Immunol Res 11:1462–1479. https://doi.org/10.1158/2326-6066.CIR-23-0045

Agbetuyi-Tayo P, Gbadebo M, Rotimi OA, Rotimi SO (2024) Advancements in biomarkers of prostate cancer: a review. Technol Cancer Res Treat 23:15330338241290029. https://doi.org/10.1177/15330338241290029

Ahmadi A, Rezaei A, Khalaj-Kondori M, Khajehdehi M (2024) A comprehensive bioinformatic analysis identifies a tumor suppressor landscape of the MEG3 lncRNA in breast cancer. Indian J Surg Oncol 15(4):752–761. https://doi.org/10.1007/s13193-024-01992-0

Albala DM, Bart AS, Badre-Hume RDT et al (2024) Prostate cancer (literature review) methods for prostate cancer risk stratification: serum-based and urine-based biomarkers for biopsy consideration. Rev Urol 23:e27–e35

Alkhathami AG, Hadi A, Alfaifi M et al (2022) Serum-based lncRNA ANRIL, TUG1, UCA1, and HIT expressions in breast cancer patients. Dis Markers 2022:9997212. https://doi.org/10.1155/2022/9997212

Amodio N, Stamato MA, Juli G et al (2018) Drugging the lncRNA MALAT1 via LNA gapmeR ASO inhibits gene expression of proteasome subunits and triggers anti-multiple myeloma activity. Leukemia 32:1948–1957. https://doi.org/10.1038/s41375-018-0067-3

Angenard G, Merdrignac A, Louis C et al (2019) Expression of long non-coding RNA ANRIL predicts a poor prognosis in intrahepatic cholangiocarcinoma. Dig Liver Dis 51:1337–1343. https://doi.org/10.1016/j.dld.2019.03.019

Arita T, Ichikawa D, Konishi H, Komatsu S, Shiozaki A, Shoda K, Kawaguchi T, Hirajima S, Nagata H, Kubota T, Fujiwara H, Okamoto K, Otsuji E (2013) Circulating long non-coding RNAs in plasma of patients with gastric cancer. Anticancer Res 33(8):3185–3193

Arriaga-Canon C, De La Rosa-Velázquez IA, González-Barrios R et al (2018) The use of long noncoding RNAs as prognostic biomarkers and therapeutic targets in prostate cancer. Oncotarget 9:20872–20890

Bai X, Lu D, Lin Y et al (2019) A seven miRNA expression based prognostic signature and its corresponding potential competing endogenous RNA network in early pancreatic cancer. Exp Ther Med 18(3):1601–1608. https://doi.org/10.3892/etm.2019.7728

Buranjiang G, Abuduwanke A, Li X, Abulizi G (2023) LncRNA HOTAIR enhances RCC2 to accelerate cervical cancer progression by sponging miR-331-3p. Clin Transl Oncol 25:1650–1660. https://doi.org/10.1007/s12094-022-03059-4

Cai J, Zuo X, Chen Z et al (2018) Prognostic value and clinical significance of long noncoding RNA CASC2 in human malignancies: a meta-analysis. Cancer Manag Res 10:1403–1412. https://doi.org/10.2147/CMAR.S161373

Cao Y, Chai W, Wang Y et al (2021) lncRNA TUG1 inhibits the cancer stem cell-like properties of temozolomide-resistant glioma cells by interacting with EZH2. Mol Med Rep 24:1–10. https://doi.org/10.3892/MMR.2021.12172

Cao S, Zhong F, Chen X et al (2025) The combination of serum lncRNA PTTG3P and mRNA PTTG1 serves as a diagnostic and prognostic marker for hepatocellular carcinoma. Mol Med Rep 31:44. https://doi.org/10.3892/mmr.2024.13409

Chen J, Lin C, Yong W et al (2015) Calycosin and genistein induce apoptosis by inactivation of HOTAIR/p-akt signaling pathway in human breast cancer MCF-7 cells. Cell Physiol Biochem 35:722–728. https://doi.org/10.1159/000369732

Chen JS, Wang YF, Zhang XQ et al (2016) H19 serves as a diagnostic biomarker and up-regulation of H19 expression contributes to poor prognosis in patients with gastric cancer. Neoplasma 63:223–230. https://doi.org/10.4149/207_150821N454

Chen T, Liu Z, Zeng W, Huang T (2019) Down-regulation of long non-coding RNA HOTAIR sensitizes breast cancer to trastuzumab. Sci Rep 9:19881. https://doi.org/10.1038/s41598-019-53699-w

Chen W, Tan X, Yang Q et al (2022) MALAT1 enhances gemcitabine resistance in non-small cell lung cancer cells by directly affecting miR-27a-5p/PBOV1 axis. Cell Signal 94:110326. https://doi.org/10.1016/j.cellsig.2022.110326

Cheng C, Qin Y, Zhi Q et al (2018) Knockdown of long non-coding RNA HOTAIR inhibits cisplatin resistance of gastric cancer cells through inhibiting the PI3K/Akt and Wnt/β-catenin signaling pathways by up-regulating miR-34a. Int J Biol Macromol 107:2620–2629. https://doi.org/10.1016/j.ijbiomac.2017.10.154

Ching T, Peplowska K, Huang S et al (2016) Pan-cancer analyses reveal long intergenic noncoding RNAs relevant to tumor diagnosis, subtyping and prognosis. EBioMedicine 7:62–72. https://doi.org/10.1016/j.ebiom.2016.03.023

Chudova D, Wilde JI, Wang ET et al (2010) Molecular classification of thyroid nodules using high-dimensionality genomic data. J Clin Endocrinol Metab 95:5296–5304. https://doi.org/10.1210/jc.2010-1087

Cui Y, Cao W, Li Q et al (2016) Evaluation of prostate cancer antigen 3 for detecting prostate cancer: a systematic review and meta-analysis. Sci Rep 6:25776. https://doi.org/10.1038/srep25776

Cui Y, Li G, Zhang X et al (2018) Increased MALAT1 expression contributes to cisplatin resistance in non-small cell lung cancer. Oncol Lett 16:4821–4828. https://doi.org/10.3892/ol.2018.9293

Dahiya V, Hans S, Kumari R, Bagchi G (2024) Prostate cancer biomarkers: from early diagnosis to precision treatment. Clin Transl Oncol 26(10):2444–2456. https://doi.org/10.1007/s12094-024-03508-2

Dana F, Mahjoub S, Eshkiki ZS et al (2025) Clinical relevance of plasma-derived exosomal long non-coding RNAs (lncRNAs) CCAT1 and XIST in colorectal cancer patients. Mol Biol Res Commun 14:157–166. https://doi.org/10.22099/mbrc.2025.51654.2061

Day JR, Jost M, Reynolds MA et al (2011) PCA3: from basic molecular science to the clinical lab. Cancer Lett 301:1–6. https://doi.org/10.1016/j.canlet.2010.10.019

Ding Y, Sun C, Li J et al (2019) The significance of long non-coding RNA HULC in predicting prognosis and metastasis of cancers: a meta-analysis. Pathol Oncol Res 25:311–318. https://doi.org/10.1007/s12253-017-0351-y

Ding Y, Feng W, Ge JK et al (2020) Serum level of long noncoding RNA B3GALT5-AS1 as a diagnostic biomarker of colorectal cancer. Future Oncol 16:827–835. https://doi.org/10.2217/fon-2019-0820

Dong L, Qi P, Xu MD et al (2015) Circulating CUDR, LSINCT-5 and PTENP1 long noncoding RNAs in sera distinguish patients with gastric cancer from healthy controls. Int J Cancer 137:1128–1135. https://doi.org/10.1002/ijc.29484

Fan Y-H, Wu M-J, Jiang Y et al (2017) Long non-coding RNA HULC as a potential prognostic biomarker in human cancers: a meta-analysis. Oncotarget 8:21410

Fang Z, Chen W, Yuan Z et al (2018) LncRNA-MALAT1 contributes to the cisplatin-resistance of lung cancer by upregulating MRP1 and MDR1 via STAT3 activation. Biomed Pharmacother 101:536–542. https://doi.org/10.1016/j.biopha.2018.02.130

Gao S, Xu X, Wang Y et al (2018) Diagnostic utility of plasma lncRNA small nucleolar RNA host gene 1 in patients with hepatocellular carcinoma. Mol Med Rep 18:3305–3313. https://doi.org/10.3892/mmr.2018.9336

Ghafouri-Fard S, Dashti S, Taheri M (2020) The role of long non-coding RNA CASC2 in the carcinogenesis process. Biomed Pharmacother 127. https://doi.org/10.1016/j.biopha.2020.110202

Golshan M, Khaleghi S, Shafiee SM et al (2021) Metformin modulates oncogenic expression of HOTAIR gene via promoter methylation and reverses epithelial–mesenchymal transition in MDA-MB-231 cells. J Cell Biochem 122:385–393. https://doi.org/10.1002/jcb.29867

Goode RR, Marshall SJ, Duff M et al (2013) Use of PCA3 in detecting prostate cancer in initial and repeat prostate biopsy patients. Prostate 73:48–53. https://doi.org/10.1002/pros.22538

Gu L, Zhang J, Shi M et al (2017a) lncRNA MEG3 had anti-cancer effects to suppress pancreatic cancer activity. Biomed Pharmacother 89:1269–1276. https://doi.org/10.1016/j.biopha.2017.02.041

Gu Y, Xiao X, Yang S (2017b) LncRNA MALAT1 acts as an oncogene in multiple myeloma through sponging miR-509-5p to modulate FOXP1 expression. Oncotarget 8(60):101984–101993. https://doi.org/10.18632/oncotarget.21957

Guo X, Wang W, Cheng X et al (2025) Diagnostic efficacy of an extracellular vesicle-derived lncRNA-based liquid biopsy signature for the early detection of early-onset gastric cancer. Gut 74:1209–1218. https://doi.org/10.1136/gutjnl-2024-333657

Gutschner T, Hämmerle M, Eissmann M et al (2013) The noncoding RNA MALAT1 is a critical regulator of the metastasis phenotype of lung cancer cells. Cancer Res 73:1180–1189. https://doi.org/10.1158/0008-5472.CAN-12-2850

Hakami MA, Hazazi A, Abdulaziz O et al (2024) HOTAIR: a key regulator of the Wnt/β-catenin signaling cascade in cancer progression and treatment. Pathol Res Pract 253:154957. https://doi.org/10.1016/j.prp.2023.154957

Han D, Gao X, Wang M et al (2016) Long noncoding RNA H19 indicates a poor prognosis of colorectal cancer and promotes tumor growth by recruiting and binding to eIF4A3. Oncotarget 7(16):22159–22173. https://doi.org/10.18632/oncotarget.8063

Han B, Peng X, Cheng D et al (2019) Delphinidin suppresses breast carcinogenesis through the HOTAIR/microRNA-34a axis. Cancer Sci 110:3089–3097. https://doi.org/10.1111/cas.14133

Han T, Zhuo M, Yuan C et al (2020) Coordinated silencing of the Sp1-mediated long noncoding RNA MEG3 by EZH2 and HDAC3 as a prognostic factor in pancreatic ductal adenocarcinoma. Cancer Biol Med 17:953–969. https://doi.org/10.20892/j.issn.2095-3941.2019.0427

Han W, Shi CT, Ma J et al (2022) Pan-cancer analysis of LncRNA XIST and its potential mechanisms in human cancers. Heliyon 8:e10786. https://doi.org/10.1016/j.heliyon.2022.e10786

Han H, Huang H, Chen AP et al (2024) High CASC expression predicts poor prognosis of lung cancer: a systematic review with meta-analysis. PLoS One 19:e0292726. https://doi.org/10.1371/journal.pone.0292726

Hashad D, Elbanna A, Ibrahim A, Khedr G (2016) Evaluation of the role of circulating long non-coding RNA H19 as a promising novel biomarker in plasma of patients with gastric cancer. J Clin Lab Anal 30:1100–1105. https://doi.org/10.1002/jcla.21987

He A, Hu R, Chen Z et al (2017) Role of long noncoding RNA UCA1 as a common molecular marker for lymph node metastasis and prognosis in various cancers: a meta-analysis. Oncotarget 8(1):1937–1943. https://doi.org/10.18632/oncotarget.12463

He X, Wang J, Chen J et al (2019) lncRNA UCA1 predicts a poor prognosis and regulates cell proliferation and migration by repressing p21 and SPRY1 expression in GC. Mol Ther Nucleic Acids 18:605–616. https://doi.org/10.1016/j.omtn.2019.09.024

He Y, Xiao B, Lei T et al (2023) LncRNA T376626 is a promising serum biomarker and promotes proliferation, migration, and invasion via binding to LAMC2 in triple-negative breast cancer. Gene 860:147227. https://doi.org/10.1016/j.gene.2023.147227

Hu X, Bao J, Wang Z et al (2016) The plasma lncRNA acting as fingerprint in non-small-cell lung cancer. Tumour Biol 37:3497–3504. https://doi.org/10.1007/s13277-015-4023-9

Hu Y, Lin J, Fang H et al (2018) Targeting the MALAT1/PARP1/LIG3 complex induces DNA damage and apoptosis in multiple myeloma. Leukemia 32:2250–2262. https://doi.org/10.1038/s41375-018-0104-2

Hu N, Chen L, Wang C, Zhao H (2019a) MALAT1 knockdown inhibits proliferation and enhances cytarabine chemosensitivity by upregulating miR-96 in acute myeloid leukemia cells. Biomed Pharmacother 112:108720. https://doi.org/10.1016/j.biopha.2019.108720

Hu X, Ding D, Zhang J, Cui J (2019b) Knockdown of lncRNA HOTAIR sensitizes breast cancer cells to ionizing radiation through activating miR-218. Biosci Rep 29:BSR20181038. https://doi.org/10.1042/BSR20181038

Huang C, Yu Z, Yang H, Lin Y (2016) Increased MALAT1 expression predicts poor prognosis in esophageal cancer patients. Biomed Pharmacother 83:8–13. https://doi.org/10.1016/j.biopha.2016.05.044

Huang JL, Liu W, Tian LH et al (2017) Upregulation of long non-coding RNA MALAT-1 confers poor prognosis and influences cell proliferation and apoptosis in acute monocytic leukemia. Oncol Rep 38:1353–1362. https://doi.org/10.3892/or.2017.5802

Huang Q, Wu J, Wang H et al (2021) LncRNA taurine upregulated gene 1 as a potential biomarker in the clinicopathology and prognosis of multiple malignant Tumors: a meta-analysis. Dis Markers 2021:8818363. https://doi.org/10.1155/2021/8818363

Ji P, Diederichs S, Wang W et al (2003) MALAT-1, a novel noncoding RNA, and thymosin β4 predict metastasis and survival in early-stage non-small cell lung cancer. Oncogene 22:8031–8041. https://doi.org/10.1038/sj.onc.1206928

Ji Y, Sun H, Liang H et al (2019) Evaluation of LncRNA ANRIL potential in hepatic cancer progression. J Environ Pathol Toxicol Oncol 38(2):119–131. https://doi.org/10.1615/JEnvironPatholToxicolOncol.2019028282

Jiang M, Wang Q, Yu G, Wan J, Liu S, Zhang Z, Le A (2020) Clinical significance of long noncoding RNA maternally expressed gene 3 in acute promyelocytic leukemia. Int J Lab Hematol 43:1–6. https://doi.org/10.1111/ijlh.13438

Jing L, Yuan W, Ruofan D et al (2015) HOTAIR enhanced aggressive biological behaviors and induced radio-resistance via inhibiting p21 in cervical cancer. Tumour Biol 36:3611–3619. https://doi.org/10.1007/s13277-014-2998-2

Kamel LM, Helmy Hassan E, Mohammed Ibrahim S et al (2024) Circulating long-non coding RNA RP11-445H22.4 as a diagnostic and prognostic biomarker of breast cancer. Zagazig Univ Med J 30:4383–4390. https://doi.org/10.21608/zumj.2024.314097.3531

Knight SB, Crosbie PA, Balata H et al (2017) Progress and prospects of early detection in lung cancer. Open Biol 7:170070. https://doi.org/10.1098/rsob.170070

Li J, Wang X, Tang J et al (2015) HULC and Linc00152 act as novel biomarkers in predicting diagnosis of hepatocellular carcinoma. Cell Physiol Biochem 37:687–696. https://doi.org/10.1159/000430387

Li J, Zhang M, An G, Ma Q (2016) LncRNA TUG1 acts as a tumor suppressor in human glioma by promoting cell apoptosis. Exp Biol Med 241:644–649. https://doi.org/10.1177/1535370215622708

Li N, Shi K, Kang X, Li W (2017a) Prognostic value of long non-coding RNA TUG1 in various tumors. Oncotarget 8(39):65659–65667. https://doi.org/10.18632/oncotarget.20025

Li P, Zhang X, Wang L et al (2017b) lncRNA HOTAIR contributes to 5FU resistance through suppressing miR-218 and activating NF-κB/TS Signaling in colorectal cancer. Mol Ther Nucleic Acids 8:356–369. https://doi.org/10.1016/j.omtn.2017.07.007

Li S, Mei Z, Hu HB, Zhang X (2018) The lncRNA MALAT1 contributes to non-small cell lung cancer development via modulating miR-124/STAT3 axis. J Cell Physiol 233:6679–6688. https://doi.org/10.1002/jcp.26325

Li J, Gao C, Liu C et al (2019a) Four lncRNAs associated with breast cancer prognosis identified by coexpression network analysis. J Cell Physiol 234:14019–14030. https://doi.org/10.1002/jcp.28089

Li Z, Qian J, Li J, Zhu C (2019b) Knockdown of lncRNA HOTAIR downregulates the drug resistance of breast cancer cells to doxorubicin via the PI3K/AKT/mTOR signaling pathway. Exp Ther Med. https://doi.org/10.3892/etm.2019.7629

Li S, Ruan X, Liu T (2020) PCAT-1 expression is associated with the prognosis in prostate cancer. https://doi.org/10.21203/rs.3.rs-112933/v1

Li Y, Ji J, Lyu J et al (2021) A novel urine exosomal lncRNA assay to improve the detection of prostate cancer at initial biopsy: a retrospective multicenter diagnostic feasibility study. Cancers (Basel) 13:4075. https://doi.org/10.3390/cancers13164075

Li Y, Miao H, Wei W et al (2023) Inhibitory effect of calycosin on breast cancer cell progression through downregulating lncRNA HOTAIR and downstream targets: HuR and IGF2BP1. Acta Biochim Biophys Sin Shanghai 55:225–236. https://doi.org/10.3724/abbs.2022197

Li HB, Wang D, Zhang Y et al (2024) Long noncoding RNA XIST: a novel independent prognostic biomarker for patients with ABC-DLBCL receiving R-CHOP treatment. Carcinogenesis 45:500–509. https://doi.org/10.1093/carcin/bgae017

Liang W, Lv T, Shi X et al (2016) Circulating long noncoding RNA GAS5 is a novel biomarker for the diagnosis of nonsmall cell lung cancer. Medicine 95:e4608. https://doi.org/10.1097/MD.0000000000004608

Liang C, Zhang J, Ge H et al (2018) Long non-coding RNA CASC2 in solid tumors: a meta-analysis. Clin Chim Acta 486:357–368. https://doi.org/10.1016/j.cca.2018.08.035

Liu H, Wang H, Wu B et al (2017a) Down-regulation of long non-coding RNA MALAT1 by RNA interference inhibits proliferation and induces apoptosis in multiple myeloma. Clin Exp Pharmacol Physiol 44:1032–1041. https://doi.org/10.1111/1440-1681.12804

Liu J, Lin J, Li Y et al (2017b) Prognostic role of lncRNA TUG1 for cancer outcome: evidence from 840 cancer patients. Oncotarget 8(30):50051–50060. https://doi.org/10.18632/oncotarget.17844

Liu N, Zhou Q, Qi YH et al (2017c) Effects of long non-coding RNA H19 and microRNA let7a expression on thyroid cancer prognosis. Exp Mol Pathol 103:71–77. https://doi.org/10.1016/j.yexmp.2017.06.004

Liu C, Jin J, Shi J et al (2019) Long noncoding RNA UCA1 as a novel biomarker of lymph node metastasis and prognosis in human cancer: a meta-analysis. Biosci Rep 39:BSR20180995. https://doi.org/10.1042/BSR20180995

Liu Y, Chen X, Chen X et al (2020) Long non-coding RNA HOTAIR knockdown enhances radiosensitivity through regulating microRNA-93/ATG12 axis in colorectal cancer. Cell Death Dis 11:175. https://doi.org/10.1038/s41419-020-2268-8

Liu J, Li Y, Zhang Q et al (2021) PVT1 expression is a predictor for poor survival of prostate cancer patients. Technol Cancer Res Treat 20:1533033820971610. https://doi.org/10.1177/1533033820971610

Lu J, Xie F, Geng L et al (2015) Investigation of serum lncRNA-uc003wbd and lncRNA-AF085935 expression profile in patients with hepatocellular carcinoma and HBV. Tumour Biol 36:3231–3236. https://doi.org/10.1007/s13277-014-2951-4

Lu R, Zhang J, Zhang W et al (2018a) Circulating HOTAIR expression predicts the clinical response to neoadjuvant chemotherapy in patients with breast cancer. Cancer Biomark 22:249–256. https://doi.org/10.3233/CBM-170874

Lu T, Wang Y, Chen D et al (2018b) Potential clinical application of lncRNAs in non-small cell lung cancer. Onco Targets Ther 11:8045–8052

Ma P-J, Guan Q-K, Meng L et al (2017) Long non-coding RNA TUG1 as a potential prognostic biomarker in human cancers: a meta-analysis. Oncotarget 8(37):62454–62462. https://doi.org/10.18632/oncotarget.19099

Masrour M, Khanmohammadi S, Habibzadeh A, Fallahtafti P (2024) LncRNA MALAT1 as diagnostic and prognostic biomarker in colorectal cancers: a systematic review and meta-analysis. PLoS One 19:e0308009. https://doi.org/10.1371/journal.pone.0308009

Mirzaei S, Paskeh MDA, Okina E et al (2022) Molecular landscape of LncRNAs in prostate cancer: a focus on pathways and therapeutic targets for intervention. J Exp Clin Cancer Res 41(1):214. https://doi.org/10.1186/s13046-022-02406-1

Moradi M-T, Hatami R, Rahimi Z (2020) Circulating CYTOR as a potential biomarker in breast. Cancer 9(1):83–90. https://doi.org/10.22088/IJMCM.BUMS.9.1.83

Pan H, Diao H, Zhong W et al (2021) A cancer cell cluster marked by LincRNA MEG3 leads pancreatic ductal adenocarcinoma metastasis. Front Oncol 11:656564. https://doi.org/10.3389/fonc.2021.656564

Panzitt K, Tschernatsch MMO, Guelly C et al (2007) Characterization of HULC, a novel gene with striking up-regulation in hepatocellular carcinoma, as noncoding RNA. Gastroenterology 132:330–342. https://doi.org/10.1053/j.gastro.2006.08.026

Peng W, Gao W, Feng J (2014) Long noncoding RNA HULC is a novel biomarker of poor prognosis in patients with pancreatic cancer. Med Oncol 31:1–7. https://doi.org/10.1007/s12032-014-0346-4

Peng S, Yang Q, Pan Y et al (2024) Expression of the long noncoding RNA CASC2 in acute myeloid leukemia and its prognostic significance. Indian J Cancer 61:728–735. https://doi.org/10.4103/ijc.IJC_1365_20

Qin Z, Yao J, Xu L et al (2020) Diagnosis accuracy of PCA3 level in patients with prostate cancer: a systematic review with meta-analysis. Int Braz J Urol 46:691–704. https://doi.org/10.1590/S1677-5538.IBJU.2019.0360

Ren K, Li Z, Li Y et al (2017) Long noncoding RNA taurine-upregulated gene 1 promotes cell proliferation and invasion in gastric cancer via negatively modulating miRNA-145-5p. Oncol Res 25:789–798. https://doi.org/10.3727/096504016X14783677992682

Ren Y, Wang YF, Zhang J et al (2019) Targeted design and identification of AC1NOD4Q to block activity of HOTAIR by abrogating the scaffold interaction with EZH2. Clin Epigenetics 11:29. https://doi.org/10.1186/s13148-019-0624-2

Roh J, Kim B, Im M et al (2023) MALAT1-regulated gene expression profiling in lung cancer cell lines. BMC Cancer 23:818. https://doi.org/10.1186/s12885-023-11347-7

Sanchez A, Lhuillier J, Grosjean G et al (2023) The long non-coding RNA ANRIL in cancers. Cancers (Basel) 15(16):4160. https://doi.org/10.3390/cancers15164160

Sayed NH, Senousy MA, Shaker OG, Kortam MA (2025) Serum lncRNA PVT1 and its targets miR-146a and SIRT1 as biomarkers for the early detection of breast cancer and its metastasis. Biotechnol Appl Biochem. https://doi.org/10.1002/bab.2769

Shang C, Guo Y, Zhang H, Xue YX (2016) Long noncoding RNA HOTAIR is a prognostic biomarker and inhibits chemosensitivity to doxorubicin in bladder transitional cell carcinoma. Cancer Chemother Pharmacol 77:507–513. https://doi.org/10.1007/s00280-016-2964-3

Si H, Chen P, Li H, Wang X (2019) Long non-coding RNA H19 regulates cell growth and metastasis via miR-138 in breast cancer am. J Transl Res 11(5):3213–3225

Sun X-H, Yang L-B, Geng X-L et al (2015) Increased expression of lncRNA HULC indicates a poor prognosis and promotes cell metastasis in osteosarcoma. Int J Clin Exp Pathol 8(3):2994–3000

Sun Z, Zhang C, Wang T et al (2019) Correlation between long non-coding RNAs (lncRNAs) H19 expression and trastuzumab resistance in breast cancer. J Cancer Res Ther 15:933–940. https://doi.org/10.4103/jcrt.JCRT_208_19

Sun Y, Jing Y, Zhang Y (2021) Serum lncRNA-ANRIL and SOX9 expression levels in glioma patients and their relationship with poor prognosis. World J Surg Oncol 19:287. https://doi.org/10.1186/s12957-021-02392-2

Svoboda M, Slyskova J, Schneiderova M et al (2014) HOTAIR long non-coding RNA is a negative prognostic factor not only in primary tumors, but also in the blood of colorectal cancer patients. Carcinogenesis 35:1510–1515. https://doi.org/10.1093/carcin/bgu055

Tan GH, Nason G, Ajib K et al (2019) Smarter screening for prostate cancer. World J Urol 37:991–999

Tan Z, Zhu K, Yin Y, Luo Z (2021) Long non coding RNA ANRIL is a potential indicator of disease progression and poor prognosis in acute myeloid leukemia. Mol Med Rep 23:112. https://doi.org/10.3892/mmr.2020.11751

Tang J, Jiang R, Deng L, Zhang X, Wang K, Sun B (2015) Circulation long non-coding RNAs act as biomarkers for predicting tumorigenesis and metastasis in hepatocellular carcinoma. Oncotarget 6(6):4505–4515. https://doi.org/10.18632/oncotarget.2934

Tantai J, Hu D, Yang Y, Geng J (2015) Combined identification of long non-coding RNA XIST and HIF1A-AS1 in serum as an effective screening for non-small cell lung cancer. Int J Clin Exp Pathol 8(7):7887–7895

Wang Q (2023) Five cuprotosis-related lncRNA signatures for prognosis prediction in acute myeloid leukaemia. Hematology 28:2231737. https://doi.org/10.1080/16078454.2023.2231737

Wang K, Guo WX, Li N et al (2015a) Serum LncRNAs profiles serve as novel potential biomarkers for the diagnosis of HBV-positive hepatocellular carcinoma. PLoS One 10:e0144934. https://doi.org/10.1371/journal.pone.0144934

Wang Y-L, Overstreet A-M, Chen M-S et al (2015b) Combined inhibition of EGFR and c-ABL suppresses the growth of triple-negative breast cancer growth through inhibition of HOTAIR. Oncotarget 6(13):11150–11161. https://doi.org/10.18632/oncotarget.3441

Wang H, Wang L, Zhang G et al (2018a) MALAT1/miR-101-3p/MCL1 axis mediates cisplatin resistance in lung cancer. Oncotarget 9(7):7501–7512. https://doi.org/10.18632/oncotarget.23483

Wang J, Yang K, Yuan W, Gao Z (2018b) Determination of serum exosomal h19 as a noninvasive biomarker for bladder cancer diagnosis and prognosis. Med Sci Monit 24:9307–9316. https://doi.org/10.12659/MSM.912018

Wang D, Zhang S, Zhao M, Chen F (2020) LncRNA MALAT1 accelerates non-small cell lung cancer progression via regulating miR-185-5p/MDM4 axis. Cancer Med 9:9138–9149. https://doi.org/10.1002/cam4.3570

Wang J, Chen X, Hu H et al (2021) PCAT-1 facilitates breast cancer progression via binding to RACK1 and enhancing oxygen-independent stability of HIF-1α. Mol Ther Nucleic Acids 24:310–324. https://doi.org/10.1016/j.omtn.2021.02.034

Wang H, Ding Y, He Y et al (2023) LncRNA UCA1 promotes pancreatic cancer cell migration by regulating mitochondrial dynamics via the MAPK pathway. Arch Biochem Biophys 748:109783. https://doi.org/10.1016/j.abb.2023.109783

Wei S, Wang K, Huang X et al (2019) Lncrna malat1 contributes to non-small cell lung cancer progression via modulating mir-200a-3p/programmed death-ligand 1 axis. Int J Immunopathol Pharmacol 33:2058738419859699. https://doi.org/10.1177/2058738419859699

Wen F, Cao Y-X, Luo Z-Y et al (2018) LncRNA MALAT1 promotes cell proliferation and imatinib resistance by sponging miR-328 in chronic myelogenous leukemia. Biochem Biophys Res Commun 507:1–8. https://doi.org/10.1016/j.bbrc.2018.09.034

Weng X, Liu H, Ruan J et al (2022) HOTAIR/miR-1277-5p/ZEB1 axis mediates hypoxia-induced oxaliplatin resistance via regulating epithelial-mesenchymal transition in colorectal cancer. Cell Death Discov 8:310. https://doi.org/10.1038/s41420-022-01096-0

Wu C, Yang L, Qi X et al (2018) Inhibition of long non-coding RNA HOTAIR enhances radiosensitivity via regulating autophagy in pancreatic cancer. Cancer Manag Res 10:5261–5271. https://doi.org/10.2147/CMAR.S174066

Xie Y, Zhang Y, Du L et al (2018) Circulating long noncoding RNA act as potential novel biomarkers for diagnosis and prognosis of non-small cell lung cancer. Mol Oncol 12:648–658. https://doi.org/10.1002/1878-0261.12188

Xin L, Liu Y-H, Martin TA, Jiang WG (2017) The era of multigene panels comes? The clinical utility of oncotype DX and MammaPrint. World J Oncol 8:34–40. https://doi.org/10.14740/wjon1019w

Xiong X, Zhu H, Chen X (2017) Low expression of long noncoding RNA CASC2 indicates a poor prognosis and promotes tumorigenesis in thyroid carcinoma. Biomed Pharmacother 93:391–397. https://doi.org/10.1016/j.biopha.2017.06.063

Xu W, Zhou G, Wang H et al (2020) Circulating lncRNA SNHG11 as a novel biomarker for early diagnosis and prognosis of colorectal cancer. Int J Cancer 146:2901–2912. https://doi.org/10.1002/ijc.32747

Xue Z, Zhu X, Teng Y (2018) Long non-coding RNA CASC2 inhibits progression and predicts favorable prognosis in epithelial ovarian cancer. Mol Med Rep 18:5173–5181. https://doi.org/10.3892/mmr.2018.9550

Xue L, Li C, Ren J, Wang Y (2021) KDM4C contributes to cytarabine resistance in acute myeloid leukemia via regulating the miR-328-3p/CCND2 axis through MALAT1. Ther Adv Chronic Dis 12:2040622321997259. https://doi.org/10.1177/2040622321997259

Yan J, Dang Y, Liu S et al (2016) LncRNA HOTAIR promotes cisplatin resistance in gastric cancer by targeting miR-126 to activate the PI3K/AKT/MRP1 genes. Tumour Biol 37:16345–16355. https://doi.org/10.1007/s13277-016-5448-5

Yan X, Zhu Y, Li F et al (2018) The value of long noncoding RNA CASC2 as a biomarker of prognosis in carcinomas: a meta-analysis. J Cancer 9:3824–3830. https://doi.org/10.7150/jca.26458

Yörüker EE, Keskin M, Kulle CB et al (2018) Diagnostic and prognostic value of circulating lncRNA h19 in gastric cancer. Biomed Rep 9:181–186. https://doi.org/10.3892/br.2018.1116

Yu X, Zheng H, Chan MTV, Wu WKK (2017) HULC: an oncogenic long non-coding RNA in human cancer. J Cell Mol Med 21:410–417

Yu L, Chen S, Bao H et al (2018) The role of LncRNA CASC2 on prognosis of malignant tumors: a meta-analysis and bioinformatics. Onco Targets Ther 11:4355–4365. https://doi.org/10.2147/OTT.S166132

Zhang L, Song X, Wang X et al (2015) Circulating DNA of HOTAIR in serum is a novel biomarker for breast cancer. Breast Cancer Res Treat 152:199–208. https://doi.org/10.1007/s10549-015-3431-2

Zhang E, He X, Yin D, Han L, Qiu M, Xu T, Xia R, Xu L, Yin R, De W (2016) Increased expression of long noncoding RNA TUG1 predicts a poor prognosis of gastric cancer and regulates cell proliferation by epigenetically silencing of p57. Cell Death Dis 7:e2109

Zhang E, Li W, Yin D et al (2016a) C-Myc-regulated long non-coding RNA H19 indicates a poor prognosis and affects cell proliferation in non-small-cell lung cancer. Tumour Biol 37:4007–4015. https://doi.org/10.1007/s13277-015-4185-5

Zhang K, Luo Z, Zhang Y et al (2016b) Circulating lncRNA H19 in plasma as a novel biomarker for breast cancer. Cancer Biomark 17:187–194. https://doi.org/10.3233/CBM-160630

Zhang D, Sun G, Zhang H et al (2017a) Long non-coding RNA ANRIL indicates a poor prognosis of cervical cancer and promotes carcinogenesis via PI3K/Akt pathways. Biomed Pharmacother 85:511–516. https://doi.org/10.1016/j.biopha.2016.11.058

Zhang K, Shi H, Xi H et al (2017b) Genome-wide lncRNA microarray profiling identifies novel circulating lncrnas for detection of gastric cancer. Theranostics 7:213–227. https://doi.org/10.7150/thno.16044

Zhang R, Xia Y, Wang Z et al (2017c) Serum long non coding RNA MALAT-1 protected by exosomes is up-regulated and promotes cell proliferation and migration in non-small cell lung cancer. Biochem Biophys Res Commun 490:406–414. https://doi.org/10.1016/j.bbrc.2017.06.055

Zhang X, Hamblin MH, Yin KJ (2017d) The long noncoding RNA Malat1: its physiological and pathophysiological functions. RNA Biol 14:1705–1714

Zhang Z, Li J-Z, Wei Z-W, Li F, Li H-M, Xiao Y, Qin Y-Q (2020a) Correlation between expression levels of lncRNA UCA1 and miR-18a with prognosis of hepatocellular cancer. Eur Rev Med Pharmacol Sci 24(7):3586–3591. https://doi.org/10.26355/eurrev_202004_20820

Zhang C, Yao K, Zhang J et al (2020b) Long noncoding RNA MALAT1 promotes colorectal cancer progression by acting as a ceRNA of miR-508-5p to regulate Rab14 expression. Biomed Res Int 2020:4157606. https://doi.org/10.1155/2020/4157606

Zhang C, Qin Y, Tang Y et al (2024) MEG3 in hematologic malignancies: from the role of disease biomarker to therapeutic target. Pharmacogenet Genomics 34:209–216. https://doi.org/10.1097/FPC.0000000000000534

Zhao Y, Guo Q, Chen J et al (2014) Role of long non-coding RNA HULC in cell proliferation, apoptosis and tumor metastasis of gastric cancer: a clinical and in vitro investigation. Oncol Rep 31:358–364. https://doi.org/10.3892/or.2013.2850

Zhao W, Sun C, Cui Z (2017) A long noncoding RNA UCA1 promotes proliferation and predicts poor prognosis in glioma. Clin Transl Oncol 19:735–741. https://doi.org/10.1007/s12094-016-1597-7

Zheng Q, Wu F, Dai WY et al (2015) Aberrant expression of UCA1 in gastric cancer and its clinical significance. Clin Transl Oncol 17:640–646. https://doi.org/10.1007/s12094-015-1290-2

Zhou X, Yin C, Dang Y et al (2015) Identification of the long non-coding RNA H19 in plasma as a novel biomarker for diagnosis of gastric cancer. Sci Rep 5:11516. https://doi.org/10.1038/srep11516

Zhou Y, Lu Y, Li R et al (2017a) Prognostic role of long non-coding RNA TUG1 expression in various cancers: a meta-analysis. Oncotarget 8(59). https://doi.org/10.18632/oncotarget.20037

Zhou Y, Sheng B, Xia Q et al (2017b) Association of long non-coding RNA H19 and microRNA-21 expression with the biological features and prognosis of non-small cell lung cancer. Cancer Gene Ther 24:317–324. https://doi.org/10.1038/cgt.2017.20

Chapter 7
Mechanistic Role of Long Non-coding RNAs in Drug and Radiotherapy Resistance of Cancer Cells

Saghar Yousefnia, Farshid Parvini, and Parisa Naji

Abstract Drug and radiotherapy resistance provide a significant challenge in the treatment of cancer. Cancer cells can develop resistance mechanisms that reduce the efficiency of the treatments, leading to disease progression and poor patient outcomes. Various factors contribute to drug and radiotherapy resistance, including genetic mutations, activation of signaling pathways, and alterations in drug metabolism. Understanding these mechanisms is crucial for developing strategies to overcome resistance and improve treatment efficacy. Recent discoveries have led to a better understanding of non-coding RNAs, which are supposed to be dysregulated in chemo/radiotherapy-resistant cancer cells. Long non-coding RNAs (lncRNAs), acting as oncogenes or tumor suppressor genes, are thought to modulate numerous signaling cascades involved in cancer pathobiology, including proliferation, metabolism, survival, epithelial-mesenchymal transition, angiogenesis, metastasis, stemness, and therapy resistance through sets of DNA, RNA, and protein interactions. The precise mechanisms of therapy resistance mediated by lncRNA are not fully understood and need to be further elucidated; however, increasing studies suggest their prominent role in cancer therapy by regulating drug resistance, tumor microenvironment interactions, gene expression, and apoptosis pathways, which makes them not only potential therapeutic targets but also a novel strategy to break down barriers of treatment resistance.

Keywords Cancer · Long non-coding RNA (lncRNA) · Oncogenic lncRNA · Tumor suppressor lncRNA · MicroRNA (miRNA) · Chemo/radio therapy resistance

S. Yousefnia (✉) · F. Parvini · P. Naji
Department of Cell and Molecular Biology, Semnan University, Semnan, Iran
e-mail: saghar_yousefnia@semnan.ac.ir

L. M. Vaschetto (ed.), *Long Non-Coding RNAs, Human Health, and Diseases*, Epigenetics and Human Health 16, https://doi.org/10.1007/978-3-032-19668-2_7

Abbreviations

5-FU	5-Fluorouracil
ABC	ATP-binding cassette
ATM	Ataxia telangiectasia mutated
ATR	Ataxia telangiectasia and Rad3 related
CAT	Catalase
CeRNAs	Competing endogenous RNAs
CircRNAs	Circular RNAs
CML	Chronic myeloid leukemia
CPS1	Carbamoyl-phosphate synthase 1
CSCs	Cancer stem cells
DDR	DNA damage response
DHFR	Dihydrofolate reductase
DNA-PKs	DNA-dependent protein kinases
DSB	Double-strand break
eIF	Eukaryotic initiation factor
EMT	Epithelial-mesenchymal transition
ESCC	Esophageal squamous cell carcinoma
EZH2	Zeste 2 Polycomb Repressive Complex 2 Subunit
FBXW7	F-box/WD repeat-containing protein 7
GPX	Glutathione peroxidase
HIF-1	Hypoxia-inducible factor
HR	Homologous recombination
IDC	Invasive ductal carcinoma
IL-24	Interleukin-24
IR	Ionizing radiation
LASP1	LIM and SH3 domain protein 1
LET	Low-linear energy transfer
LncRNAs	Long ncRNAs
MDR1	Multidrug resistance protein-1
MiRNAs	MicroRNAs
MREs	MicroRNA response elements
MTX	Methotrexate
NcRNAs	Noncoding RNAs
NHEJ	Non-homologous end joining
NPC	Nasopharyngeal carcinoma
NSCLC	Non-small cell lung cancer
Onco-lncRNA	Oncogenic lncRNA
PD-L1	Programmed death-ligand 1
PIKKs	Phosphatidylinositol 3-kinases-related kinases
PRC2	Polycomb repressive complex 2
PTBP1	Polypyrimidine tract-binding protein 1
PTEN	Phosphatase and Tensin Homolog

RBPs	RNA-binding proteins
ROS	Reactive oxygen species
RUNX3	Runt-related transcription factor 3
SiRNAs	Small-interfering RNAs
SNPs	Single-nucleotide polymorphisms
SnRNAs	Small nucleolar RNAs
SOCS3	Suppressor of cytokine signaling 3
SOD	Superoxide dismutase
SRSF2	Serine and Arginine Rich Splicing Factor 2
SSB	Single-strand break
THF	Tetrahydrofolate
TOP1	Topoisomerase I
TS-lncRNA	Tumor suppressor lncRNA
ULK1	Unraveling the complexity of UNC-51-like kinase.

7.1 Introduction

The incidence of newly diagnosed cancer cases and deaths is increasing each year. GLOBOCAN 2022 data reveals a significant global cancer burden, with 20 million new diagnoses and 9.7 million fatalities. It indicates a further increase in new cancer cases, reaching an estimated 35 million by the year 2050 (Ren et al. 2025). Although the mortality rate has declined due to earlier diagnosis and therapeutic advancements, the overall cancer burden remains substantial. This ongoing challenge underscores the urgent need to better understand the molecular mechanisms underlying tumor initiation and progression to develop more effective diagnostic and therapeutic strategies. Tumor development and malignancy are driven by genetic and epigenetic alterations as well as potential structural and/or functional single-nucleotide polymorphisms (SNPs)) Asadi et al. 2023; Shirvani et al. 2025; Yousefnia 2024a). A large number of oncogenes and tumor suppressor genes promote the malignant characteristics of cancer through dysregulation of recognized molecular mechanisms of signaling pathways, including MAPK, PI3K, Notch, Hedgehog, and Wnt/β-catenin pathways (Bou Antoun and Chioni 2023; Yousefnia et al. 2020). Recent research efforts have increasingly focused on identifying molecular biomarkers and therapeutic targets in cancer. These include strategies such as oncogene silencing, tumor suppressor gene modulation, targeting cancer cell pathways, nanomedicine-based delivery systems, and retroviral therapies (Abdolmaleki et al. 2022; Nasri et al. 2023; Rahimi et al. 2023; Saharkhiz et al. 2023). The increasing cancer burden, despite therapeutic progress, necessitates a focus on molecular drivers and novel biomarkers, such as noncoding RNAs (ncRNAs).

Aberrant expression of ncRNAs contributes to malignant properties, including proliferation, migration, invasion, epithelial-mesenchymal transition (EMT), and chemo/radiotherapy resistance (Yousefnia 2022, 2024b, 2025). Roughly three-quarters of the genome has been defined as non-coding RNAs, yet they are being discovered as valuable molecules in not only the initiation and progression regulation of a wide range of cancers but also in detection and treatment (Márton et al. 2025; Yan and Bu 2021). NcRNAs are broadly categorized based on their size into small ncRNAs (≤200 nucleotides) and long ncRNAs (lncRNAs, >200 nucleotides). Among the small ncRNAs, microRNAs (miRNAs) are the most extensively studied and play crucial roles in gene regulation. Other important regulatory ncRNAs include lncRNAs and circular RNAs (circRNAs), which modulate gene expression at epigenetic, transcriptional, and post-transcriptional levels. These regulatory ncRNAs are increasingly recognized for their involvement in cancer initiation, progression, and metastasis through their effects on oncogenes, tumor suppressors, and cellular signaling pathways (Abbasi et al. 2025; Amin et al. 2019; Márton et al. 2025; Yan and Bu 2021). Furthermore, circRNAs are implicated in the modulation of diverse signaling pathways and participate in a wide array of biological, physiological, cellular, and disease-related processes. CircRNAs demonstrate the capacity to interact with a variety of biomolecules, including RNA, proteins, and DNA, thereby influencing a broad spectrum of cellular functions such as transcription, translation, and alternative splicing. Through the presence of multiple microRNA response elements (MREs), circRNAs can act as molecular sponges that sequester specific miRNAs and modulate gene expression. In addition, many circRNAs serve as versatile molecular scaffolds or decoys for RNA-binding proteins (RBPs), mediating their localization and activity in both physiological and pathological contexts. These interactions enable circRNAs to modulate protein-protein interaction networks and facilitate the recruitment of transcription factors and modifying enzymes such as DNA methyltransferase and histone acetyltransferase to chromatin and promote chromatin remodeling (Yousefnia 2025).

This chapter summarizes the biogenesis of lncRNAs and discusses the biological roles of lncRNAs in various types of cancer. Also, the following highlights the regulatory roles of lncRNAs as miRNA sponges and/or protein decoys and chromatin modification regulators in the expression of genes related to chemo/radiotherapy resistance. Understanding the molecular mechanistic roles of lncRNAs in cellular pathways and malignancy, especially chemo/radiotherapy resistance ability, provides significant potential tools for both diagnostic and therapeutic approaches in cancer treatment.

7.2 Long Non-coding RNAs and Their Biological Activity in Cancer

lncRNAs constitute a vast and diverse component of the human transcriptome, exhibiting remarkable heterogeneity in their genomic origins, biogenesis, and cellular functions (Fig. 7.1). Advances in genome annotation projects such as Ensembl/GENCODE have identified over 35,000 lncRNA genes and more than 190,000 transcript loci, highlighting their extensive representation across the genome (https://www.gencodegenes.org/human/release_47.html).

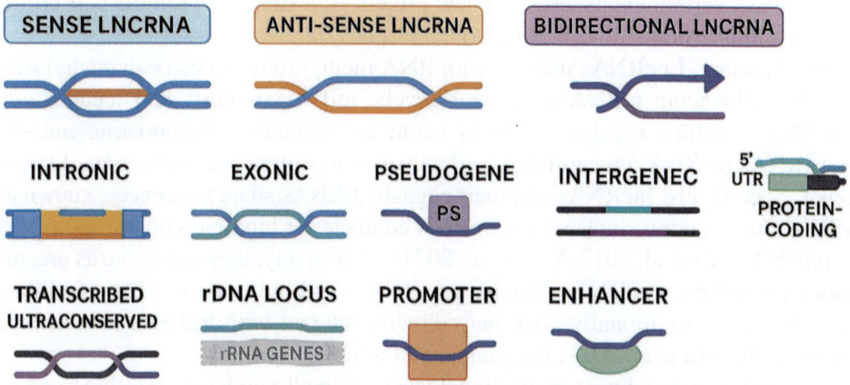

Fig. 7.1 Different types of lncRNAs based on their genomic localization

LncRNAs are now recognized as key regulators of gene expression that interact with DNA, RNA, and proteins to form complex regulatory networks. Through these multifaceted molecular interactions, alterations in lncRNA expression can disrupt cellular homeostasis and contribute to various pathological states, including cancer. Consequently, lncRNAs have emerged as both promising diagnostic biomarkers and potential therapeutic targets in diverse diseases and malignancies (Chen et al. 2022).

They are implicated in the progression of various cancers through dysregulation of gene expression at epigenetic, transcriptional and post-transcriptional, and translational and post-translational levels (Shah et al. 2022; Yousefnia 2024b).

LncRNAs can act as epigenetic regulators by interacting with histone-modifying enzymes. These interactions can either activate or repress gene transcription through two primary mechanisms: (a) by directly recruiting protein complexes (histone methyltransferases/demethylases and histone acetyltransferase/deacetylases) that modify histones, thereby altering the chromatin structure and accessibility for transcription; and (b) by acting as structural scaffolds, bringing together multiple histone-modifying enzymes to a specific genomic location, facilitating coordinated modification of histones. Furthermore, lncRNAs can modulate gene expression by recruiting enzymes involved in DNA methylation. They can either recruit DNA methyltransferases or typically repress transcription, or demethylases, which activate transcription (Dakal et al. 2025; Li et al. 2021).

At transcriptional and post-transcriptional levels, lncRNAs can interact with transcription factors, either activating or repressing the expression of target genes (Li et al. 2021). Also, they interact with splicing factors and proteins to modulate alternative splicing of mRNA and mRNA expression. For example, lncRNA SNHG1 interacts with the polypyrimidine tract-binding protein 1 (PTBP1) protein, a splicing factor, leading to alternative splicing that favors cancer cell proliferation. LncRNA SNHG1 acts as a scaffold and has a specific three-dimensional structure that allows it to bind to and interact with the PTBP1 protein. This “scaffolding” brings the PTBP1 protein to specific splicing locations on a target pre-mRNA, resulting in alternative splicing of the target genes. This can lead to the inclusion or exclusion of certain exons, creating new protein isoforms that promote cell proliferation and invasion (Liu et al. 2021a).

Furthermore, LncRNAs interact with RNA methyltransferases or demethylases, thereby influencing mRNA expression levels, mRNA stability, and localization. LncRNAs exhibit a regulatory role by acting as "sponges" or competing endogenous RNAs (ceRNAs) for miRNAs, effectively controlling the expression of target genes. Specifically, lncRNAs and their cognate mRNAs share sequences, known as MREs. Consequently, lncRNA and mRNA compete for binding with the same pool of miRNAs (Xu et al. 2017; Yang et al. 2021). Conversely, they can serve as precursors for miRNAs or siRNAs. Finally, LncRNAs can bind to proteins and facilitate mRNA decay. Additionally, they may directly interact with mRNAs to influence their stability (Li et al. 2021; Pisignano and Ladomery 2021).

In translational and post-translational levels, a small number of lncRNAs can be translated into peptides under certain circumstances. Moreover, they directly influence mRNA translation. Specifically, lncRNAs bind to the eukaryotic initiation factor (eIF) complex, which plays a crucial role in initiating protein synthesis (Karakas and Ozpolat 2021). Furthermore, lncRNAs interact with proteins involved in post-translational modifications, such as phosphorylation, acetylation, and ubiquitination, thereby modulating protein function and stability (Li et al. 2021).

LncRNAs play a prominent role in the tumorigenesis and progression of various cancers in different stages, including proliferation, migration, invasion, EMT, angiogenesis, and metastasis through acting as an oncogenic lncRNA (onco-lncRNA) or tumor suppressor lncRNA (TS-lncRNA) (Márton et al. 2025; Roy et al. 2025; Sweef et al. 2022) (Tables 7.1 and 7.2). In addition, numerous lncRNAs are dysregulated in cancers with chemotherapy and radiotherapy resistance ability, which is considered one of the most critical reasons for unsuccessful treatment, recurrence, and malignancy (Giuliani et al. 2023; Márton et al. 2025; Roy et al. 2025; Shah et al. 2022).

7.3 Chemotherapy and Their Targets

Chemotherapy is usually considered an effective method for cancer treatment. All chemotherapy agents exert their effects by interfering with the cell cycle through one or more different processes. Chemotherapy drugs are divided into three categories based on their mechanisms of function: alkylating agents, antimetabolites, and natural products (Table 7.1). Alkylating agents exert their cytotoxic effects by transferring alkyl groups to various cellular components (DNA, RNA, and protein). Alkylation of nuclear DNA provides the primary interaction that leads to cell death (Katzung 2021; Karati et al. 2022). Antimetabolites are structural analogs of naturally produced metabolites involved in the synthesis of DNA and RNA. They act by replacing these metabolites or by competing for the active sites of key enzymes. For example, methotrexate (MTX), a folic acid analog, binds to dihydrofolate reductase (DHFR), inhibiting the formation of tetrahydrofolate (THF). THF is essential for

Table 7.1 Chemotherapy agents, targets and mechanistic function in different type of cancer

Chemotherapy agents	Targets and the mechanistic function	Examples	Type of cancer
Alkylating agents	Disrupting cellular function through the formation of covalent bonds on important molecules in proteins, DNA, and RNA by the chemical structure	Cisplatin and carboplatin	Lung, breast, bladder, esophagus, head and neck, and ovarian cancer
		Oxaliplatin	Colorectal, esophagus, and pancreatic cancer
		Chlorambucil	Non-Hodgkin's lymphoma and chronic lymphocytic leukemia
		Cyclophosphamide	Breast cancer, ovarian cancer, non-Hodgkin's lymphoma, chronic lymphocytic leukemia, sarcoma, neuroblastoma, and Wilms tumor
		Mechlorethamine	Hodgkin's and non-Hodgkin's lymphoma
		Bendamustine	Non-Hodgkin's lymphoma and chronic lymphocytic leukemia
		Melphalan	Myeloma, breast cancer, and ovarian cancer
		Thiotepa	Breast, ovarian, and bladder cancer
		Busulfan	Chronic myeloid leukemia
		Carmustine	Glioma, Hodgkin's and non-Hodgkin's lymphoma
		Lomustine	Glioma
		Altretamine	Ovarian cancer
		Temozolomide	Glioma and melanoma
		Procarbazine	Glioma, Hodgkin's and non-Hodgkin's lymphoma
		Dacarbazine	Sarcoma and Hodgkin's and non-Hodgkin's lymphoma

(continued)

Table 7.1 (continued)

Chemotherapy agents	Targets and the mechanistic function	Examples	Type of cancer
Antimetabolites	Structural analogs of naturally produced metabolites involved in DNA and RNA synthesis. They either replace a metabolite that naturally participates in DNA or RNA or compete with the catalytic site of a key enzyme	5-fluorouracil	Colorectal, breast, esophagus, head and neck, and hepatocellular carcinoma
		Methotrexate	Breast cancer, head and neck, sarcoma, central nervous system lymphoma, non-Hodgkin's lymphoma, and bladder cancer
		Pemetrexed	Lung cancer
		Mercaptopurine	Acute myeloid leukemia
		Gemcitabine	Pancreatic, bladder, breast, lung, ovarian cancer, non-Hodgkin's lymphoma, and sarcoma
		Capecitabine	Breast cancer, colorectal, esophagus, hepatocellular, and pancreatic cancer
		Cytarabine	Acute myeloid leukemia, acute lymphocytic leukemia, and chronic myeloid leukemia
		Fludarabine	Non-Hodgkin's lymphoma and chronic lymphocytic leukemia
		Cladribine	Non-Hodgkin's lymphoma and chronic lymphocytic leukemia
		6-thioguanine	Acute myeloid leukemia and acute lymphocytic leukemia
Anti-cancer antibiotics	Interference with a specific DNA sequence, creating free radicals that cause strand breaks	Bleomycin	Hodgkin's and non-Hodgkin's lymphoma and head and neck cancer
		Anthracyclines	Breast cancer, Hodgkin's and non-Hodgkin's lymphoma, sarcoma, ovarian, lung, thyroid cancer, neuroblastoma, and Wilms tumor
		Mitomycin	Bladder, stomach, breast, lung, head, and neck cancer (in combination with radiotherapy)
Topoisomerase inhibitors	Topoisomerases are enzymes that regulate the three-dimensional structure of DNA. Topoisomerase I and II are the enzymes responsible for unwinding the twisting of DNA during DNA replication	Topoisomerase I inhibitor: Irinotecan	Colorectal, esophagus, and lung cancer
		Topoisomerase I inhibitor: Topotecan	Lung and ovarian cancer
		Topoisomerase II inhibitor: Etoposide	Lung, stomach cancer, and non-Hodgkin's lymphoma

Tubulin-binding drugs	Vinca alkaloids bind to tubulins and prevent the formation of microtubules, which are important during mitosis for cell shape, intracellular transport, and axonal function, while taxoids prevent the depolymerization of microtubules, thereby inhibiting their normal function	Vinca alkaloids: Vincristine	Acute lymphocytic leukemia, Hodgkin's and non-Hodgkin's lymphoma, sarcoma, neuroblastoma, and Wilms tumor
		Vinca alkaloids: Vinblastine	Hodgkin's and non-Hodgkin's lymphoma, breast cancer, and sarcoma
		Vinca alkaloids: Vinorelbine	Lung, breast, and ovarian cancer
		Taxoids: Docetaxel	Breast, lung, prostate, stomach, head and neck, ovarian, and bladder cancer
		Taxoids: Paclitaxel	Breast, lung, ovarian, esophagus, prostate, bladder, and head and neck cancer

Based on data from Katzung (2021), Chiorcea-Paquim and Oliveira-Brett (2023) and Karati et al. (2022)

Table 7.2 LncRNAs, their Onco/TS roles, and the mechanisms of function in drug resistances of various cancers

LncRNA	Onco/TS	Direct targets and the mechanism of function	Drug resistance	Cancer	References
SNHG6	Onco	Targets miR-26a-5p/ULK1 axis and induces autophagy	5-FU ↑	Colorectal cancer	Wang et al. (2019b)
FEZF1-AS1	Onco	Targets ATG5 and induces autophagy	5-FU ↑	Gastric cancer	Gui et al. (2021)
H19	Onco	Activates PI3K/Akt/mTOR pathway	5-FU and methotrexate ↑	Choriocarcinoma	Yu et al. (2019)
EIF3J-DT	Onco	Interacts with and stabilizes ATG14 mRNA and competes with miR-188-3p to induce autophagy	5-FU and oxaliplatin ↑	Gastric cancer	Luo et al. (2021)
MALAT1	Onco	Targets miR-23b-3p/ ATG12 and induces autophagy	5-FU and cisplatin	Gastric cancer	YiRen et al. (2017)
AC078883.3	TS	Targets miR-19a/PTEN and inhibits the PTEN/AKT pathway	Cisplatin ↓	Non-small cell lung cancer	Xing et al. (2019)
CASC2	TS	Targets miR-21/PTEN pathway	Cisplatin ↓	Cervical cancer	Feng et al. (2017)
FOXD2-AS1	Onco	Targets miR185-5p/SIX1 axis	Cisplatin ↑	Non-small cell lung cancer	Ge et al. (2019)
GAS5	TS	Targets miR-21 and regulates phosphorylation of Akt and PI3K/Akt pathway	Cisplatin ↓	Cervical cancer	Wen et al. (2017)
HOXA11-AS	Onco	Targets miR-214-3p/PIM1 axis and promoting proliferation	Cisplatin ↑	Oral squamous cell carcinoma	Wang et al. (2019c)
Linc00312	TS	Induces apoptosis via regulating Bcl-2 and caspase-3 expression	Cisplatin ↓	Ovarian cancer	Zhang et al. (2018)
Linc00473	TS	Targets ZBTB7A-mediated LINC00473-C/EBPβ-IL24 pathway	Cisplatin ↓	Osteosarcoma	Zhang et al. (2017b)

MEG3	TS	Decreases the extracellular vesicle-mediated transfer of miR-214	Cisplatin ↓	Ovarian cancer	Zhang et al. (2017a)
MALAT1	Onco	Activates STAT3 and increases MRP1 and MDR1 expression	Cisplatin ↑	Non-small cell lung cancer	Fang et al. (2018)
NCK1-AS1	Onco	Targets miR-134-5p/MSH2 axis and increases DNA repair, and inhibits apoptosis	Cisplatin ↑	Cervical cancer	Zhang et al. (2019b)
NRAL	Onco	Targets miR-340-5p/Nrf2 and inhibits apoptosis	Cisplatin ↑	Hepatocellular carcinoma	Wu et al. (2019a)
OIP5-AS1	Onco	Targets miR-340-5p/LPAATβ and activates / PI3K/Akt/mTOR pathway	Cisplatin ↑	Osteosarcoma	Song et al. (2019)
TATDN1	Onco	Targets miR-451/ TRIM66 axia and promotes proliferation, adhesion, and invasion	Cisplatin ↑	Lung cancer	Wang et al. (2019a)
SCAT7	Onco	Regulates ATR signaling pathway and targets and degrades topoisomerase I (TOP1)	Cisplatin ↑	Lung adenoma carcinoma	Statello et al. (2021)
PANDAR	Onco	Interacts with SRSF2 splicing factor and inhibitsp53-mediated transcription of BAX and PUMA	Cisplatin ↑	Ovarian cancer	Wang et al. (2018a)
XIST	Onco	Targets miR-520/Bax	Cisplatin ↑	NSCLC	Liu et al. (2021b)
GAS5	TS	Targets miR-21/PTEN	Cisplatin ↓	NSCLC	Cao et al. (2017)
ERVH48–1	Onco	miR-4784/ WNT2B	Docetaxel ↑	Prostate cancer	Chen et al. (2023)
GBCDRlnc1	Onco	Interacts with PGK1 and reduces the levels of the ATG5-ATG12 and inhibits autophagy	Doxorubicin ↑	Gallbladder cancer	Cai et al. (2019)
SNHG12	Onco	Targets miR-320a/Mcl-1	Doxorubicin ↑	Osteosarcoma	Tang et al. (2021)
XIST	Onco	Targets miR-30a-5p/ROR1 axis promotes proliferation and drug resistance	Doxorubicin ↑	Colorectal cancer	Zhang et al. (2019a)

(continued)

Table 7.2 (continued)

LncRNA	Onco/TS	Direct targets and the mechanism of function	Drug resistance	Cancer	References
NONHSAT101069	Onco	Targets miR-129-5p/Twist1 axis and induces EMT	Epirubicin ↑	Breast cancer	Yao et al. (2019)
Linc00460	Onco	Targets miR-769-5p/EGFR axis	Gefitinib ↑	Non-small cell lung cancer	Ma et al. (2019)
Linc00665	Onco	Targets EZH2 and PI3K/Akt pathway	Gefitinib ↑	Non-small cell lung cancer	Liu et al. (2019)
GAS5	TS	Targets miR-221/SOCS3 pathway, reverses EMT and tumor stem cell self-renewal	Gemcitabine ↓	Pancreatic cancer	Liu et al. (2018a)
Linc00346	Onco	Targets miR-188-3p/ BRD4As and promotes cell proliferation and drug resistance	Gemcitabine ↑	Pancreatic cancer	Shi et al. (2019)
MEG3	TS	Targets snail and inhibits EMT, sphere-forming ability	Gemcitabine ↓	Pancreatic cancer	Ma et al. (2018)
PVT1	Onco	Targets miR-152 and activate c-MET/PI3K/Akt pathway	Gemcitabine ↑	Osteosarcoma	Sun et al. (2019)
UCA1	Onco	Targets miR-16 to increase MDR1 expression	Imatinib ↑	CML	Xiao et al. (2017)
KCNQ1OT1	Onco	Targets miR-7-5p/ ABCC1 axis	Oxaliplatin ↑	Hepatocellular carcinoma	Hu et al. (2018)
NR2F1-AS1	Onco	Targets miR-363/ ABCC1	Oxaliplatin ↑	Hepatocellular carcinoma	Huang et al. (2018)
UPK1A-AS1	Onco	Facilitating the interaction between Ku70 and Ku80, enhancing NHEJ DNA repair	Oxaliplatin ↑	Pancreatic cancer	Zhang et al. (2022)
Linc00518	TS	Targets miR-216b-5p	Paclitaxel ↑	Prostate cancer	He et al. (2019)
Linc00672	TS	Interacts with p53 and suppresses LASP1 expression	Paclitaxel ↓	Endometrial cancer	Li et al. (2017a)

Linc01118	Onco	Targets miR-134/ABCC1 axis, inhibits apoptosis	Paclitaxel ↑	Epithelial ovarian cancer	Shi and Wang (2018)
UCA1	Onco	Targets miR-129/ABCB1 axis	Paclitaxel ↑	Ovarian cancer	Wang et al. (2018b)
NEAT1	Onco	Targets miR-335/c-met and activates c-met/Akt pathway	Sorafenib and doxorubicin ↑	Hepatocellular carcinoma	Chen and Xia (2019)
UCA1	Onco	Targets miR-18a-HIFα feedback regulatory loop, targets Wnt/β-catenin axis and mTOR signaling pathway, regulates EZH2/p21 axis and PI3K/Akt pathway	Tamoxifen ↑	Breast cancer	Malhotra et al. (2017)
HOTAIR	Onco	Targets PRC2 and silencing of estrogen receptor-related genes	Tamoxifen ↑	Breast cancer	Shi et al. (2024)
CASC2	TS	Targets miR-18a/PTEN and decreases pAkt expression	TMZ ↓	Glioma	Li et al. (2018)
Lnc-TALC	Onco	Targets miR-20b-3p/c-met axis	TMZ ↑	Glioblastoma	Wu et al. (2019b)
RP11-838 N2.4	TS	Targets miR-10a/ EphA8, activates apoptosis and inhibits TGF-β	TMZ ↓	Glioblastoma	Liu et al. (2016b)
SBF2-AS1	Onco	Targets miR-151a-3p/ZEB1 axis and activates EMT	TMZ ↑	Glioblastoma	Rezaei et al. (2021)
TUSC7	TS	Targets miR-10a/MDR1 axis	TMZ ↓	Glioblastoma	Shang et al. (2018)
GAS5	TS	Targets miR-21/PTEN axis	Trastuzumab ↓	Breast cancer	Li et al. (2016)

one-carbon transfer reactions required for the de novo synthesis of thymidylate, purine nucleotides, and the amino acids serine and methionine. Consequently, MTX disrupts the synthesis of DNA, RNA, and proteins, which impairs cell proliferation and can lead to cell death (Tiwari 2012). Natural product-derived agents include anticancer antibiotics, topoisomerase inhibitors, and tubulin-binding drugs. Screening of microbial products has led to the discovery of several growth-inhibiting compounds that have been approved as chemotherapeutic drugs. A large number of these antibiotics interfere with DNA replication by intercalating between specific DNA bases and blocking the synthesis of DNA, RNA, or both. All anticancer antibiotics currently used in clinical practice are originated from various soil *Streptomyces* species (Gao et al. 2020). Topoisomerase inhibitors also interfere with the function of topoisomerase I and II enzymes, which play a role in unwinding DNA during DNA replication (Yakkala et al. 2023). Tubulin-binding drugs are another class of approved chemotherapeutic agents. Vinca alkaloids inhibit the polymerization of tubulin, disrupting the formation of microtubules, especially those involved in the formation of the mitotic spindle apparatus (Wang et al. 2023).

7.4 Radiotherapy and Their Targets

Radiotherapy is a physical treatment that uses ionizing radiation (IR) to destroy cancer cells by depositing energy within tissues and inducing molecular damage. Approximately 50% of all cancer patients receive radiotherapy during their disease course, and it contributes to about 40% of curative cancer treatments. Radiotherapy is often used in combination with surgery, chemotherapy, or immunotherapy (Baskar et al. 2012; Gong et al. 2021).

X-rays and gamma rays are the most commonly used forms of IR in cancer therapy. They are low-linear energy transfer (LET) electromagnetic radiations that differ in their sources: X-rays are produced by devices such as linear accelerators, while gamma rays originate from the decay of radioactive isotopes like Cobalt-60, Radium, and Cesium (Abshire and Lang 2018). The primary goal of radiotherapy is to maximize the radiation dose delivered to tumor cells while minimizing exposure to surrounding healthy tissues. The biological target of radiotherapy is cellular DNA, which can be damaged through two main mechanisms: direct ionization of the DNA molecule and indirect effects mediated by free radicals generated from the radiolysis of water. These interactions result in single-strand breaks (SSBs) or double-strand breaks (DSBs). Consequently, radiation-induced cell death may occur through apoptosis, mitotic catastrophe, necrosis, senescence, or autophagy (Gong et al. 2021).

7.5 Chemotherapy and Radiotherapy Resistance in Cancer

Therapy resistance is classified as intrinsic or acquired. Intrinsic resistance characterizes a lack of treatment response, existing initially before the treatment started or occurring after a short time, though acquired resistance occurs after a certain duration, first responds to the treatment, but progresses later (Liu et al. 2021c; Yousefnia et al. 2020).

7.5.1 Molecular Mechanisms of Chemotherapy Resistance

The most important factor that induces drug resistance is tumor heterogeneity, which can be mediated by a subset of cancer cells exhibiting stem cell-like properties. These cells are implicated in various aspects of cancer development and treatment response, including tumor growth, resistance to chemotherapeutic agents and radiotherapy, and the recurrence of disease. Cancer stem cells (CSCs) are characterized by their aggressive and malignant behavior. They show self-renewal, migration and invasion, metastasis, angiogenesis, EMT, and chemo/radiotherapy resistance ability (Dakal et al. 2024).

Therapeutic resistance in cancer cells and CSCs arises from two primary mechanisms: pump and non-pump resistance. Non-pump resistance is associated with the activation of BCL2, a protein that inhibits apoptosis. Conversely, pump resistance involves the overexpression of ATP-binding cassette (ABC) transporters. ABC transporters actively efflux chemotherapy drugs from cells, thereby decreasing their intracellular concentration and diminishing their effectiveness. Additionally, survivin, a protein known to suppress apoptosis, is frequently overexpressed in both cancer cells and CSCs. This overexpression contributes to reduced cell death and enhanced drug resistance (Abdullah and Chow 2013).

7.5.2 Molecular Mechanisms of Radiotherapy Resistance

Cancer cells and CSCs possess adaptive mechanisms to counteract the damaging effects of reactive oxygen species (ROS) generated during radiotherapy. One such mechanism involves the overexpression of ROS scavenging factors. These factors are encoded by genes like catalase (CAT), superoxide dismutase (SOD), and glutathione peroxidase (GPX), which effectively neutralize ROS, thereby protecting cancer cells from radiation-induced oxidative stress (Asadi et al. 2023; Yousefnia et al. 2020). Likewise, radiotherapy resistance is controlled by genomic instability as well as by immune, vascular, and pro-fibrotic stromal components of the tumor microenvironment (Wu et al. 2023).

Moreover, cancer cells activate intricate survival mechanisms, potentially leading to tumor recurrence and radiotherapy resistance. The cellular response to DNA damage, particularly DSBs, involves a network of proteins. Rad24p, in conjunction with Ddc1p and Mec3p, forms a complex that initiates cell cycle arrest. Additionally, Rad24p interacts with Rfc2p or Rfc5p, facilitating the recruitment of the Rad24p-Ddc1p-Mec3p complex and triggering downstream signaling cascades via kinases like Rad53p. DSB repair can occur through two primary pathways; non-homologous end joining (NHEJ) and homologous recombination (HR). NHEJ is mediated by the Ku protein, which binds to DSB ends, forming a protective scaffold and allowing access for other NHEJ factors. HR, predominantly activated in response to SSBs, involves multiple proteins, including the MRN complex. The ataxia telangiectasia mutated (ATM) and ataxia telangiectasia and Rad3 related (ATR) signaling pathways play a crucial role in activating HR (Liu et al. 2021c). Furthermore, exposure to IR triggers the activation of phosphatidylinositol 3-kinases-related kinases (PIKKs), a family of enzymes including ATM, ATR, and DNA-dependent protein kinases (DNA-PKs). These PIKKs play a crucial role in amplifying and transducing DNA damage signals. Subsequently, components involved in DNA repair are recruited to the site of damage, initiating the repair process. Cellular resistance to IR is directly correlated with the efficiency of DSB repair mechanisms. Cells exhibiting robust DSB repair pathways demonstrate increased radioresistance (Liu et al. 2021c). Also, alterations in genes that regulate the DNA damage response (DDR), such as p53, can lead to increased radioresistance (Wu et al. 2023).

On the other hand, the tumor microenvironment plays a crucial role in their ability to develop radiotherapy resistance. Hypoxia is a major cause of radioresistance. In a low-oxygen environment, the production of the damaging ROS is reduced, making the cells more resistant. Hypoxia also promotes the expression of genes that contribute to radioresistance, such as hypoxia-inducible factor (HIF-1) (Jain et al. 2024).

Many lncRNAs have been shown to play a role in therapy resistance. Although the precise mechanisms of drug/radiotherapy resistance mediated by lncRNAs are not fully understood and need to be further elucidated, an increasing amount of evidence has revealed that lncRNA dysregulation brings chemo/radiotherapy resistance through the targets.

7.6 Role of lncRNAs in Drug Resistance in Cancer

Increasing in vitro and in vivo evidence indicates that lncRNAs are potentially involved in resistance to several chemotherapy agents in various cancers (Table 7.2).

7.6.1 *Role of Onco-lncRNAs as miRNA Sponges in Cancer Drug Resistance*

Numerous studies validate that lncRNAs play roles in the cancer treatment barrier. The role of onco-lncRNAs as miRNA sponges in cancer drug resistance is a compelling area of research (Chen et al. 2023; Liu et al. 2021b; Xiao et al. 2017). Some onco-lncRNAs can enhance drug resistance by modulating the expression of ABC transporters, which are involved in drug efflux. By sponging specific miRNAs, these lncRNAs prevent miRNAs from downregulating ABC transporter genes. For example, imatinib resistance in CML cells is often attributed to drug efflux mediated by multidrug resistance protein-1 (MDR1). It has been shown that overexpression of lncRNA UCA1 elevates MDR1 levels, consequently promoting imatinib resistance in CML cells. Mechanistically, UCA1 functions as a ceRNA for MDR1 by directly binding to miR-16, a microRNA known to regulate MDR1 expression. These findings suggest that targeting the UCA1-MDR1 axis may represent a novel therapeutic strategy for enhancing imatinib efficacy in CML patients (Xiao et al. 2017). In most contexts, autophagy acts as a cytoprotective mechanism that enables cancer cells to survive therapeutic stress. Consequently, modulation of autophagy is considered a key mechanism by which cancer cells acquire therapy resistance (Sui et al. 2013). For example, lncRNA SNHG6 has been shown to enhance resistance to 5-fluorouracil (5-FU) by promoting autophagy. Mechanistically, SNHG6 acts as a "sponge" for miR-26a-5p, a microRNA that targets Unraveling the complexity of UNC-51-like kinase 1 (ULK1), a key regulator of autophagy (Wang et al. 2019b). The protein ULK1 plays a crucial role in initiating autophagy. It achieves this by assembling into a complex composed of four subunits: ULK1, along with the proteins ATG13, ATG101, and FIP200. This tetrameric complex is essential for the initiation of starvation-induced autophagy (Zou et al. 2022). By sequestering miR-26a-5p, SNHG6 increases ULK1 expression, thereby inducing autophagy and promoting 5-FU resistance. These findings suggest that targeting the SNHG6/miR-26a-5p/ULK1 axis could be a potential strategy to overcome chemoresistance in colorectal cancer (Wang et al. 2019b).

Some onco-lncRNAs can regulate chemotherapy resistance through modulating proliferation and apoptosis pathways. For example, using bioinformatic analysis and experimental validation, researchers have identified lncRNA ERVH48–1 as a key lncRNA contributing to docetaxel resistance. ERVH48–1 is found to act as a ceRNA, sponging miR-4784 and subsequently upregulating WNT2B expression. This activates the Wnt/-catenin signaling pathway, ultimately promoting cell proliferation, invasion, and migration, and inhibiting apoptosis in docetaxel-resistant prostate cancer cells (Chen et al. 2023). Also, inhibiting apoptosis allows cancer cells to survive during chemotherapy. For example, in non-small cell lung cancer (NSCLC), lncRNA XIST acts as a sponge for miR-520, leading to the upregulation of BAX, a miR-520 target gene and a pro-apoptotic protein. This interaction contributes to cisplatin resistance by modulating the p53 signaling pathway (Liu et al. 2021b).

Moreover, researchers have found that FOXD2-AS1 and SIX1 are upregulated, while miR185-5p is downregulated, in cisplatin-resistant NSCLC cells. Silencing FOXD2-AS1 enhances chemosensitivity by inhibiting cell proliferation, migration, invasion, and drug resistance-associated protein expression. Mechanistically, FOXD2-AS1 sequesters miR185-5p, thereby relieving the repression of its target gene SIX1 (Ge et al. 2019). The SIX1 transcription factor complex plays a significant role in various facets of cancer development. It influences both tumorigenesis and metastasis of cancer cells by modulating the expression of critical genes involved in cell cycle control, oncogenic activity, and other pro-tumorigenic factors (Blevins et al. 2015). These results suggest that the FOXD2-AS1/miR185-5p/SIX1 axis plays a crucial role in chemoresistance and could serve as a potential therapeutic target for cisplatin-resistant NSCLC.

7.6.2 Role of Onco-lncRNAs as Protein Decoys in Drug Resistance of Cancer Cells

One notable mechanism is the function of lncRNAs as protein decoys. It can disrupt normal gene regulation and signaling pathways, thereby facilitating cancer cell survival and resistance to chemotherapy. LncRNAs can act as decoys by binding to and inhibiting the activity of proteins involved in apoptosis, DNA repair, and drug metabolism (Fang et al. 2018; Gui et al. 2021; Luo et al. 2021; Shi et al. 2024; Statello et al. 2021; Wang et al. 2018a; Zhang et al. 2022).

It has been reported that lncRNA SCAT7 plays a critical role in the cellular response to DNA damage. Researchers have found that SCAT7 expression increases when lung cancer cells are exposed to DNA-damaging drugs, and its presence is crucial for cell survival under these conditions. Silencing SCAT7 reduces the proliferation of cisplatin-resistant lung adenoma carcinoma by disrupting cell cycle checkpoints and DNA repair pathways. Specifically, SCAT7 regulates the ATR signaling pathway, which promotes HR repair. Additionally, SCAT7 participates in the degradation of topoisomerase I (TOP1), an enzyme involved in DNA replication. Depletion of SCAT7 leads to accumulation of TOP1-DNA complexes, resulting in increased DNA damage. These findings highlight SCAT7 as a critical component of the DNA damage response pathway and suggest its potential as a therapeutic target for treating drug-resistant cancers (Statello et al. 2021). In addition, elevated FEZF1-AS1 expression has been observed in gastric cancer cells. Mechanistically, FEZF1-AS1 enhances multidrug resistance by directly targeting and upregulating the autophagy-related gene *ATG5*. This interaction ultimately leads to increased autophagic activity in gastric cancer cells, contributing to chemoresistance. Silencing FEZF1-AS1 suppresses tumor growth and sensitizes gastric cancer cells to 5-FU treatment, highlighting its potential as a therapeutic target for overcoming chemoresistance in gastric cancer (Gui et al. 2021). In breast cancer, lncRNA HOTAIR is overexpressed and contributes to tamoxifen resistance by reprogramming chromatin states through its interaction with polycomb repressive complex 2

(PRC2), leading to the silencing of estrogen receptor-related genes (Shi et al. 2024). LncRNA MALAT1 has been associated with resistance to cisplatin in NSCLC by modulating the expression of multidrug resistance proteins such as MRP1 and MDR1 through phosphorylation and activation of STAT3, a nuclear transcription factor (Fang et al. 2018). In pancreatic cancer, lncRNA UPK1A-AS1 has been shown to promote oxaliplatin resistance by facilitating the interaction between Ku70 and Ku80, enhancing non-homologous end joining DNA repair mechanisms, thus allowing cancer cells to survive DNA-damaging chemotherapies (Zhang et al. 2022). It has also been reported that lncRNA PANDAR interacts with the splicing factor serine−/arginine-rich splicing factor 2 (SRSF2) through the 5′-CCAG-3′ sequence, leading to the downregulation of p53 and inhibition of p53 phosphorylation. Consequently, it reduces transcription of pro-apoptotic genes such as BAX and PUMA. By acting as a scaffold for SRSF2, PANDAR impedes the normal apoptotic response to DNA damage—thereby promoting cell survival and contributing to cisplatin resistance in ovarian cancer cells (Wang et al. 2018a).

On the other hand, researchers have found that lncRNAs can interact with RNA directly, rather than proteins, to stabilize mRNA. For example, it has been reported that EIF3J-DT is highly expressed in drug-resistant gastric cancer cells, where it promotes autophagy. EIF3J-DT achieves this by directly binding to and stabilizing the mRNA of ATG14—a key autophagy protein—while also blocking its degradation by competitively binding with miR-188-3p. Elevated levels of both EIF3J-DT and ATG14 were observed in chemotherapy-resistant gastric cancer patients, correlating with poor prognosis. Therefore, EIF3J-DT emerges as a potential therapeutic target for overcoming chemoresistance and improving outcomes in gastric cancer treatment (Luo et al. 2021).

7.6.3 Role of TS-lncRNAs as miRNA Sponges in Drug Resistance of Cancer Cells

The ability of TS-lncRNAs to act as miRNA sponges is crucial for modulating drug resistance in cancer cells. By sequestering specific miRNAs, these lncRNAs can stimulate TSGs, thereby activating pathways that enhance cell apoptosis and sensitivity to chemotherapy. For instance, research has demonstrated that lncRNA GAS5 acts as a tumor suppressor by inhibiting cell proliferation, migration, EMT, and gemcitabine resistance. GAS5 achieves these effects by directly targeting and suppressing miR-221 expression, leading to increased expression of suppressor of cytokine signaling 3 (SOCS3), a known suppressor of cytokine signaling. This mechanism highlights the GAS5/miR-221/SOCS3 pathway as a potential therapeutic target for overcoming chemoresistance and metastasis in pancreatic cancer (Liu et al. 2018a). Furthermore, lncRNA GAS5 acts as a decoy for miR-21, a known oncomiRNA, thereby preventing miR-21 from suppressing its target gene, Phosphatase and Tensin Homolog (PTEN), in NSCLC. This interaction leads to the activation of the PTEN/PI3K/AKT pathway, promoting apoptosis and sensitizing cancer cells to cisplatin

(Cao et al. 2017). In addition, the study of miR-21 and its interaction with PTEN has established a crucial role in promoting EMT and drug resistance in cancerous cells. Recent research suggests that PTEN activates a long non-coding RNA derived from its pseudogene, PTENP1, which subsequently inhibits miR-21. The existence of an interplay between PTEN, miR-21, and PTENP1 during the DDR in breast cancer, hepatocellular carcinoma, and OSCC has been proposed (Gupta et al. 2024). Moreover, increased expression of lncRNA MT1JP has been associated with decreased proliferation and drug resistance ability of several cancers, such as gastric cancer, intrahepatic cholangiocarcinoma, hepatocellular carcinoma, and breast cancer. MT1JP functions as a ceRNA by sponging miR-92a-3p, miR-214-3p, and miR-24-3p, leading to the upregulation of its target genes, such as PTEN, F-box/WD repeat-containing protein 7 (FBXW7), and Runt-related transcription factor 3 (RUNX3), subsequently disrupting oncogenic pathways including PI3K/AKT, Wnt/β-catenin, and p53. These disruptions promote apoptosis and inhibit tumor progression as well as promote drug sensitivity (He et al. 2025).

In addition, it has been shown that Linc00312 expression was lower in cisplatin-resistant ovarian cancer cells compared to sensitive cells. Manipulating Linc00312 levels demonstrated its influence on cisplatin sensitivity. Increasing Linc00312 enhances the effectiveness of cisplatin in resistant cells, while decreasing it reduces sensitivity in susceptible cells. This effect was mediated through the Bcl-2/Caspase-3 pathway, promoting apoptosis (programmed cell death) in cancer cells. Before, it has been proposed that Linc00312 exhibits tumor-suppressing activity by binding to and inhibiting miR-197-3p. This interaction leads to decreased invasion and migration of cancer cells. Furthermore, Linc00312 may interact with other miRNAs, potentially relieving their repressive effects on target protein expression. This upregulation of target proteins subsequently activates the Bcl-2/Caspase-3 apoptotic signaling pathway, promoting apoptosis in ovarian cancer cells (Zhang et al. 2018).

7.6.4 *Role of TS-lncRNAs as Protein Decoys in Drug Resistance of Cancer Cells*

Beyond their well-characterized role as miRNA sponges, TS-lncRNAs can also function as protein decoys, sequestering key regulatory proteins and thereby modulating various cellular pathways that contribute to chemoresistance. For example, researchers have found that MEG3 expression is significantly downregulated in pancreatic cancer (Ma et al. 2018). Further experiments demonstrate that reducing MEG3 levels promotes cancer cell proliferation, migration, invasion, and stemness while also decreasing sensitivity to the chemotherapy drug gemcitabine. Conversely, increasing MEG3 expression has the opposite effect. These findings indicate that MEG3 functions as a tumor suppressor in pancreatic cancer. It is suggested that MEG3 plays a crucial role in modulating epigenetic modifications and gene expression at both transcriptional and post-transcriptional levels. Given this regulatory function, Snail—a protein involved in cell migration and invasion—may be a target of MEG3 regulation (Ma et al. 2018). Moreover, LINC00672 acts as a tumor suppressor by interacting with the p53 protein and other factors to suppress the expression of LIM and SH3 domain protein 1 (LASP1), a gene potentially involved in tumor progression in endometrial cancer. Overexpressing LINC00672 in endometrial cancer cells reduces LASP1 levels, slows cancer cell growth, and increases sensitivity to paclitaxel (Li et al. 2017a). LASP1, an actin-binding protein that acts as an oncogene and cellular structural remodeler, promotes migration and invasion in many cancers. Also, LASP1 can affect cell cycle regulators like cyclin A2 and cyclin B1 and activate signaling pathways such as the FAK-AKT pathway (Butt et al. 2022). Moreover, researchers have found that elevated ZBTB7A levels inhibit cisplatin-induced apoptosis by suppressing the expression of lncRNA LINC00473 (Zhang et al. 2017b). Mechanistically, ZBTB7A, a transcriptional repressor, directly binds to and represses the LINC00473 promoter. LINC00473, in turn, interacts with the transcription factor C/EBPβ, facilitating its binding to the interleukin-24 (IL-24) promoter, which ultimately reduces chemoresistance. Consequently, it has been proposed that LINC00473 acts as a tumor suppressor lncRNA through its interaction with C/EBPβ and sensitizes osteosarcoma to cisplatin through the expression of IL-24 (Zhang et al. 2017b). IL-24 is a cytokine that can enhance the effectiveness of chemotherapy by inducing apoptosis in a wide variety of cancer cells. It can increase the expression of pro-apoptotic protein BAX and decrease anti-apoptotic protein BCL-2. Also, IL-24 can reverse MDR by downregulating the expression of genes such as MDR1. It can also disrupt signaling pathways like the GTP-RhoA-ERK pathway and AKT-mTOR pathway, which are crucial for drug resistance, cell growth, and survival (Aravind et al. 2025).

These findings suggest that targeting the ZBTB7A-mediated LINC00473-C/EBPβ-IL24 pathway may be a promising therapeutic strategy for overcoming cisplatin resistance in osteosarcoma.

Understanding specific interaction mechanisms involving TS-lncRNAs, their cognate protein partners, and miRNAs is crucial for elucidating the molecular mechanisms underlying chemoresistance. These interactions can provide valuable insights into how TS-lncRNAs contribute to drug resistance in cancer cells. By targeting these molecular pathways, we can develop therapeutic strategies designed to disrupt the interactions of TS-lncRNAs with specific proteins and miRNAs, thereby interfering with their role in disease progression. Such approaches may offer potential targets to overcome drug resistance and improve the effectiveness of cancer treatments.

7.7 Role of lncRNAs in Radiotherapy Resistance in Cancer

There are several onco/TS lncRNAs upregulated and downregulated in different types of radiotherapy-resistant cancer cells (Table 7.3).

Table 7.3 LncRNAs, their Onco/TS roles, and the mechanisms of function in radiotherapy resistances in various cancers

LncRNA	Onco/ TS-lncRNA	Direct targets and the mechanism of function	Cancer	Reference
CCAT1	Onco	Targets miR-148 and inhibits apoptosis	Breast cancer	Podralska et al. (2020)
HOTAIR	Onco	Targets miR-218 and inhibits apoptosis and promotes cell cycle and DSB repair	Breast cancer	Hu et al. (2019)
LINC02582	Onco	Targets miR-200c/USP7/CHK1 and promotes DSB repair	Breast cancer	Podralska et al. (2020)
lincRNA-ROR	Onco	Targets miR-145 and inhibits apoptosis	Colorectal cancer	Yang et al. (2017)
MALAT1	Onco	Targets miR-1/SLUG and inhibits apoptosis	Nasopharyngeal carcinoma	Podralska et al. (2020)
NEAT1	Onco	Targets miR-204/ZEB1 and inhibits apoptosis and promotes EMT	Nasopharyngeal carcinoma	Wang et al. (2017)
NEAT1	Onco	Targets miR-193b/cyclin D1 and promotes cell cycle and inhibits apoptosis	Cervical cancer	Han et al. (2017)
PCAT6	Onco	Targets miR-185/TPD52 and promotes cell cycle and inhibiting apoptosis	Breast cancer	Shi et al. (2020)
SBF2-AS1	Onco	Targets miR-302a/MBNL3 and inhibits apoptosis	Non-small cell lung cancer	Yu et al. (2020)
Lnc-RI	Onco	Targets miR-193a-3p/RAD51 and promotes DSB repair	–	Shen et al. (2018)

(continued)

Table 7.3 (continued)

LncRNA	Onco/ TS-lncRNA	Direct targets and the mechanism of function	Cancer	Reference
NORAD	Onco	Targets miR-199a-5p/ EEPD1 and promotes DNA repair	Esophageal squamous cell carcinoma	Sun et al. (2021)
BOKAS	Onco	Targets WISP1 and promotes Wnt/β-catenin	Esophageal squamous cell carcinoma	Zhang et al. (2015)
FAM83H-AS1	Onco	Targets HuR and promotes DSB repair	Ovarian cancer	Dou et al. (2019)
HOTAIR	Onco	Targets p21, promotes cell cycle and inhibits apoptosis	Cervical cancer	Jing et al. (2015)
HOTAIR	Onco	Targets WIF-1 and promotes Wnt/β-catenin signaling pathway	Pancreatic cancer	Jiang et al. (2016)
HOTAIR	Onco	Targets LC3-II, ATG7 and inhibits autophagy and apoptosis	Pancreatic cancer	Wu et al. (2018)
HOTAIR	Onco	Targets EZH2 and recruits it to the MYC promoter and expresses KU70, KU80, DNA-PKs, ATM and promotes DNA repair	Invasive ductal breast cancer	Qian et al. (2020)
HOTAIR	Onco	Targets ATR and inhibits apoptosis and promotes cell cycle	Colorectal cancer	Hu et al. (2024)
lincRNA-ROR	Onco	Targets p53 and inhibits apoptosis	Colorectal cancer	Podralska et al. (2020)
HULC	Onco	Targets LC3B-II, Beclin-1 and inhibits apoptosis and autophagy	Prostate cancer	Podralska et al. (2020)
LINP1	Onco	Targets Ku80, DNA-PKcs and promotes DSB repair and inhibits apoptosis	Cervical cancer	Wang et al. (2018c)

(continued)

Table 7.3 (continued)

LncRNA	Onco/ TS-lncRNA	Direct targets and the mechanism of function	Cancer	Reference
lincRNA-p21	Onco	Targets HIF-1α and inhibits autophagy	Hepatoma and glioma	Shen et al. (2017)
MALAT1	Onco	Targets Cks1 and inhibits apoptosis	Esophageal squamous-cell carcinoma	Li et al. (2017b)
POU6F2-AS2	Onco	Targets YBX1 and promotes DSB repair	Esophageal squamous-cell carcinoma	Liu et al. (2016a)
PVT1	Onco	Targets ATM, p53, CHK2, HIF1 and inhibits apoptosis and promotes DSB repair	Nasopharyngeal carcinoma	He et al. (2018)
TP73-AS1	Onco	Targets PTEN, AKT, and inhibits apoptosis	Hepatocellular carcinoma	Podralska et al. (2020)
TUG1	Onco	Targets HMGB1 and inhibits apoptosis	Breast cancer	Jiang et al. (2017)
UCA1	Onco	Targets AKT, FAK, FGR, AMPKα1, and promotes cell cycle	Gastric cancer	Podralska et al. (2020)
UCA1	Onco	Targets HK2 and promotes glycolysis	Cervical cancer	Podralska et al. (2020)
MEG3	TS	Targets miR-182 and inhibits DSB repair	Thyroid cancer	Liu et al. (2018b)
NEAT1	TS	Targets miR-101/ EMP2 axis and promotes apoptosis and inhibits EMT	Nasopharyngeal carcinoma	Podralska et al. (2020)
OIP5-AS1	TS	Targets miR-369/ DYRK1A promotes apoptosis	Colorectal cancer	Podralska et al. (2020)
LOC401312	TS	Targets some specific miRNAs, sponging CPS1 and suppresses DNA repair	Non-small cell lung cancer	Cao et al. (2025)

(continued)

Table 7.3 (continued)

LncRNA	Onco/ TS-lncRNA	Direct targets and the mechanism of function	Cancer	Reference
lincRNA-p21	TS	Targets β-catenin and inhibits Wnt/β-catenin signaling pathway	Gastric cancer	Chen et al. (2019)
linc00312	TS	Targets DNA-PKcs protein and inhibits DNA repair	Nasopharyngeal carcinoma	Guo et al. (2021)
LINC-PINT	TS	Targets DNA-PKcs protein and ATM/ATR-Chk1/Chk2 signaling pathway and suppresses DNA repair	Nasopharyngeal cancer	Wang et al. (2021)

7.7.1 *Role of Onco-lncRNAs as miRNA Sponges in Radiotherapy Resistance*

There is some evidence showing the oncogenic roles of lncRNAs in radiotherapy resistance as miRNA sponges in cancer cells. The study has demonstrated that lnc-RI plays a critical role in DSB repair. Researchers have found that lnc-RI expression is inversely correlated with micronucleus formation, a marker of DSBs. Knockdown of lnc-RI leads to reduced HR repair efficiency, the primary pathway for repairing DSBs. Mechanistically, lnc-RI stabilizes RAD51—an HR repair, and increases the efficiency of this system repair by acting as a ceRNA for miR-193a-3p, which targets *RAD51* mRNA. These findings establish a novel regulatory mechanism for DSB repair involving lnc-RI and highlights its importance in preserving genomic integrity and promoting radiotherapy resistance (Shen et al. 2018).

Research investigates the role of the lncRNA NORAD (activated by DNA Damage) in the radioresistance of esophageal squamous cell carcinoma (ESCC). Researchers have found that NORAD expression is upregulated in radioresistant ESCC cells and contributes to their resistance by promoting DNA repair mechanisms. NORAD achieves this by suppressing miR-199a-5p, a microRNA that targets EEPD1, a protein involved in the ATR/Chk1 signaling pathway crucial for DDR. Furthermore, NORAD knockdown increases programmed death-ligand 1 (PD-L1) ubiquitination, enhancing sensitivity to both radiation and anti-PD-1 immunotherapy. These findings suggest that targeting NORAD could be a promising strategy to overcome radioresistance and improve treatment outcomes for ESCC patients (Sun et al. 2021).

7.7.2 Role of Onco-lncRNAs as Protein Decoy in Radiotherapy Resistance

There is evidence showing the oncogenic roles of lncRNAs in radiotherapy resistance as protein decoys in cancer cells. Researchers have observed that HOTAIR expression is elevated in invasive ductal carcinoma (IDC) tissue and that its overexpression promotes breast cancer cell proliferation by promoting the cell cycle and inhibiting apoptosis. Mechanistically, HOTAIR promotes DNA damage repair by upregulating key repair factors, including KU70, KU80, DNA-PKcs, and ATM. This upregulation occurs through HOTAIR-mediated recruitment of the Enhancer of Zeste 2 Polycomb Repressive Complex 2 Subunit (EZH2) transcription factor to the MYC promoter, which enhances MYC transcription and subsequently increases the expression of these DNA repair proteins. In conclusion, it has been shown that HOTAIR is an oncogenic lncRNA, a crucial modulator of breast cancer prognosis, and a potential biomarker for radiotherapy response (Qian et al. 2020). In addition, it has been demonstrated that HOTAIR expression is significantly elevated in colorectal tumor tissues, particularly in radioresistant samples. High expression level of HOTAIR correlates with advanced disease stage and poor prognosis. Silencing HOTAIR in colorectal cells suppresses proliferation, promotes apoptosis, and enhances radiosensitivity. Mechanistically, HOTAIR interacts with the DNA damage response mediator, ATR, promoting its function and downstream signaling of the Chk1/Chk2/p53/p21 axis. These findings suggest that HOTAIR promotes radiotherapy resistance in colorectal cancer by modulating the ATR-mediated DNA damage response pathway (Hu et al. 2024).

Research demonstrates that lncRNA PVT1 is overexpressed in nasopharyngeal carcinoma (NPC) compared to normal tissue. Knockdown of PVT1 in NPC cells results in decreased proliferation, increased apoptosis, and enhanced radiosensitivity. Mechanistically, PVT1 knockdown appears to induce apoptosis by impacting the DNA damage repair pathway following radiation therapy. Reducing the expression of the PVT1 gene leads to a significant decrease in the phosphorylation of key proteins involved in DNA repair: ATM, p53, and Chk2. This suggests that suppressing PVT1 impairs the ability of nasopharyngeal carcinoma cells to repair radiation-induced DNA damage, thereby increasing their susceptibility to radiotherapy. These findings suggest that PVT1 may serve as an onco-lncRNA, a prognostic biomarker, and a potential therapeutic target for NPC. Targeting PVT1 could offer a novel strategy for improving NPC treatment outcomes and enhancing the efficacy of radiotherapy (He et al. 2018).

7.7.3 Role of TS-lncRNAs as miRNA Sponges in Radiotherapy Resistance

Researchers have identified LOC401312 as an lncRNA that enhances radiosensitivity in non-small cell lung cancer cells. LOC401312 has been reported to upregulate carbamoyl-phosphate synthase 1 (CPS1), an enzyme involved in pyrimidine biosynthesis, by targeting specific overlapping miRNAs. Mechanistically, CPS1 suppresses ATM kinase phosphorylation, a key step in DNA damage response. This study establishes LOC401312-CPS1-ATM, a novel regulatory axis that governs radiation sensitivity. By targeting this axis, it may be possible to improve radiotherapy efficacy in NSCLC (Cao et al. 2025).

7.7.4 Role of TS-lncRNAs as Protein Decoy in Radiotherapy Resistance

There is evidence regarding the tumor suppressive roles of lncRNAs in radiotherapy resistance as protein decoys in cancer cells. Research demonstrates that higher linc00312 expression correlates with improved survival and better radiotherapy response in NPC patients. Mechanistically, linc00312 directly binds to DNA-PKcs, a key protein involved in DNA repair, thereby hindering its function and subsequently inhibiting the DNA damage response pathways. This disruption of DNA repair mechanisms enhances the sensitivity of NPC to ionizing radiation. The findings indicate that linc00312 functions as a TS-lncRNA; thus, its overexpression in cancer cells may enhance their sensitivity to radiotherapy, suggesting linc00312-based approaches could potentially improve radiotherapy efficacy in patients with NPC (Guo et al. 2021).

Recently, LINC-PINT expression has been shown to be significantly lower in NPC tissues than in normal tissues, with low expression correlating with poorer prognosis in radiotherapy patients. LINC-PINT inhibits malignant phenotypes and enhances radiosensitivity. Mechanistically, LINC-PINT responds to DNA damage by suppressing DNA repair through the ATM/ATR-Chk1/Chk2 signaling pathways. It also increases radiosensitivity by interacting with—and thereby negatively regulating—DNA-PKcs, a key protein involved in DNA repair. Therefore, LINC-PINT emerges as a potential therapeutic target for overcoming radioresistance in NPC (Wang et al. 2021).

7.8 Application of lncRNAs in Cancer-Targeted Therapy

Due to their critical roles as key regulators in cancer biology, lncRNAs are recognized as diagnostic biomarkers and potential targets for cancer therapy (Beylerli et al. 2022). They can be implicated in cancer treatment through silencing

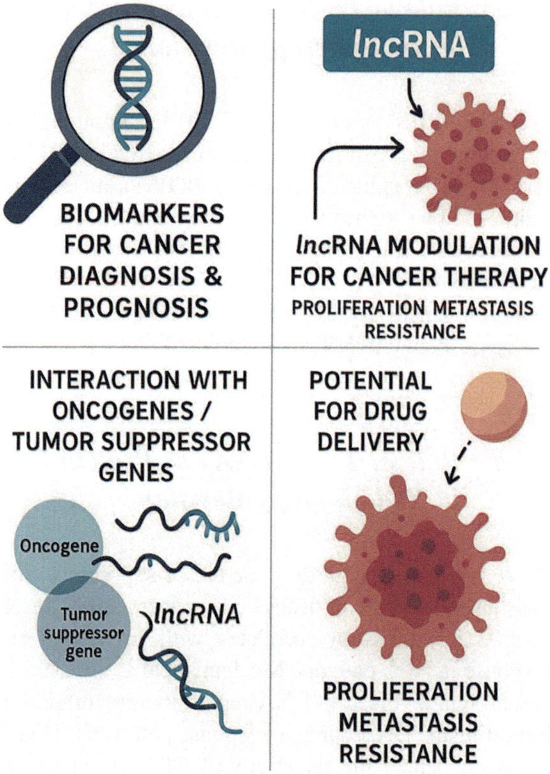

Fig. 7.2 Application of lncRNAs in cancer-targeted therapy. LncRNAs can be implicated as potential biomarkers for the diagnosis and prognosis of different types of cancer. Also, lncRNAs can be applied as drugs in drug delivery systems in cancer cells and inhibit proliferation, metastasis, and drug resistance ability of cancer cells through modulating oncogenes and tumor suppressor genes

onco-lncRNAs using antisense oligonucleotides, siRNAs, and CRISPR-based approaches, as well as restoring TS-lncRNAs via vector-mediated overexpression and synthetic lncRNA mimics (Williams and Pickard 2016). TS-lncRNAs can be delivered into cancer cells using various drug delivery systems and nanoparticle-based platforms, such as lipid nanoparticles, polymeric nanoparticles, and inorganic nanocarriers. These nanoparticle systems can also deliver siRNAs, antisense oligonucleotides, or CRISPR components to target onco-lncRNAs (Davodabadi et al. 2024). The ability to selectively target specific cells, their roles in gene regulation, and their potential to overcome drug resistance make lncRNAs attractive tools in personalized cancer therapy and other therapeutic applications (Arun et al. 2018; Chen et al. 2022). Figure 7.2 summarizes lncRNAs' applications in cancer-targeted therapy.

7.9 Conclusion

Multidisciplinary treatment is the most effective approach to overcoming cancer. However, chemotherapy and radiotherapy—traditional standard methods—result in only partial or complete response in some patients. LncRNAs are involved in important regulatory functions owing to their diverse cellular and subcellular interactomes. Many in vitro studies have established lncRNA signatures in disease pathobiology, proving them to be drivers of numerous diseases. These ncRNAs control cancer cell proliferation, apoptosis, migration, invasion, and stemness, making them not only potential therapeutic targets but also novel strategies to break down barriers of treatment resistance.

Acknowledgments Not applicable.

Funding Not applicable.

Compliance with Ethical Standards The manuscript does not require ethics standards.

Conflict of Interest The authors do not have any conflicts of interest to disclose.

Ethical Approval The manuscript does not require ethics committee approval.

References

Abbasi BA, Suravajhala R, Sundararajan V, Valadi J, Suravajhala P (2025) Non-coding RNAs and their integrated biological networks. Prog Mol Biol Transl Sci 214:179–194

Abdolmaleki A, Karimian A, Asadi A, Ghanimi HA, Akram M (2022) Application of novel molecular biology in cancer therapy basic & clinical cancer research

Abdullah LN, Chow EK-H (2013) Mechanisms of chemoresistance in cancer stem cells. Clin Transl Med 2:3

Abshire D, Lang MK (2018) The evolution of radiation therapy in treating cancer. In: Seminars in oncology nursing, vol 2. Elsevier, pp 151–157

Amin N, McGrath A, Chen Y-PP (2019) Evaluation of deep learning in non-coding RNA classification. Nat Mach Intell 1:246–256

Aravind A, Shaji V, Ahmed M, Dagamajalu S, Vijayakumar M, Prasad TSK, Raju R (2025) An assembled molecular signaling map of interleukin-24: a resource to decipher its multifunctional immunoregulatory role in pathophysiological conditions. Front Immunol 16:1608101

Arun G, Diermeier SD, Spector DL (2018) Therapeutic targeting of long non-coding RNAs in cancer. Trends Mol Med 24:257–277

Asadi S, Abkar M, Zamanzadeh Z, Taghipour Kamalabad S, Sedghi M, Yousefnia S (2023) Association of SOD2 rs2758339, rs5746136 and rs2842980 polymorphisms with increased risk of breast cancer: a haplotype-based case–control study genes &. Genomics 45:1165–1178

Baskar R, Lee KA, Yeo R, Yeoh K-W (2012) Cancer and radiation therapy: current advances and future directions. Int J Med Sci 9:193

Beylerli O, Gareev I, Sufianov A, Ilyasova T, Guang Y (2022) Long noncoding RNAs as promising biomarkers in cancer. Non-coding RNA Res 7:66–70

Blevins MA, Towers CG, Patrick AN, Zhao R, Ford HL (2015) The SIX1-EYA transcriptional complex as a therapeutic target in cancer. Expert Opin Ther Targets 19:213–225. https://doi.org/10.1517/14728222.2014.978860

Bou Antoun N, Chioni A-M (2023) Dysregulated signalling pathways driving anticancer drug resistance. Int J Mol Sci 24:12222

Butt E, Howard CM, Raman D (2022) LASP1 in cellular signaling and gene expression: more than just a cytoskeletal regulator. Cells 11:3817

Cai Q et al (2019) Long non-coding RNA GBCDRlnc1 induces chemoresistance of gallbladder cancer cells by activating autophagy. Mol Cancer 18:82

Cao L, Chen J, Ou B, Liu C, Zou Y (2017) GAS5 knockdown reduces the chemo-sensitivity of non-small cell lung cancer (NSCLC) cell to cisplatin (DDP) through regulating miR-21/PTEN axis. Biomed Pharmacother 93:570–579

Cao Z et al (2025) Long non-coding RNA LOC401312 induces radiosensitivity through upregulation of CPS1 in non-small cell lung cancer. Int J Mol Sci 26:5865

Chen S, Xia X (2019) Long noncoding RNA NEAT1 suppresses sorafenib sensitivity of hepatocellular carcinoma cells via regulating miR-335–c-met. J Cell Physiol 234:14999–15009

Chen L, Yuan D, Yang Y, Ren M (2019) LincRNA-p21 enhances the sensitivity of radiotherapy for gastric cancer by targeting the β-catenin signaling pathway. J Cell Biochem 120:6178–6187

Chen B, Dragomir MP, Yang C, Li Q, Horst D, Calin GA (2022) Targeting non-coding RNAs to overcome cancer therapy resistance. Signal Transduct Target Ther 7:121

Chen B, Xu K, Zhang Y, Xu P, Li C, Liu J, Xu Y (2023) LncRNA ERVH48-1 contributes to the drug resistance of prostate cancer and proliferation through sponging of miR-4784 to the activation of the Wnt/β-catenin pathway. Cancers 15:1902

Chiorcea-Paquim A-M, Oliveira-Brett AM (2023) Electrochemistry of chemotherapeutic alkylating agents and their interaction with DNA. J Pharm Biomed Anal 222:115036

Dakal TC et al (2024) Intricate relationship between cancer stemness, metastasis, and drug resistance. MedComm 5:e710

Dakal TC, Philip RR, Bhushan R, Sonar PV, Rajagopal S, Kumar A (2025) Genetic and epigenetic regulation of non-coding RNAs: implications in cancer metastasis, stemness and drug resistance. Pathol Res Pract 266:155728

Davodabadi F et al (2024) Nanotherapeutic approaches for delivery of long non-coding RNAs: an updated review with emphasis on cancer. Nanoscale 16:3881–3914

Dou Q, Xu Y, Zhu Y, Hu Y, Yan Y, Yan H (2019) LncRNA FAM83H-AS1 contributes to the radioresistance, proliferation, and metastasis in ovarian cancer through stabilizing HuR protein. Eur J Pharmacol 852:134–141

Fang Z, Chen W, Yuan Z, Liu X, Jiang H (2018) LncRNA-MALAT1 contributes to the cisplatin-resistance of lung cancer by upregulating MRP1 and MDR1 via STAT3 activation. Biomed Pharmacother 101:536–542

Feng Y et al (2017) Modulation of CASC2/miR-21/PTEN pathway sensitizes cervical cancer to cisplatin. Arch Biochem Biophys 623:20–30

Gao Y et al (2020) Antibiotics for cancer treatment: a double-edged sword. J Cancer 11:5135

Ge P, Cao L, Yao Y-J, Jing R-J, Wang W, Li H-J (2019) lncRNA FOXD2-AS1 confers cisplatin resistance of non-small-cell lung cancer via regulation of miR185-5p–SIX1 axis. OncoTargets Ther 12:6105–6117

Giuliani B, Tordonato C, Nicassio F (2023) Mechanisms of long non-coding RNA in breast cancer. Int J Mol Sci 24:4538

Gong L, Zhang Y, Liu C, Zhang M, Han S (2021) Application of radiosensitizers in cancer radiotherapy. Int J Nanomed 16:1083–1102

Gui Z, Zhao Z, Sun Q, Shao G, Huang J, Zhao W, Kuang Y (2021) LncRNA FEZF1-AS1 promotes multi-drug resistance of gastric cancer cells via upregulating ATG5. Front Cell Dev Biol 9:749129. https://doi.org/10.3389/fcell.2021.749129

Guo Z et al (2021) LncRNA linc00312 suppresses radiotherapy resistance by targeting DNA-PKcs and impairing DNA damage repair in nasopharyngeal carcinoma. Cell Death Dis 12:69

Gupta S, Silveira DA, Lorenzoni PR, Mombach JCM, Hashimoto RF (2024) LncRNA PTENP1/miR-21/PTEN axis modulates EMT and drug resistance in cancer: dynamic boolean modeling for cell fates in DNA damage response. Int J Mol Sci 25:8264
Han D, Wang J, Cheng G (2017) LncRNA NEAT1 enhances the radio-resistance of cervical cancer via miR-193b-3p/CCND1 axis. Oncotarget 9:2395
He Y et al (2018) Long non-coding RNA PVT1 predicts poor prognosis and induces radioresistance by regulating DNA repair and cell apoptosis in nasopharyngeal carcinoma. Cell Death Dis 9:235
He J, Sun M, Geng H, Tian S (2019) Long non-coding RNA Linc00518 promotes paclitaxel resistance of the human prostate cancer by sequestering miR-216b-5p. Biol Cell 111:39–50
He H et al (2025) MT1JP: a pivotal tumor-suppressing LncRNA and its role in cancer progression and therapeutic potential current drug targets. Curr Drug Targets 26:394–409
Hu H, Yang L, Li L, Zeng C (2018) Long non-coding RNA KCNQ1OT1 modulates oxaliplatin resistance in hepatocellular carcinoma through miR-7-5p/ABCC1 axis. Biochem Biophys Res Commun 503:2400–2406
Hu X, Ding D, Zhang J, Cui J (2019) Knockdown of lncRNA HOTAIR sensitizes breast cancer cells to ionizing radiation through activating miR-218. Biosci Rep 39:BSR20181038
Hu H, Yang H, Fan S, Jia X, Zhao Y, Li H (2024) LncRNA HOTAIR promotes DNA damage repair and radioresistance by targeting ATR in colorectal cancer. Oncol Res 32:1335
Huang H, Chen J, Ding CM, Jin X, Jia ZM, Peng J (2018) Lnc RNA NR 2F1-AS 1 regulates hepatocellular carcinoma oxaliplatin resistance by targeting ABCC 1 via miR-363. J Cell Mol Med 22:3238–3245
Jain SM et al (2024) Understanding the molecular mechanism responsible for developing therapeutic radiation-induced radioresistance of rectal cancer and improving the clinical outcomes of radiotherapy—a review. Cancer Biol Ther 25:2317999
Jiang Y et al (2016) The long non-coding RNA HOTAIR affects the radiosensitivity of pancreatic ductal adenocarcinoma by regulating the expression of Wnt inhibitory factor 1. Tumour Biol 37:3957–3967
Jiang H, Hu X, Zhang H, Li W (2017) Down-regulation of LncRNA TUG1 enhances radiosensitivity in bladder cancer via suppressing HMGB1 expression. Radiat Oncol 12:65
Jing L, Yuan W, Ruofan D, Jinjin Y, Haifeng Q (2015) HOTAIR enhanced aggressive biological behaviors and induced radio-resistance via inhibiting p21 in cervical cancer. Tumour Biol 36:3611–3619
Karakas D, Ozpolat B (2021) The role of LncRNAs in translation non-coding. RNA 7:16
Karati D, Mahadik KR, Trivedi P, Kumar D (2022) Alkylating agents, the road less traversed, changing anticancer therapy. Anti-Cancer Agents Med Chem Anti-Cancer Agents 22:1478–1495
Katzung BG (2021) Basic & clinical pharmacology. EGC
Li W, Zhai L, Wang H, Liu C, Zhang J, Chen W, Wei Q (2016) Downregulation of LncRNA GAS5 causes trastuzumab resistance in breast cancer. Oncotarget 7:27778
Li W et al (2017a) Long non-coding RNA LINC00672 contributes to p53 protein-mediated gene suppression and promotes endometrial cancer chemosensitivity. J Biol Chem 292:5801–5813
Li Z, Zhou Y, Tu B, Bu Y, Liu A, Kong J (2017b) Long noncoding RNA MALAT 1 affects the efficacy of radiotherapy for esophageal squamous cell carcinoma by regulating Cks1 expression. J Oral Pathol Med 46:583–590
Li L, Zhang H, Wang X, Wang J, Wei H (2018) Long non-coding RNA CASC2 enhanced cisplatin-induced viability inhibition of non-small cell lung cancer cells by regulating the PTEN/PI3K/Akt pathway through down-regulation of miR-18a and miR-21. RSC Adv 8:15923–15932
Li G, Deng L, Huang N, Sun F (2021) The biological roles of lncRNAs and future prospects in clinical application. Diseases 9:8
Liu J, Sun X, Zhu H, Qin Q, Yang X, Sun X (2016a) Long noncoding RNA POU6F2-AS2 is associated with oesophageal squamous cell carcinoma. J Biochem 160:195–204
Liu Y et al (2016b) Long noncoding RNA RP11-838N2. 4 enhances the cytotoxic effects of temozolomide by inhibiting the functions of miR-10a in glioblastoma cell lines. Oncotarget 7:43835

Liu B et al (2018a) lncRNA GAS5 reverses EMT and tumor stem cell-mediated gemcitabine resistance and metastasis by targeting miR-221/SOCS3 in pancreatic cancer. Mol Ther Nucleic Acids 13:472–482

Liu Y, Yue P, Zhou T, Zhang F, Wang H, Chen X (2018b) LncRNA MEG3 enhances 131I sensitivity in thyroid carcinoma via sponging miR-182 biomedicine &. Pharmacotherapy 105:1232–1239

Liu X et al (2019) LINC00665 induces acquired resistance to gefitinib through recruiting EZH2 and activating PI3K/AKT pathway in NSCLC. Mol Ther Nucleic acids 16:155–161

Liu F, Feng X-X, Zhu S-L, Lin L, Huang H-Y, Zhang B-Y, Huang J-L (2021a) Long non-coding RNA SNHG1 regulates rheumatoid synovial invasion and proliferation by interaction with PTBP1. Int Immunopharmacol 90:107182

Liu TT, Li R, Liu X, Zhou XJ, Huo C, Li JP, Qu YQ (2021b) LncRNA XIST acts as a MicroRNA-520 sponge to regulate the cisplatin resistance in NSCLC cells by mediating BAX through CeRNA network. Int J Med Sci 18:419–431. https://doi.org/10.7150/ijms.49730

Liu YP, Zheng CC, Huang YN, He ML, Xu WW, Li B (2021c) Molecular mechanisms of chemo- and radiotherapy resistance and the potential implications for cancer treatment. MedComm 2:315–340

Luo Y et al (2021) Long noncoding RNA (lncRNA) EIF3J-DT induces chemoresistance of gastric cancer via autophagy activation. Autophagy 17:4083–4101

Ma L et al (2018) Long non-coding RNA MEG3 functions as a tumour suppressor and has prognostic predictive value in human pancreatic cancer. Oncol Rep 39:1132–1140

Ma G, Zhu J, Liu F, Yang Y (2019) Long noncoding RNA LINC00460 promotes the gefitinib resistance of nonsmall cell lung cancer through epidermal growth factor receptor by sponging miR-769-5p. DNA Cell Biol 38:176–183

Malhotra A, Jain M, Prakash H, Vasquez KM, Jain A (2017) The regulatory roles of long non-coding RNAs in the development of chemoresistance in breast cancer. Oncotarget 8:110671–110684. https://doi.org/10.18632/oncotarget.22577

Márton É et al (2025) Non-coding RNAs in cancer: structure, function, and clinical application. Cancers 17:579

Nasri N, Saharkhiz S, Dini G, Yousefnia S (2023) Thermo-and pH-responsive targeted lipid-coated mesoporous nano silica platform for dual delivery of paclitaxel and gemcitabine to overcome HER2-positive breast cancer. Int J Pharm 648:123606

Pisignano G, Ladomery M (2021) Post-transcriptional regulation through long non-coding rnas (lncrnas), vol 7. MDPI

Podralska M, Ciesielska S, Kluiver J, van den Berg A, Dzikiewicz-Krawczyk A, Slezak-Prochazka I (2020) Non-coding RNAs in cancer radiosensitivity: microRNAs and lncRNAs as regulators of radiation-induced signaling pathways. Cancers 12:1662

Qian L et al (2020) lncRNA HOTAIR promotes DNA repair and radioresistance of breast cancer via EZH2. DNA Cell Biol 39:2166–2173

Rahimi G, Yousefnia S, Angnes L, Negahdary M (2023) Design a PEGylated nanocarrier containing lemongrass essential oil (LEO), a drug delivery system: application as a cytotoxic agent against breast cancer cells. J Drug Delivery Sci Technol 80:104183

Ren J et al (2025) Cancer chemoprevention: signaling pathways and strategic approaches signal transduction and targeted. Therapy 10:113

Rezaei O, Tamizkar KH, Sharifi G, Taheri M, Ghafouri-Fard S (2021) Emerging role of long non-coding RNAs in the pathobiology of glioblastoma. Front Oncol 10:625884

Roy D et al (2025) LncRNAs in oncogenic microenvironment: from threat to therapy. Front Cell Dev Biol 12:1423279

Saharkhiz S, Nasri N, Dini G, Yousefnia S (2023) Development of a new smart theranostic anti-PSMA-aptamer conjugated cationic-lipid coated mesoporous silica platform for targeted delivery of paclitaxel and CdSe/ZnS quantum dots to LNCaP cell line. J Drug Delivery Sci Technol 88:104964

Shah IM, Dar MA, Bhat KA, Dar TA, Ahmad F, Ahmad SM (2022) Long non-coding RNAs: biogenesis, mechanism of action and role in different biological and pathological processes. In: Recent advances in noncoding RNAs. IntechOpen

Shang C, Tang W, Pan C, Hu X, Hong Y (2018) Long non-coding RNA TUSC7 inhibits temozolomide resistance by targeting miR-10a in glioblastoma. Cancer Chemother Pharmacol 81:671–678

Shen Y, Liu Y, Sun T, Yang W (2017) LincRNA-p21 knockdown enhances radiosensitivity of hypoxic tumor cells by reducing autophagy through HIF-1/Akt/mTOR/P70S6K pathway. Exp Cell Res 358:188–198

Shen L, Wang Q, Liu R, Chen Z, Zhang X, Zhou P, Wang Z (2018) LncRNA lnc-RI regulates homologous recombination repair of DNA double-strand breaks by stabilizing RAD51 mRNA as a competitive endogenous RNA. Nucleic Acids Res 46:717–729

Shi C, Wang M (2018) LINC01118 modulates paclitaxel resistance of epithelial ovarian cancer by regulating miR-134/ABCC1 Medical science monitor. Int Med J Exp Clin Res 24:8831

Shi W et al (2019) Long non-coding RNA LINC00346 promotes pancreatic cancer growth and gemcitabine resistance by sponging miR-188-3p to derepress BRD4 expression. J Exp Clin Cancer Res 38:60

Shi R, Wu P, Liu M, Chen B, Cong L (2020) Knockdown of lncRNA PCAT6 enhances radiosensitivity in triple-negative breast cancer cells by regulating miR-185-5p/TPD52 axis. OncoTargets Ther 13:3025–3037

Shi Y et al (2024) Long non-coding RNAs in drug resistance across the top five cancers: update on their roles and mechanisms. Heliyon 10:e27207. https://doi.org/10.1016/j.heliyon.2024.e27207

Shirvani M, Yadollahi A, Zamanzadeh Z, Abkar M, Yousefnia S (2025) Association of MT1A rs11640851, rs8052394 and rs11076161 polymorphisms with the risk of developing breast cancer: a haplotype-based case-control study and in silico analysis Indian J Clin Biochem 1–13. https://doi.org/10.21203/rs.3.rs-3867462/v1

Song L et al (2019) Long noncoding RNA OIP5-AS1 causes cisplatin resistance in osteosarcoma through inducing the LPAATβ/PI3K/AKT/mTOR signaling pathway by sponging the miR-340-5p. J Cell Biochem 120:9656–9666

Statello L, Ali MM, Reischl S, Mahale S, Kosalai ST, Huarte M, Kanduri C (2021) The DNA damage inducible lncRNA SCAT7 regulates genomic integrity and topoisomerase 1 turnover in lung adenocarcinoma. NAR Cancer 3:zcab002. https://doi.org/10.1093/narcan/zcab002

Sui X et al (2013) Autophagy and chemotherapy resistance: a promising therapeutic target for cancer treatment. Cell Death Dis 4:e838. https://doi.org/10.1038/cddis.2013.350

Sun Z-Y, Jian Y-K, Zhu H-Y, Li B (2019) lncRNAPVT1 targets miR-152 to enhance chemoresistance of osteosarcoma to gemcitabine through activating c-MET/PI3K/AKT pathway. Pathol Res Pract 215:555–563

Sun Y et al (2021) Radiation induces NORAD expression to promote ESCC radiotherapy resistance via EEPD1/ATR/Chk1 signalling and by inhibiting pri-miR-199a1 processing and the exosomal transfer of miR-199a-5p. J Exp Clin Cancer Res 40:306

Sweef O, Yang C, Wang Z (2022) The oncogenic and tumor suppressive long non-coding RNA–microRNA–messenger RNA regulatory axes identified by analyzing multiple platform omics data from cr (VI)-transformed cells and their implications in lung cancer. Biomedicines 10:2334

Tang Z, Lu Y, Chen Y, Zhang J, Chen Z, Wang Q (2021) Research progress of MicroRNA in chemotherapy resistance of osteosarcoma. Technol Cancer Res Treat 20:15330338211034262

Tiwari M (2012) Antimetabolites: established cancer therapy. J Cancer Res Ther 8:510

Wang Y et al (2017) Long non-coding RNA NEAT1 regulates epithelial membrane protein 2 expression to repress nasopharyngeal carcinoma migration and irradiation-resistance through miR-101-3p as a competing endogenous RNA mechanism. Oncotarget 8:70156

Wang H et al (2018a) The cisplatin-induced lncRNA PANDAR dictates the chemoresistance of ovarian cancer via regulating SFRS2-mediated p53 phosphorylation. Cell Death Dis 9:1103

Wang J, Ye C, Liu J, Hu Y (2018b) UCA1 confers paclitaxel resistance to ovarian cancer through miR-129/ABCB1 axis. Biochem Biophys Res Commun 501:1034–1040

Wang X, Liu H, Shi L, Yu X, Gu Y, Sun X (2018c) LINP1 facilitates DNA damage repair through non-homologous end joining (NHEJ) pathway and subsequently decreases the sensitivity of cervical cancer cells to ionizing radiation. Cell Cycle 17:439–447

Wang L, Shang X, Feng Q (2019a) LncRNA TATDN1 contributes to the cisplatin resistance of non-small cell lung cancer through TATDN1/miR-451/TRIM66 axis. Cancer Biol Ther 20:261–271

Wang X et al (2019b) LncRNA SNHG6 promotes chemoresistance through ULK1-induced autophagy by sponging miR-26a-5p in colorectal cancer cells. Cancer Cell Int 19:234

Wang X, Li H, Shi J (2019c) LncRNA HOXA11-AS promotes proliferation and cisplatin resistance of oral squamous cell carcinoma by suppression of miR-214-3p expression. BioMed Res Int 2019:8645153

Wang Y-H et al (2021) LINC-PINT impedes DNA repair and enhances radiotherapeutic response by targeting DNA-PKcs in nasopharyngeal cancer. Cell Death Dis 12:454

Wang X, Gigant B, Zheng X, Chen Q (2023) Microtubule-targeting agents for cancer treatment: seven binding sites and three strategies. MedComm–Oncology 2:e46

Wen Q et al (2017) Long noncoding RNA GAS5, which acts as a tumor suppressor via microRNA 21, regulates cisplatin resistance expression in cervical cancer. Int J Gynecol Cancer 27:1096–1108

Williams GT, Pickard MR (2016) Long non-coding RNAs: new opportunities and old challenges in cancer therapy. Transl Cancer Res 5. https://doi.org/10.21037/tcr.2016.09.04

Wu C, Yang L, Qi X, Wang T, Li M, Xu K (2018) Inhibition of long non-coding RNA HOTAIR enhances radiosensitivity via regulating autophagy in pancreatic cancer. Cancer Manag Res 10:5261–5271

Wu L-L, Cai W-P, Lei X, Shi K-Q, Lin X-Y, Shi L (2019a) NRAL mediates cisplatin resistance in hepatocellular carcinoma via miR-340-5p/Nrf2 axis. J Cell Commun Signal 13:99–112

Wu P et al (2019b) Lnc-TALC promotes O6-methylguanine-DNA methyltransferase expression via regulating the c-Met pathway by competitively binding with miR-20b-3p. Nat Commun 10:2045

Wu Y, Song Y, Wang R, Wang T (2023) Molecular mechanisms of tumor resistance to radiotherapy. Mol Cancer 22:96

Xiao Y, Jiao C, Lin Y, Chen M, Zhang J, Wang J, Zhang Z (2017) lncRNA UCA1 contributes to imatinib resistance by acting as a ceRNA against miR-16 in chronic myeloid leukemia cells. DNA Cell Biol 36:18–25. https://doi.org/10.1089/dna.2016.3533

Xing S, Qu Y, Li C, Huang A, Tong S, Wu C, Fan K (2019) Deregulation of lncRNA-AC078883. 3 and microRNA-19a is involved in the development of chemoresistance to cisplatin via modulating signaling pathway of PTEN/AKT. J Cell Physiol 234:22657–22665

Xu S, Kong D, Chen Q, Ping Y, Pang D (2017) Oncogenic long noncoding RNA landscape in breast cancer. Mol Cancer 16:129

Yakkala PA, Penumallu NR, Shafi S, Kamal A (2023) Prospects of topoisomerase inhibitors as promising anti-cancer agents. Pharmaceuticals 16:1456

Yan H, Bu P (2021) Non-coding RNA in cancer. Essays Biochem 65:625–639

Yang P, Yang Y, An W, Xu J, Zhang G, Jie J, Zhang Q (2017) The long noncoding RNA-ROR promotes the resistance of radiotherapy for human colorectal cancer cells by targeting the p53/miR-145 pathway. J Gastroenterol Hepatol 32:837–845

Yang J, Qi M, Fei X, Wang X, Wang K (2021) LncRNA H19: a novel oncogene in multiple cancers. Int J Biol Sci 17:3188

Yao N et al (2019) Long non-coding RNA NONHSAT101069 promotes epirubicin resistance, migration, and invasion of breast cancer cells through NONHSAT101069/miR-129-5p/Twist1 axis. Oncogene 38:7216–7233

YiRen H et al (2017) Long noncoding RNA MALAT1 regulates autophagy associated chemoresistance via miR-23b-3p sequestration in gastric cancer. Mol Cancer 16:174

Yousefnia S (2022) A comprehensive review on miR-153: mechanistic and controversial roles of miR-153 in tumorigenicity of cancer cells. Front Oncol 12:985897

Yousefnia S (2024a) Breast cancer, subtypes, risk factors, and treatment. In: The Palgrave encyclopedia of disability. Springer, pp 1–14

Yousefnia S (2024b) A comprehensive review on lncRNA LOXL1-AS1: molecular mechanistic pathways of lncRNA LOXL1-AS1 in tumorigenicity of cancer cells Frontiers in. Oncology 14:1384342

Yousefnia S (2025) Mechanistic role of circRNAs in malignant characterizations of cancer stem cells. J Cancer Biomol Ther 2:19–28

Yousefnia S, Seyed Forootan F, Seyed Forootan S, Nasr Esfahani MH, Gure AO, Ghaedi K (2020) Mechanistic pathways of malignancy in breast cancer stem cells. Front Oncol 10:452

Yu S, Wu C, Tan Q, Liu H (2019) Long noncoding RNA H19 promotes chemotherapy resistance in choriocarcinoma cells. J Cell Biochem 120:15131–15144

Yu Z, Wang G, Zhang C, Liu Y, Chen W, Wang H, Liu H (2020) LncRNA SBF2-AS1 affects the radiosensitivity of non-small cell lung cancer via modulating microRNA-302a/MBNL3 axis. Cell Cycle 19:300–316

Zhang H et al (2015) Targeting WISP1 to sensitize esophageal squamous cell carcinoma to irradiation. Oncotarget 6:6218

Zhang J, Liu J, Xu X, Li L (2017a) Curcumin suppresses cisplatin resistance development partly via modulating extracellular vesicle-mediated transfer of MEG3 and miR-214 in ovarian cancer. Cancer Chemother Pharmacol 79:479–487

Zhang L et al (2017b) ZBTB7A enhances osteosarcoma chemoresistance by transcriptionally repressing lncRNALINC00473-IL24 activity. Neoplasia 19:908–918

Zhang C, Wang M, Shi C, Shi F, Pei C (2018) Long non-coding RNA Linc00312 modulates the sensitivity of ovarian cancer to cisplatin via the Bcl-2/Caspase-3 signaling pathway. Biosci Trends 12:309–316

Zhang R et al (2019a) Atractylenolide II reverses the influence of lncRNA XIST/miR-30a-5p/ROR1 axis on chemo-resistance of colorectal cancer cells. J Cell Mol Med 23:3151–3165

Zhang WY, Liu YJ, He Y, Chen P (2019b) Suppression of long noncoding RNA NCK1-AS1 increases chemosensitivity to cisplatin in cervical cancer. J Cell Physiol 234:4302–4313

Zhang X et al (2022) Cancer-associated fibroblast-induced lncRNA UPK1A-AS1 confers platinum resistance in pancreatic cancer via efficient double-strand break repair. Oncogene 41:2372–2389

Zou L et al (2022) Autophagy and beyond: Unraveling the complexity of UNC-51-like kinase 1 (ULK1) from biological functions to therapeutic implications. Acta Pharm Sin B 12:3743–3782

Chapter 8
Long Non-coding RNAs in Gastrointestinal Diseases

Sibin Nambidi, Antara Banerjee, and Surajit Pathak

Abstract Long non-coding RNAs (lncRNAs) have emerged as essential regulators of gastrointestinal (GI) diseases, influencing gene expression through chromatin remodeling, transcriptional modulation, and post-transcriptional control involving RNA-binding proteins and competing endogenous RNA networks. Their dysregulation contributes to the pathogenesis of major GI disorders, including colorectal, gastric, and esophageal cancers, inflammatory bowel disease, liver diseases, and pancreatic cancer. In GI cancer, lncRNAs such as HOTAIR, MALAT1, H19, and CCAT1 function as oncogenic drivers or tumor suppressors by modulating key signaling pathways, including NF-κB, Wnt/β-catenin, PI3K/AKT/mTOR, TGF-β, JAK/STAT, and Notch. In inflammatory and metabolic GI conditions, lncRNAs influence epithelial integrity, immune responses, and fibrosis, highlighting their diverse biological roles. Given their tissue-specific expression and stability, lncRNAs hold potential as diagnostic and prognostic biomarkers, as well as therapeutic targets. Emerging technologies such as antisense oligonucleotides and exosome-mediated delivery systems expand their potential for clinical translation. However, challenges remain, including optimization of delivery strategies and minimization of off-target effects. Future research integrating multi-omics approaches and lncRNA–microbiota interactions may advance precision medicine in gastroenterology.

Keywords Long non-coding RNAs · Gastrointestinal diseases · Biomarkers · Therapeutic targets · Precision medicine

S. Nambidi · A. Banerjee · S. Pathak (✉)
Medical Biotechnology lab, Faculty of Allied Health Sciences, Chettinad Academy of Research and Education (CARE), Chettinad Hospital and Research Institute, Chennai, India
e-mail: drsurajitpathak@care.edu.in

L. M. Vaschetto (ed.), *Long Non-Coding RNAs, Human Health, and Diseases*, Epigenetics and Human Health 16, https://doi.org/10.1007/978-3-032-19668-2_8

Abbreviations

LncRNAs	Long non-coding RNAs
GI	Gastrointestinal
NF-κB	Nuclear Factor kappa B
Wnt/β catenin	Wnt signaling pathway/β-catenin
PI3K/AKT	Phosphoinositide 3-kinase/Protein kinase B
DNA	Deoxyribonucleic Acid
RNAPII	RNA Polymerase II
RBP	RNA-Binding Protein
ceRNA	Competing Endogenous RNA
CRC	Colorectal Cancer
HOTAIR	HOX Transcript Antisense RNA
MYC	MYC Proto-Oncogene
EMT	Epithelial-Mesenchymal Transition
GC	Gastric Cancer
MAPK	Mitogen-Activated Protein Kinase
EC	Esophageal Cancer
STAT3	Signal Transducer and Activator of Transcription 3
CD	Crohn's Disease
UC	Ulcerative Colitis
HSC	Hepatic Stellate Cell
NAFLD	Non-Alcoholic Fatty Liver Disease
NASH	Non-Alcoholic Steatohepatitis
HBV	Hepatitis B Virus
T2DM	Type 2 Diabetes Mellitus
PC	Pancreatic Cancer
YAP	Yes-Associated Protein
TGF-β	Transforming Growth Factor Beta
JAK	Janus Kinase
mTOR	Mammalian Target of Rapamycin
CEA	Carcinoembryonic Antigen
CA19-9	Carbohydrate Antigen 19-9
ASOs	Antisense Oligonucleotides
RNAi	RNA Interference
CRISPR/Cas9	Clustered Regularly Interspaced Short Palindromic Repeats/ CRISPR-associated Protein 9

8.1 Introduction

Long non-coding RNAs (lncRNAs) represent a class of RNA higher than 200 nucleotides in length that do not code for a protein. Out of the approximately 270,000 lncRNAs identified in the human genome, only a fraction has been studied thoroughly. This is primarily due to the enormous potential biological functions of lncRNAs that have yet to be explored, making them crucial for numerous complicated processes with many diseases. Recently, lncRNAs have been reported to control transcriptional regulation and other cellular processes, which may help in various therapies (Potemkin and Clarkson 2023). Classifying lncRNAs is challenging due to their structural diversity, poor conservation, and heterogeneous functionality, lacking the clear coding frameworks that define protein-coding genes. These RNA molecules are usually specific to a cell type, interact with chromatin, and organize spatially within the nucleus to control gene expression, which is important during the development and differentiation of cells. Besides, lncRNAs also play a role in translation, metabolism, and signaling pathways in the cytoplasm. The specific mechanisms through which lncRNAs operate are still largely unknown and pose a significant area for future exploration (Mattick et al. 2023; Statello et al. 2021).

Gastrointestinal diseases (GID) include various conditions that affect the digestive system and its organs, such as the liver and the pancreas, the biliary tract, and the gallbladder (Fig. 8.1). They range from mild to severe disorders affecting and contributing to the death rates of a high number of people, along with increasing the

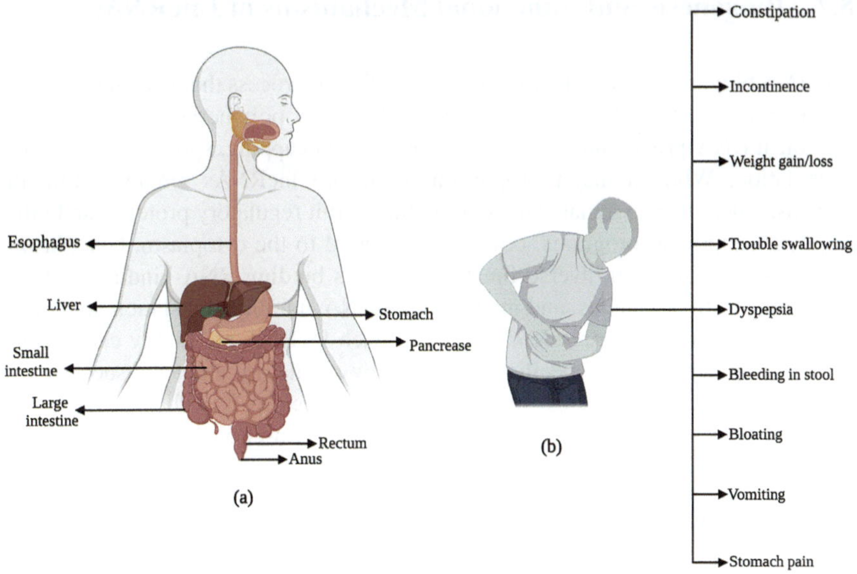

Fig. 8.1 (**a**) Illustration of the GI tract, including major organs such as the oesophagus, liver, stomach, pancreas, small intestine, large intestine, rectum, and anus. (**b**) Common symptoms associated with GID. (Adapted from Livesey 2014)

burden of chronic illness. Some of the major health threats that have serious complications include GIDs in liver diseases, pancreatitis, infections, and inflammatory and functional gastrointestinal disorders (Khan et al. 2023). GIDs fall into two categories: functional, which present with normal structures but impaired motility, and structural, which comprise motility disorders with physical abnormalities, including pancreatitis, hepatitis, peptic ulcers, cholelithiasis, complicated GERD, some neoplasms, diverticular diseases, hemorrhoids, and inflammatory bowel diseases like Crohn's disease and ulcerative colitis (Jabłońska and Mrowiec 2023). GIDs are investigated with clinical, imaging, endoscopic, and serological methods, infections being some of the vaguest and most expensive to elucidate. Substances found in tissues or fluids of the body, known as biomarkers, can further aid in distinguishing infectious/non-infectious disorders of the gastrointestinal system. Certain genetic biomarkers, like lncRNAs and microRNAs also possess significant advantages in this area (Ranjbar et al. 2022).

Consequently, lncRNAs have emerged as key regulatory molecules in GID, offering insights into disease mechanisms and potential targets for diagnosis and therapy (Lashgarian et al. 2020). lncRNAs play a vital role in maintaining intestinal barrier function and homeostasis. They contribute to intestinal stability by influencing epithelial regeneration, immune responses, and gut microbiota balance. Due to their distinct expression patterns in specific tissues and conditions, lncRNAs are considered promising biomarkers for a range of diseases (Lu et al. 2023).

8.2 Biogenesis and Functional Mechanisms of LncRNAs

LncRNA biogenesis is a tightly regulated, multi-step process that resembles mRNA production. Most lncRNAs are transcribed by RNA polymerase II and undergo canonical RNA-processing events, including 5′ m^7G capping, splicing, and 3′ polyadenylation. While a major proportion of mature lncRNAs remain within the nucleus, where they associate with chromatin, recruit regulatory proteins, and influence transcriptional programs, others are exported to the cytoplasm. Cytoplasmic lncRNAs participate in diverse functions such as binding RNA-binding proteins, interacting with organelles, and modulating mRNA stability, translation, and cellular homeostasis. The expression and fate of each lncRNA are highly cell-specific and developmental stage-specific, controlled by regulatory elements such as promoters, enhancers, and nuclear structures like paraspeckles, as well as by ribonucleases that help in lncRNA maturation and turnover (Lu et al. 2023; Majeed Shah et al. 2022).

8.2.1 LncRNAs in Pre-transcriptional Regulation

LncRNAs regulate gene expression at the pre-transcriptional level mainly by reshaping chromatin structure and altering the accessibility of gene promoters. By interacting with chromatin-modifying complexes, these lncRNAs can recruit enzymes that add or remove histone marks (particularly on histone H3), leading to changes in nucleosome positioning and the overall openness of chromatin. When chromatin is relaxed, transcription factors and RNA polymerase II can easily access the promoter; when it is compact, transcription is suppressed. Through these interactions, lncRNAs act as guides or scaffolds, directing chromatin-remodeling machinery to specific genomic regions and setting the stage for whether transcription will be activated or silenced (Liu et al. 2019). Furthermore, lncRNA serves as miRNA sponges that influence chromatin structure and, consequently, gene transcription by interfering with histone modification patterns, thus forming regulatory networks (Bure et al. 2022). Moreover, lncRNA interacts with histone modification complexes such as PRC1 and PRC2. They act *in cis* or *in trans* to recruit these complexes to specific genomic regions, leading to histone modifications like H3K27 methylation, which suppress transcription (Majeed Shah et al. 2022). Recent studies have demonstrated that lncRNAs transcribed from rRNA gene promoters are capable of regulating DNA methylation and rDNA transcription by generating stable triple-helix (triplex) DNA structures. These triplexes serve as a scaffold to recruit epigenetic modifiers, such as DNA methyltransferases (DNMTs), primarily DNMT3B, which facilitates de novo methylation at the rDNA *locus*, resulting in transcriptional repression or alteration of rDNA, thus playing an essential role in cellular homeostasis. Importantly, dysregulated triplex-mediated control of DNA methylation has been significantly associated with cancer and other disease pathologies (Huang et al. 2022; Kuo et al. 2019).

Furthermore, lncRNAs are involved in regulating gene expression by modulating the functions of DNMTs in a wide variety of biological processes, such as muscle regeneration, neural differentiation, adipogenesis, neuropsychiatric disorders, cardiovascular diseases, osteoarthritis, and various cancers (Huang et al. 2022). LncRNAs also facilitate the recruitment of RNA polymerase II to specific gene promoters, regulating initiation, and contributing to nuclear architecture by organizing chromatin into topologically associated domains (TADs) (Herman et al. 2022). Their interactions with transcription factors and other non-coding RNAs, along with their roles in various disease pathways, underpin their critical role in regulating gene expression and therapeutic potential (Gao et al. 2020a, b). Consequently, lncRNAs play critical roles in gene regulation by modulating chromatin structure, pre-mRNA splicing, and post-transcriptional processes, including mRNA stability, translation, protein stability, and ceRNA network formation, thereby bridging nuclear and cytoplasmic functions (Figs. 8.2 and 8.3).

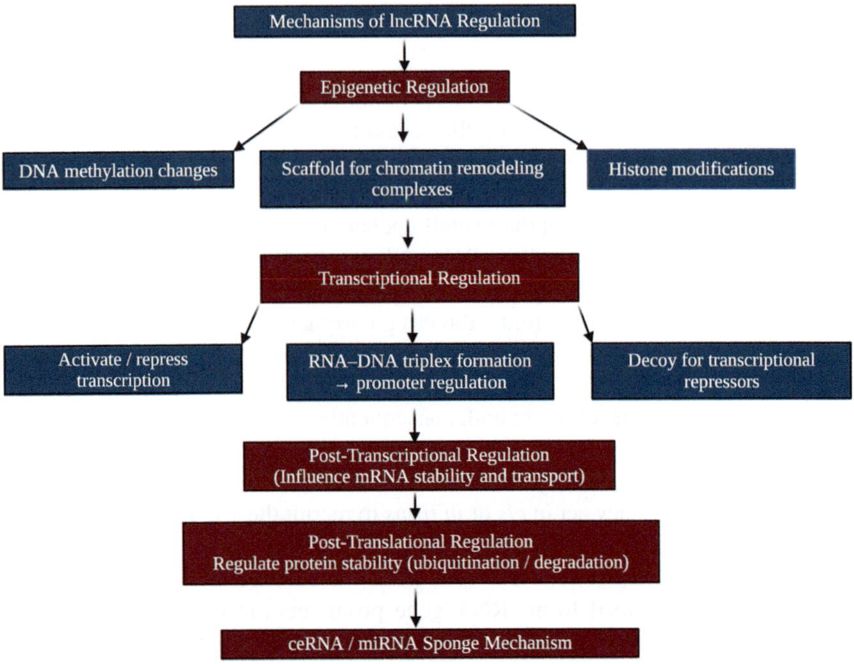

Fig. 8.2 Mechanisms of lncRNA regulation. (Adapted from Li et al. 2021)

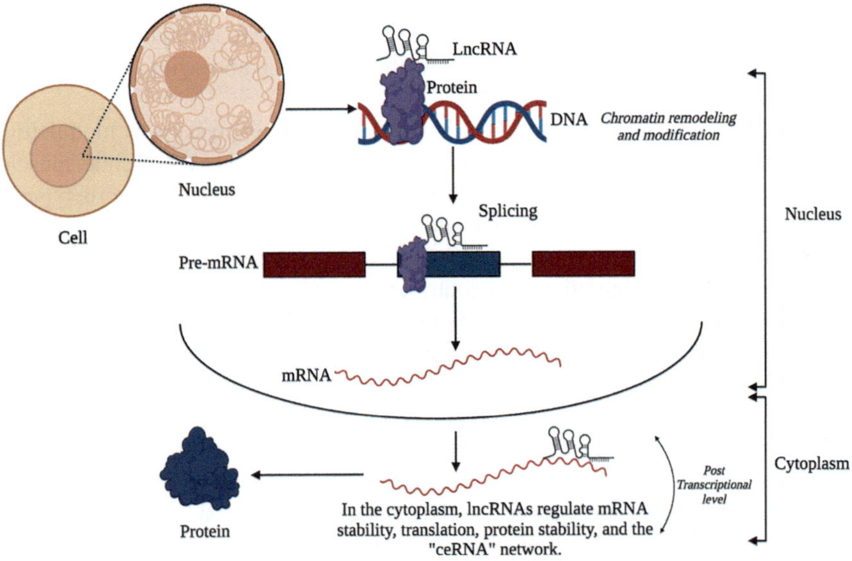

Fig. 8.3 LncRNAs play critical roles in gene regulation by modulating chromatin remodelling, pre-mRNA splicing, and post-transcriptional processes such as mRNA stability, translation, protein stability, and the ceRNA network, bridging nuclear and cytoplasmic functions. (Adapted from Gao et al. 2020a, b)

8.2.2 LncRNAs in Transcriptional Regulation

LncRNA takes an active part in transcriptional control by influencing the activity of RNA Polymerase II (RNAPII) and chromatin organization. LncRNAs that act *in trans* include 7SK and B2; these inhibit RNAPII elongation or DNA binding. LncRNAs that act *in cis* include HOTTIP; these tether epigenetic modifiers to certain *loci* during transcription. Moreover, transcriptional interference (TI) occurs when lncRNA transcription alters chromatin structure or overlaps with regulatory elements, ultimately reducing gene expression (Kornienko et al. 2013).

8.2.3 LncRNAs in Post-transcriptional Regulation

LncRNAs exert important post-transcriptional regulatory functions primarily through their interactions with RNA-binding proteins (RBPs). By forming ribonucleoprotein complexes, lncRNAs can modulate several aspects of mRNA processing, including alternative splicing, mRNA stability, nuclear export, and translation. These regulatory effects determine the lifespan of specific mRNAs and influence the efficiency with which they are translated into proteins. In pathological conditions such as cancer, dysregulated lncRNA–RBP interactions have been shown to alter gene expression programs that drive proliferation, metastasis, and resistance to therapy. Elucidating these molecular interactions is therefore essential for identifying new regulatory mechanisms and developing targeted therapeutic interventions (He et al. 2019). They can also act as scaffold or adaptor molecules, bringing together splicing machinery in the nucleus or processing stress granules in the cytoplasm. Furthermore, lncRNAs are involved in the interactions between RNA-binding proteins and miRNAs, modulating mRNA decay and translation efficiency, and also modulating protein production. By dynamically modulating these processes, lncRNAs have critical functions in maintaining normal cellular function, and dysregulation in lncRNA expression can lead to the development of various diseases (Yoon et al. 2013). Some lncRNAs are acutely processed into smaller non-coding RNAs, while others serve as endogenous RNAs (ceRNAs), which can bind to miRNAs through common sequence motifs, limiting the availability of those miRNAs. These lead to an indirect upregulation of genes that depend on miRNAs. It has also been shown that lncRNAs interact with protein-Staufen 1 (STAU1) to regulate the stability of mRNA or the degradation through the build-up of RNA-protein complexes or double-stranded RNA motifs. All these activities showcase the mechanistic versatility that lncRNAs possess to regulate gene expression post-transcriptionally (Dykes and Emanueli 2017). Figure 8.2 is a model schematizing the mechanisms of regulation via lncRNAs, which act at multiple levels, including epigenetic, transcriptional, post-transcriptional, and post-translational levels, and by acting as ceRNAs/miRNA sponges.

8.3 Role of LncRNAs in Specific Gastrointestinal (GI) Diseases

LncRNAs contribute to key regulatory processes involved in gastrointestinal diseases, influencing gene expression and cellular behavior. However, many of their specific functional roles remain insufficiently understood.

8.3.1 Colorectal Cancer (CRC)

lncRNAs silence different processes, such as splicing, degradation, transcription, and translation. lncRNAs can be oncogenic or tumor suppressors in cancer progression. LncRNA overexpression in CRC cells is often associated with poor patient prognosis and metastasis, which positions their catalytic use as prognostic biomarkers and therapeutic targets (Siddiqui et al. 2019). Some lncRNAs associated with tumorigenesis influence key genes and cellular processes. For instance, the lncRNA HOTAIR interacts with PRC2, altering the expression of target genes; MALAT1 stimulates cancer cell proliferation and invasion through AKAP-9; and CCAT1 is involved in the activation of the MYC target and cellular proliferation, invasion, and metastasis. Likewise, CCAT2 interacts with TCF7L2 and modulates MYC protein expression, whereas UCA1 promotes tumorigenesis by inhibiting apoptosis. Other lncRNAs like CASC11, CRNDE, and GAS5 are implicated in metastasis and tumor progression by either interacting with several proteins or activating certain pathways (Siddiqui et al. 2019). Their association with poor clinical outcomes in CRC highlights the potential of these lncRNAs to serve as prognostic biomarkers and therapeutic targets. Furthermore, the Consensus Molecular Subtypes (CMS) of colorectal cancer display distinct lncRNA expression patterns, each influencing tumor behavior and treatment responsiveness in different ways (Siddiqui et al. 2019). Such CMS1 subtype (MSI immune) lncRNAs include HOTAIR and LINK-A, involved in immune activation and hypermutation, while those in the CMS2 (canonical) subtype, like CCAT1 and CRNDE, possess regulatory roles over the WNT and MYC signaling pathways. A number of lncRNAs, including H19, which links CMS4 to epithelial-mesenchymal transition and metastasis, contribute not only to the molecular landscape of CRC but also represent possible therapeutic targets and biomarkers (Shakhpazyan et al. 2023).

LncRNAs also influence CRC progression at each disease stage. In the early stages (Stage 0–I), when the tumor remains localized, lncRNAs help regulate gene expression programs that support initial tumor development (Saini et al. 2025). As CRC advances, additional lncRNAs become dysregulated and promote invasion, metastasis, and ultimately distant spread. The tumor advances in Stage II and Stage III, in which lncRNAs allow metastases. At Stage IV, where the cancer metastasizes to distant sites, lncRNAs take part in the invasiveness regulated by epithelial-to-mesenchymal transition, which further assists in the spread of the tumor. Thus, lncRNAs serve as crucial regulators in CRC progression and can serve as possible markers for diagnosis and treatment (Sung and Hong 2023).

8.3.2 Gastric Cancer

LncRNAs play crucial roles in gastric cancer (GC) by regulating various cellular processes. Many lncRNAs are frequently dysregulated in GC and can be easily detected in tissues and body fluids, thus offering potential for future clinical applications. LncRNAs affect GC using pathways like mRNA stability, microRNA sponging, and epigenetic alterations. They may also serve as prognostic markers, circulating biomarkers, and indicators of drug resistance, providing potential for novel therapeutic strategies and biomarker identification in GC (Gao et al. 2020a, b). HOXA11-AS and DANCR lncRNAs promote gastric cancer cell migration by regulating pathways such as the miR-124-3p/ITGB3 axis and EZH2, both of which contribute to epithelial-mesenchymal transition. Similarly, XIST and AK025387 enhance the invasive capacity of gastric cancer cells by activating key signaling pathways, including JAK/STAT3 and MAPK (You et al. 2021). Upregulation of LINC00152 and LINC00355 represents a poor prognostic factor for overall survival in GC, possessing potential as novel biomarkers and therapeutic targets in GC (Silva et al. 2023). In recent years, lncRNAs have been established as potentially useful markers for cancer diagnosis through liquid biopsies. While lncRNAs in plasma/serum and gastric juice have been studied for gastric cancer, other fluids such as saliva and urine have been studied for various cancers of the head and neck and urinary system. However, the elucidation of the potential roles of lncRNAs in liquid biopsies as biomarkers for gastric cancer has remained largely unexplored (Yuan et al. 2020). LncRNAs have been shown to play an important role in drug resistance in gastric cancer by regulating major pathways and drug resistance genes. Tumorigenic lncRNAs such as PCAT-1, MALAT-1, and DANCR enforce resistive traits toward platinum-based therapies such as Adriamycin, 5-fluorouracil, and Taxol-modulating apoptosis, cell cycle progression, and MDR gene expression (Meng et al. 2024). For instance, PCAT-1 inactivates PTEN, whereas DANCR activates MDR genes and inhibits apoptosis. Tumor suppressor lncRNAs like CRAL and CASC2 prevent drug resistance by encouraging DNA damage and apoptosis through regulating specific microRNAs and tumor suppressor genes. In summary, lncRNAs modulate cell survival and chemotherapeutic responses in GC (Wei et al. 2020).

8.3.3 Oesophageal Cancer

lncRNAs are vital in the development and progression of esophageal cancer (EC) by regulating distinct pathways underlying tumor growth, migration, invasion, and apoptosis. Oncogenic lncRNAs, such as H19, MALAT1, UCA1, or PVT1, stimulate EC progression through modulation of several signaling cascades, including ATM-Chk2, STAT3-EZH2-β-catenin, and cell-cycle regulation, while other lncRNAs, such as CASC9 and MIAT, facilitate tumor proliferation or metastasis (Han et al. 2022). Both MIR22HG and VESTAR are recently discovered lncRNAs that may be useful as diagnostic and therapeutic targets against cancer (Han et al. 2022). In addition to driving tumor advancement, lncRNAs also impact therapeutic responses in EC. For example, TUSC7 and linc00261 potentiate chemosensitivity by regulating the EGFR/AKT pathway and enhancing apoptosis. In contrast, NORAD and CCAT2 promote radioresistance by supporting genome integrity and regulating cell-cycle checkpoints. In opposition, however, MAGI2-AS3 and GAS5 increase the susceptibility of EC cells to radiation by initiating apoptosis and inhibiting oncogenic signaling (Yang and Chen 2022).

Various lncRNAs, which are present in oesophageal squamous-cell carcinoma (ESCC), modulate gene expression and thus influence tumor progression through epigenetic mechanisms. For example, Double-homeobox A pseudogene 10 is an lncRNA that recruits EZH2 and leads to P21 silencing, thereby promoting cellular proliferation and metastasis (Yan et al. 2024). FOXP4-AS1 enhances FOXP4 expression through the recruitment of MLL2, stimulating β-catenin-mediated cell growth and migration. The TMPO-AS1 activates TMPO gene transcription through the action of p300, while LINC01419 contributes to resistance against 5-fluorouracil treatment by methylation of the GSTP1 promoter. Lymphoid enhancer binding factor 1-AS1 promotes cell proliferation by degrading p53 (Yan et al. 2024). In EC, several lncRNAs associate with specific proteins to act in a variety of cellular processes. LINC01503 binds EBP-1 and ERK2, impeding important signaling pathways, such as ERK/MAPK and PI3K/AKT, promoting cell proliferation and invasion. SPRY4-IT1 stimulates Snail to high levels and nuclear localization to drive epithelial-mesenchymal transition, functioning as an oncogene in ESCC. Additionally, EZR-AS1 interacts with SMYD3 and the RNA polymerase II complex, facilitating transcription of EZR and promoting cell migration. These interactions point to the role of lncRNAs in prescribing the modulation of proteins and regulation of cellular functions in EC (Su et al. 2018).

8.3.4 Inflammatory Bowel Disease (IBD)

Inflammatory bowel disease (IBD), which includes Crohn's disease (CD) and ulcerative colitis (UC), is characterized by chronic inflammation of the gastrointestinal tract, with early disruption of intestinal barrier integrity. CD typically presents with patchy, transmural lesions that can affect any part of the digestive tract, whereas UC is marked by continuous mucosal inflammation confined to the colon (Hu et al. 2024). This pathology weakens intestinal barrier permeability, letting through toxins and bacteria, which might worsen systemic inflammation. The disturbance of the intestinal barrier occurs during both active and inactive phases of the disease in IBD patients, which involves epithelial cell apoptosis and destroyed tight junctions. lncRNAs are key modulators of the intestinal barrier through regulating epithelial regeneration and immunity, representing a valid target for IBD therapy (Hu et al. 2024). For example, lncRNAs such as H19, PLncRNA1, and SPRY4-IT1 affect intestinal tight junction proteins that provide effective gut barrier integrity. Certain lncRNAs, including BC012900 and CRNDE, modulate epithelial cell apoptosis, which directly affects intestinal barrier stability. Other lncRNAs, such as CCAT1, contribute to barrier regulation through alternative mechanisms, including the modulation of epithelial proliferation and inflammatory signaling. They also modulate immune responses in IBD, with lncRNAs like NEAT1, IFNG-AS1, and LINC01882 involved in this aspect. Other lncRNAs show great promise as potential diagnostic biomarkers for IBD, including KIF9-AS1, LINC01272, and DIO3OS (Hu et al. 2024).

LncRNAs are not only involved in the regulation of inflammation and modulation of immune responses but are also critical in maintaining the integrity of the intestinal epithelial barrier. Different studies previously showed that aberrant expression of lncRNAs contributes to the progression of IBD and mucosal alterations in experimental colitis, thus highlighting the potential of lncRNAs to inform the mechanisms of disease (Ortega Moreno et al. 2024). For example, H19 correlates with epithelial repair, while NEAT1 and MALAT1 were associated with immune-cell activation and inflammatory signaling pathways (Bitaraf et al. 2025). Due to their numerous actions, lncRNAs are being explored to function as potential diagnostic biomarkers as well as therapeutic targets in IBD, thus opening new avenues for personalized disease-management strategies. Their complex involvement in IBD pathophysiology requires fundamental studies to elucidate their actual molecular mechanisms and therapeutic potential (Ortega Moreno et al. 2024).

8.3.5 Liver Diseases

The liver is vital for detoxification and nutrient metabolism, which are essential for good health. The onset of liver diseases is usually insidious and begins with vague or no symptoms, often progressing asymptomatically in early stages. As the disease advances, however, it passes through distinct stages in which significant impairment of the liver and the translational well-being of the host occur (Wazir et al. 2023). LncRNAs play a crucial role in promoting hepatic fibrosis, cirrhosis, and hepatocellular carcinoma (HCC). Some lncRNAs, such as MEG3, GAS5, Gm5091, NR-002155.1, and HIF1A-AS1, act as inhibitors of liver fibrosis, thus possessing properties such as downregulation in expression levels in fibrotic liver tissue. They inhibit HSC activation and delay the progression of fibrosis when overexpressed. Strategies to upregulate the expression of these lncRNAs, therefore, might give a unique therapeutic approach for liver fibrosis (He et al. 2020). Multiple lncRNAs are involved in both chronic liver diseases and HCC, either promoting inflammation, fibrosis, or tumorigenesis triggered via pathways such as TGF-β1, TNF signaling, or ECM deposition. Interestingly, in NAFLD and NASH, NEAT1, MALAT1, and PVT1 are overexpressed; the risk of HCC for some variants, including rs7763881 in HULC and rs619586 in MALAT1, reduces among HBV carriers. The promoter or tumor suppressive role of lncRNAs, such as HULC, HOTAIR, and MALAT1, when deregulated, might either enhance or inhibit cancer cell proliferation, invasion, and apoptosis. Other lncRNAs, like GAS5 or MEG3, are associated with a better prognosis, while AFAP-AS1 and ZEB1-AS1 are suggestive of a relatively poor prognosis and chemoresistance, asserting their importance in developing therapeutic regimes against liver diseases (Kim et al. 2020).

8.3.6 Pancreatic Cancer

lncRNAs play a major role in the pathogenesis of T2DM and PC, two interconnected diseases. Specifically, MEG3, MALAT1, UCA1, HOTAIR, PVT1, and H19 lncRNAs modulate insulin resistance (IR), tumor growth, and alteration of metabolism. For example, the MEG3 gene regulates insulin production and inhibits tumor growth through the PI3K/AKT pathway, whereas, in contrast, MALAT1 and UCA1 promote PC progression via EMT and stem cell properties. HOTAIR also promotes glucose uptake in PC cells, connecting IR with the enhancement of tumor growth. Another example is PVT1 and H19, which could affect IR by being involved in inducing hyperglycemia and consequently possibly in EMT, which in turn enables T2DM complications and PC metastasis. On this basis, emerging evidence has called for more clinical and experimental research to investigate the complex interrelationship therein, promoting the opportunities for diagnostic and therapeutic strategies in diabetes-associated PC (Bin Wang et al. 2023). LncRNAs such as LINC01559 enhance YAP transcriptional activity, while LINC01133 promotes pancreatic cancer cell proliferation. Other lncRNAs, including LINC01197 and LINC00460, modulate the Wnt/β-catenin pathway and have been proposed as potential biomarkers. Small nucleolar RNA host genes (SNHGs), including SNHG14, SNHG15, and SNHG7, have been implicated in pancreatic cancer, with SNHG15 in particular identified as a marker of poor prognosis. Antisense lncRNAs, such as siAFAP1-AS1 and ADPGK-AS1, induce apoptosis and cell migration, suggesting their potential as therapeutic targets (Lou et al. 2019; Ramya Devi et al. 2021; Weng et al. 2018). Other lncRNAs, such as PVT1, MEG3, and XIST, have emerged as paths through which the tumor is driven in terms of progression and metastasis as possible markers and therapeutic agents on PC (Ramya Devi et al. 2021). lncRNAs exert their influence on chemoresistance development in PC through pathways with respect to miRNA sponging. Reduced expression of GAS5 in chemo-resistant cells affects the Hippo pathway and DCSKS3 expression via miR-181c-5p and miR-221. Linc-ROR promotes autophagy by binding to miR-124, which leads to PTBP1-mediated switching of PKM1 to the PKM2 isoform. In contrast, TUG1 contributes to chemoresistance through activation of the ERK pathway, a process that can be reversed using ERK inhibitors such as Scheme 772984. Serum lncRNAs such as MALAT1, HOXA transcript at the distal tip, and PVT1 correlate with poor gemcitabine response and shorter survival, highlighting their potential as prognostic biomarkers (Wang et al. 2020).

8.4 Mechanisms by Which LncRNAs Contribute to GI Diseases

LncRNAs exhibit diverse mechanistic roles in gastrointestinal diseases, acting as regulators of gene expression and modulating multiple signaling pathways that govern inflammation, proliferation, differentiation, and tumor progression. The NF-κB, Wnt/β-catenin, MAPK/ERK, PI3K/AKT, TGF-β, JAK/STAT, and Notch signaling pathways mainly modulate inflammation, cell proliferation, immune responses, as well as tumor progression. The NF-κB signaling pathway plays a central role in regulating inflammation and immune responses and is tightly controlled by multiple lncRNAs. Certain lncRNAs directly modulate key molecular events within this pathway, thereby influencing NF-κB activation status (Pourmehran et al. 2024). For instance, the lncRNA NKILA functions as a tumor suppressor by inhibiting the phosphorylation of IκBα, which prevents NF-κB nuclear translocation and subsequent transcriptional activation of pro-inflammatory and pro-tumorigenic target genes. In contrast, HOTAIR can enhance NF-κB activity, and this is often associated with the promotion of cancer. Additional lncRNAs, including MALAT1 and Carlr, affect NF-κB signaling in inflammation and immune disorders. These lncRNAs are vital for maintaining an equilibrium in NF-κB activity and stable targets for therapy (Li et al. 2024a, b, c; Gupta et al. 2020; Shin et al. 2024). Acting directly on NF-κB proteins, signaling intermediates of NF-κB or through modulation of the expression of genes in the NF-κB pathway has made lncRNAs integral components of NF-κB activation. These interactions influence macrophage-mediated inflammatory responses, which play a role in the development and progression of GIDs (Gao et al. 2023).

LncRNAs associated with the Wnt pathway are significantly altered in oesophageal, gastric, liver, colorectal, and pancreatic cancers, making them possible biomarkers. EWSAT1, LINC01606, and GASL1 show diagnostic capacities, while MEG3 and LINC01226 correlate with survival outcomes, emphasizing their importance in cancer diagnosis and prognosis (Hu et al. 2018). The Wnt/β-catenin signaling pathway is a key regulator of cancer development, affected by many lncRNAs in different cancers. In CRC, tumor suppressive lncRNAs such as BC032913 and lincRNA-p21 inhibit Wnt/β-catenin activity by either stabilizing the inactive form of β-catenin, thus impeding β-catenin nuclear translocation, or directly targeting β-catenin post-transcriptionally, and suppressing tumor growth and metastasis (Hu et al. 2018). On the other hand, NEAT1 and HOTAIR support the activation of the Wnt/β-catenin pathway in glioma, gastric cancer, and non-small cell lung cancer (NSCLC) by promoting proliferation, migration, and increased resistance to therapies. This includes their interaction with EZH2 and miRNA modulation, showing the dual role of lncRNAs as a tumor suppressor or an oncogenic feature with the potential to serve as future therapeutic targets (Hu et al. 2018). The lncRNA GASL1 is downregulated in gastric carcinoma and acts as a tumor suppressor by inactivation of the Wnt/β-catenin signaling pathway, preventing tumor growth. In comparison, lnc-CHAF1B-2 is overexpressed in gastric cancer and activates the Wnt/β-catenin pathway that promotes cancer cell proliferation, invasion, and metastasis (Tian et al. 2025). Consequently, lncRNAs can potentially modulate tumor progression, aggressiveness, and therapeutic sensitivity through interactions with key Wnt/β-catenin signaling components, such as β-catenin and TCF; these lncRNAs show distinct tissue- and cell-type-specific expression patterns in various cancers and lead to specific differences in signaling that mediate cancer pathology and mechanisms of resistance (Hu et al. 2018).

The MAPK signaling pathway, a key regulator of cell proliferation, is modulated by several classes of non-coding RNAs (ncRNAs) in liver cancer, particularly lncRNAs, circRNAs, and small nucleolar RNAs (snoRNAs). The ARA lncRNA inhibits cell growth and enhances cell death by modulating MAPK (ERK) signaling, while URHC, which is upregulated in hepatocellular carcinoma, contributes to cell growth and apoptosis resistance through ERK/MAPK pathways by targeting the zipper-containing kinase. On the whole, these lncRNAs modulate MAPK pathways and collectively influence cancer cell proliferation, survival, and metastasis (Li et al. 2024a, b, c). LINC00152 is one example of an lncRNA that suppresses MAPK/ERK signaling, leading to reduced tumor cell proliferation, enhanced apoptosis, and an overall inhibition of cancer progression. The significance of those regulatory activities illustrates lncRNAs as key modulators of MAPK/ERK pathways involved in gastrointestinal disease development and progression (Shi and Sun 2020).

The PI3K/AKT/mTOR pathway plays a central role in normal cellular physiology, regulating protein synthesis, nutrient utilization, metabolism, cell growth, and proliferation functions largely mediated through mTOR complex 1 (mTORC1). In GI cancers, dysregulation of this pathway is frequently influenced by aberrant lncRNA expression. Many lncRNAs act as competing endogenous RNAs (ceRNAs), sequestering specific miRNAs and thereby modifying the expression of PI3K/AKT/mTOR pathway components (Riquelme et al. 2023). Conversely, proteins within this pathway can regulate the transcription or stability of particular lncRNAs, creating reciprocal regulatory loops. These altered lncRNA–PI3K/AKT interactions promote malignant behaviors such as increased proliferation, migration, invasion, apoptosis resistance, and reduced responsiveness to therapy (Li et al. 2024a, b, c). Several lncRNAs have been specifically linked to the activation of the PI3K/AKT pathway in GI tumors. In gastric cancer, AK023391, UCA1,

LOC101928316, and HOTTIP enhance tumor formation by upregulating PI3K/AKT signaling. Other lncRNAs, including CRNDE, XLOC_006753, and HAGLROS, increase PI3K and AKT phosphorylation, thereby driving aggressive phenotypes and promoting disease progression (Hashemi et al. 2024; Riquelme et al. 2023). Importantly, alteration of these pathways also contributes to chemoresistance in several GI malignancies (Li et al. 2024a, b, c).

LncRNAs have also been found to be involved in the TGF-β signaling pathway, a multifunctional cytokine that plays an essential role in regulating several cellular behaviors. TGF-β signaling plays an important role in the initiation of tumors and their progression in GID (Han et al. 2022). Current studies suggest that lncRNAs seem to act as specific mediators of TGF-β1, regulating approaches for the promotion of inflammatory diseases and offering an attractive target for new gene-based therapeutic strategies (Tang et al. 2022). Moreover, the JAK/STAT signaling pathway is important for gene expression, cell survival, and the progression of various malignancies, including gastrointestinal diseases. Abnormal activation of such pathways drives tumorigenesis due to other genetic alterations, regulating processes, including immune responses to apoptosis, proliferation, and angiogenesis. LncRNAs and miRNAs play a significant role in modulating JAK/STAT activity in gastrointestinal diseases. For instance, lncRNAs PVT1, CCAT1, and CCAT5 are known to support colorectal cancer progression by upregulating the activity of STAT3, while miR-34a and miR-365a-3p exert inhibitory effects on proliferation and metastasis in colorectal cancer by targeting this pathway. These interactions underscore the important role of JAK/STAT signaling and its regulation by non-coding RNAs in GI diseases (Ghasemian et al. 2023).

The Notch signaling pathway is a key pathway involved in a whole range of cellular processes from cell fate decision to differentiation. Some lncRNAs have been shown to be positively or negatively correlated with this pathway and with miRNAs or mRNAs involved in Notch-related processes in the physiological state. In cancer, especially CRC, dysregulated lncRNAs generate signals to influence the activation of the Notch signaling pathway, while activation of Notch itself may, in turn, regulate the expression of those or other lncRNAs. Certain lncRNAs have also been found to modulate the Notch signaling pathway and exert their action in cancers including gastrointestinal cancers (Emam et al. 2022). FOXD2-AS1 promotes colorectal cancer progression through Notch and epithelial-to-mesenchymal transition, while its knockdown leads to a decrease in tumor growth (Yang et al. 2017). In gastric cancer, ANRIL promotes survival and migration of tumor cells through miR-99a and BMI1, thus impacting both the Notch and mTOR signaling pathways (Liu et al. 2018). LincRNA-p21 exerts a suppressive effect on Notch signaling and epithelial-to-mesenchymal transition in liver cancer (Miyamoto and Weinmaster 2009; Emam et al. 2022; Ghafouri-Fard et al. 2021). An overview of representative lncRNAs, their associated gastric malignancies, biological functions, and regulatory pathways is provided in Table 8.1, and an overview of lncRNA mechanisms of action in GI diseases and cancers is represented in Fig. 8.4.

Table 8.1 Representative lncRNAs associated with gastrointestinal cancers, their functional roles, and key signaling pathways

lncRNA	Associated GI disease	Function	Pathway involved	Refrences
CRNDE	Colorectal and gastric cancer	Promotes proliferation, migration, invasion	PI3K/AKT/mTOR	Zhang et al. (2018)
UCA1	Gastric cancer	Enhances tumorigenesis, apoptosis resistance	PI3K/AKT	Dai et al. (2020)
HOTTIP	Gastric cancer	Promotes invasion and metastasis	PI3K/AKT	Xin et al. (2023)
GASL1	Gastric cancer	Tumor suppressor, inhibits tumor growth	Wnt/β-catenin (inactivated)	Peng et al. (2019)
lnc-CHAF1B-2	Gastric cancer	Promotes proliferation, invasion, metastasis	Wnt/β-catenin (activated)	Tian et al. (2025)
HAGLROS	Gastric cancer	Induces malignant phenotype	PI3K/AKT	Chen et al. (2018)
XLOC_006753	Gastric cancer	Enhances proliferation and invasion	PI3K/AKT	Zeng et al. (2018)

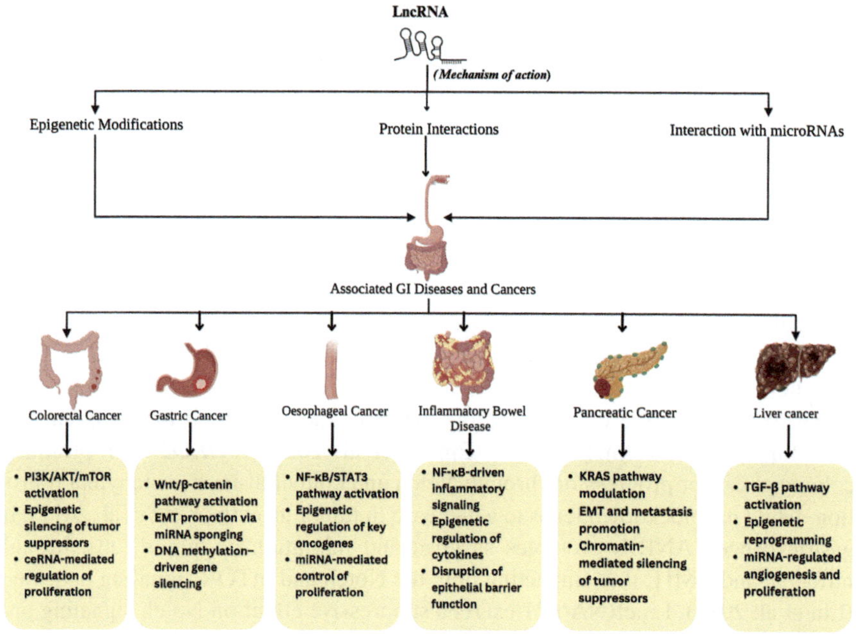

Fig. 8.4 Overview of the mechanisms of action of lncRNAs in GI diseases and cancers, including their roles in modulating epigenetic modifications, protein interactions, and miRNA crosstalk. These roles impact key signaling pathways and processes across various cancers and inflammatory conditions. (Adapted from Li et al. 2024a, b, c; Gao et al. 2023; and Hu et al. 2018). The roles of lncRNAs in cancer pathways are detailed in Chap. 6 by Aoki et al.

8.5 Clinical Implications of LncRNAs in GI Diseases

The most well-studied lncRNAs, including HOTAIR, H19, MALAT1, ZFAS1, and ST7-AS1, are known to act as key regulators of GID and are critical drivers of cell proliferation, disease progression, and therapy resistance. Thus, lncRNAs present opportunities for designing novel prognostic, diagnostic, and therapeutic strategies in GI cancers. Advanced molecular techniques such as ASOs, RNAi, CRISPR/Cas9, viral-mediated delivery of shRNAs targeting lncRNAs, and small-molecule inhibitors could modulate their production (Sayedeh Azimeh Hosseini et al. 2023).

In GI cancers, the use of traditional markers, CEA and CA19-9, is prevalent; however, their implications for diagnosis are rather limited due to the lack of specificity and low sensitivity. Several important lncRNAs, including SNHG5, PVT1, and NORAD, have emerged as pivotal contributors toward developing chemotherapy resistance in GI cancers. However, lncRNA-based therapy holds the potential of revolutionizing outcomes, thereby paving the path for future investigations (Liu et al. 2022). Owing to their stability in the bloodstream and ability for early detection, lncRNAs are considered promising, additional options as non-invasive biomarkers for gastrointestinal cancers as opposed to the classical diagnostic assessment (Badowski et al. 2022).

The advancement of liquid biopsy applications for lncRNAs is hindered by several methodological limitations. A major issue is the absence of unified protocols for their isolation and detection, which leads to variability across studies. In addition, circulating lncRNAs are present at low concentrations in biological fluids, making their measurement technically demanding. Obtaining exosomal lncRNAs of sufficient purity for dependable analysis remains difficult. These challenges, together with concerns about the sensitivity and reproducibility of current detection platforms, continue to restrict the clinical use of lncRNA-based liquid biopsies. Exosomal lncRNAs from blood represent potential candidates for liquid biopsy-based early and less-invasive diagnosis of GI cancers. In addition, the exosomal transfer of such lncRNAs has shown enormous anti-tumor efficacy (Sun et al. 2023). Biologically and analytically, limited sensitivity and specificity, tumor heterogeneity, and lncRNAs from non-tumor sources are factors that contribute to a lack of reliable interpretation of results (Ma et al. 2024; Lone et al. 2022). Furthermore, lncRNAs cannot mimic the complexity of cancer biology, and non-tumor-derived lncRNAs, especially age-related clonal haematopoiesis, inherently elevate the probability of false-positive results. These barriers point to the need for methodological improvement and biological insight before liquid lncRNA is used clinically (Ma et al. 2024; Tellez-Gabriel and Heymann 2019).

The present challenges with the clinical use of lncRNAs for diagnosis and therapy include the complicated lncRNA action mechanisms, a lack of standardization in lncRNA detection methodologies, and a lack of large-scale clinical testing to assess the precision and reliability of lncRNAs as biomarkers. Furthermore, delivering therapeutic lncRNAs to target cells and tissues will require more research, in addition to issues with lncRNA stability and degradation in biological samples (Eldakhakhny et al. 2024). LncRNAs are under investigation as diagnostic biomarkers, therapeutic targets, and prognostic markers for cancer, but to date, clinical implementation is still lacking, with PCA3 (prostate cancer antigen 3) for prostate cancer being the only lncRNA to be FDA approved for a clinical test (Lemos et al. 2019).

8.6 Conclusion and Future Perspectives

LncRNAs have emerged as key regulators in gastrointestinal diseases, influencing a wide range of biological processes, including gene expression, chromatin remodeling, and intracellular signaling. Their involvement in the initiation and progression of colorectal cancer, gastric cancer, inflammatory bowel disease, liver disease, and pancreatic cancer underscores their functional versatility and growing clinical relevance. Through modulation of major pathways such as NF-κB, Wnt/β-catenin, and PI3K/AKT, and through intricate interactions with other non-coding RNAs and proteins, lncRNAs form complex regulatory networks that shape disease behavior. Despite significant progress, several challenges continue to limit the translation of lncRNA research into clinical practice. Critical gaps remain in understanding the precise mechanisms of many lncRNAs, and efficient, targeted delivery systems for therapeutic applications are still under development. Future work should prioritize the discovery and characterization of unannotated lncRNAs, integrate multi-omic approaches to map their functions, and explore their interactions with the gut microbiota to refine biomarker discovery. Emerging technologies including antisense oligonucleotides and exosome-based delivery systems offer promising avenues for therapeutic development and may eventually enable personalized interventions for gastrointestinal diseases. Addressing current limitations will be essential for harnessing the full potential of lncRNA biology and advancing more accurate diagnostic tools and targeted therapeutic strategies.

Acknowledgement We want to thank the Chettinad Academy of Research and Education, Chettinad Hospital, and Research Institute for providing the facilities.

Authors' Contributions Sibin Nambidi, Antara Banerjee, and Surajit Pathak did the book chapter preparation and editing; the chapter was reviewed and edited by Antara Banerjee and Surajit Pathak.

Funding Not Applicable.

Compliance with Ethical Standards

Conflict of Interest All authors declare that they have no conflicts of interest.

Ethical Approval This chapter is a review of previously published accounts; as such, no animal or human studies were performed.

References

Badowski C, He B, Garmire LX (2022) Blood-derived lncRNAs as biomarkers for cancer diagnosis: the good, the bad and the beauty. Npj Precis Oncol 6(1):1–18. https://doi.org/10.1038/s41698-022-00283-7

Bitaraf A, Zafarani A, Jahandideh P, Hakak-Zargar B, Haghi A, Asgaritarghi G, Babashah S (2025) MALAT1 as a molecular driver of tumor progression, immune evasion, and resistance to therapy. Mol Cancer 24(1):245. https://doi.org/10.1186/s12943-025-02415-6

Bure IV, Nemtsova MV, Kuznetsova EB (2022) Histone modifications and non-coding RNAs: mutual epigenetic regulation and role in pathogenesis. Int J Mol Sci 23(10):5801. https://doi.org/10.3390/ijms23105801

Chen J, Wu P, Xia R, Yang J, Huo X, Gu D et al (2018) STAT3-induced lncRNA HAGLROS overexpression contributes to the malignant progression of gastric cancer cells via mTOR signal-mediated inhibition of autophagy. Mol Cancer 17(1):6. https://doi.org/10.1186/s12943-017-0756-y

Dai Q, Zhang T, Pan J, Li C (2020) LncRNA UCA1 promotes cisplatin resistance in gastric cancer via recruiting EZH2 and activating PI3K/AKT pathway. J Cancer 11(13):3882–3892. https://doi.org/10.7150/jca.43446

Dykes IM, Emanueli C (2017) Transcriptional and post-transcriptional gene regulation by long non-coding RNA. Genomics Proteomics Bioinformatics 15(3):177–186. https://doi.org/10.1016/j.gpb.2016.12.005

Eldakhakhny B, Sutaih AM, Siddiqui MA, Aqeeli YM, Awan AZ, Alsayegh MY et al (2024) Exploring the role of noncoding RNAs in cancer diagnosis, prognosis, and precision medicine. Non-coding RNA Res 9(4):1315–1323. https://doi.org/10.1016/j.ncrna.2024.06.015

Emam O, Wasfey EF, Hamdy NM (2022) Notch-associated lncRNAs profiling circuiting epigenetic modification in colorectal cancer. Cancer Cell Int 22(1):316. https://doi.org/10.1186/s12935-022-02736-2

Gao N, Li Y, Li J, Gao Z, Yang Z, Li Y, Liu H, Fan T (2020a) Long non-coding RNAs: the regulatory mechanisms, research strategies, and future directions in cancers. Front Oncol 10:598817. https://doi.org/10.3389/fonc.2020.598817

Gao Y, Wang J-W, Ren J-Y, Guo M, Guo C-W, Ning S-W, Yu S (2020b) Long noncoding RNAs in gastric cancer: from molecular dissection to clinical application. World J Gastroenterol 26(24):3401–3412. https://doi.org/10.3748/wjg.v26.i24.3401

Gao J, Cao B, Zhao R, Li H, Xu Q, Wei B (2023) Critical signaling transduction pathways and intestinal barrier: implications for pathophysiology and therapeutics. Pharmaceuticals 16(9):1216–1216. https://doi.org/10.3390/ph16091216

Ghafouri-Fard S, Glassy MC, Abak A, Hussen BM, Niazi V, Taheri M (2021) The interaction between miRNAs/lncRNAs and notch pathway in human disorders. Biomed Pharmacother 138:111496. https://doi.org/10.1016/j.biopha.2021.111496

Ghasemian A, Omear HA, Mansoori Y, Mansouri P, Deng X, Darbeheshti F, Zarenezhad E, Kohansal M, Pezeshki B, Wang Z, Tang H (2023) Long non-coding RNAs and JAK/STAT signaling pathway regulation in colorectal cancer development. Front Genet 14:1297093. https://doi.org/10.3389/fgene.2023.1297093

Gupta SC, Awasthee N, Rai V, Chava S, Gunda V, Challagundla KB (2020) Long non-coding RNAs and nuclear factor-κB crosstalk in cancer and other human diseases. Biochim Biophys Acta Rev Cancer 1873(1):188316. https://doi.org/10.1016/j.bbcan.2019.188316

Han Y, Zhao G, Shi X, Wang Y, Wen X, Zhang L, Guo X (2022) The emerging role of long non-coding RNAs in esophageal cancer: functions in tumorigenesis and clinical implications. Front Pharmacol 13:885075. https://doi.org/10.3389/fphar.2022.885075

Hashemi M, Mohandesi Khosroshahi E, Asadi S, Tanha M, Ghatei Mohseni F, Abdolmohammad Sagha R et al (2024) Emerging roles of non-coding RNAs in modulating the PI3K/Akt pathway in cancer. Non-coding RNA Res 10:1–15. https://doi.org/10.1016/j.ncrna.2024.08.002

He R-Z, Luo D-X, Mo Y-Y (2019) Emerging roles of lncRNAs in the post-transcriptional regulation in cancer. Gene Dis 6(1):6–15. https://doi.org/10.1016/j.gendis.2019.01.003

He Z, Yang D, Fan X, Zhang M, Li Y, Gu X, Yang M (2020) The roles and mechanisms of lncRNAs in liver fibrosis. Int J Mol Sci 21(4):1482. https://doi.org/10.3390/ijms21041482

Herman AB, Tsitsipatis D, Gorospe M (2022) Integrated lncRNA function upon genomic and epigenomic regulation. Mol Cell 82(12):2252–2266. https://doi.org/10.1016/j.molcel.2022.05.027

Hosseini SA, Haddadi MH, Fathizadeh H, Nemati F, Aznaveh HM, Taraj F, Aghabozorgizadeh AA, Gandomkar G, Bazazzadeh E (2023) Long non-coding RNAs and gastric cancer: an update of potential biomarkers and therapeutic applications. Biomed Pharmacother 163:114407–114407. https://doi.org/10.1016/j.biopha.2023.114407

Hu X, Hou P, Li T, Quan H, Minteer SD, Lin T, Liu J, Zhao Y-P, Zheng J (2018) The roles of Wnt/β-catenin signaling pathway related lncRNAs in cancer. Int J Biol Sci 14(14):2003–2011. https://doi.org/10.7150/ijbs.27977

Hu Y, Lu Y, Fang Y, Zhang Q, Zheng Z, Zheng X, Ye X, Chen Y, Ding J, Yang J (2024) Role of long non-coding RNA in inflammatory bowel disease. Front Immunol 15:1406538. https://doi.org/10.3389/fimmu.2024.1406538

Huang W, Li H, Yu Q, Xiao W, Wang D (2022) LncRNA-mediated DNA methylation: an emerging mechanism in cancer and beyond. J Exp Clin Cancer Res 41(1):100. https://doi.org/10.1186/s13046-022-02319-z

Jabłońska B, Mrowiec S (2023) Gastrointestinal disease: new diagnostic and therapeutic approaches. Biomedicine 11(5):1420–1420. https://doi.org/10.3390/biomedicines11051420

Khan WY, Jabeen F, Khursheed N, Malla BA, Ali A, Ganie SA (2023) Pharmacogenomics in gastroenterology. Elsevier EBooks, pp 271–297. https://doi.org/10.1016/b978-0-443-15336-5.00011-7

Kim Y-A, Park K-K, Lee S-J (2020) LncRNAs act as a Link between chronic liver disease and hepatocellular carcinoma. Int J Mol Sci 21(8):2883. https://doi.org/10.3390/ijms21082883

Kornienko AE, Guenzl PM, Barlow DP, Pauler FM (2013) Gene regulation by the act of long non-coding RNA transcription. BMC Biol 11(1):59. https://doi.org/10.1186/1741-7007-11-59

Kuo C-C, Hänzelmann S, Sentürk Cetin N, Frank S, Zajzon B, Derks J-P et al (2019) Detection of RNA–DNA binding sites in long noncoding RNAs. Nucleic Acids Res 47(6):e32–e32. https://doi.org/10.1093/nar/gkz037

Lashgarian HE, Karkhane M, Marzban A, Yazdi M, Shahzamani K (2020) Emerging involvement of long non-coding RNAs in gastrointestinal associated inflammatory disorders. Comp Immunol Microbiol Infect Dis 69:101428. https://doi.org/10.1016/j.cimid.2020.101428

Lemos AEG, Matos A d R, Ferreira LB, Gimba ERP (2019) The long non-coding RNA PCA3: an update of its functions and clinical applications as a biomarker in prostate cancer. Oncotarget 10(61):6589–6603. https://doi.org/10.18632/oncotarget.27284

Li L, Wei H, Zhang YW, Zhao S, Che G, Wang Y, Chen L (2021) Differential expression of long non-coding RNAs as diagnostic markers for lung cancer and other malignant tumors. Aging (Albany NY) 13(20):23842–23867. https://doi.org/10.18632/aging.203523

Li P, Ma X, Gu X (2024a) The essential roles of lncRNAs/PI3K/AKT axis in gastrointestinal tumors. Front Cell Dev Biol 12:1442193. https://doi.org/10.3389/fcell.2024.1442193

Li P, Ma X, Huang D (2024b) Role of the lncRNA/Wnt signaling pathway in digestive system cancer: a literature review. Eur J Med Res 29(1):447. https://doi.org/10.1186/s40001-024-02033-w

Li Q, Geng S, Luo H, Wang W, Mo Y-Q, Luo Q, Wang L, Song G-B, Sheng J-P, Xu B (2024c) Signaling pathways involved in colorectal cancer: pathogenesis and targeted therapy. Signal Transduct Target Ther 9(1):266. https://doi.org/10.1038/s41392-024-01953-7

Liu P, Zhang M, Niu Q, Zhang F, Yang Y, Jiang X (2018) Knockdown of long non-coding RNA ANRIL inhibits tumorigenesis in human gastric cancer cells via microRNA-99a-mediated down-regulation of BMI1. Braz J Med Biol Res 51(10):e6839. https://doi.org/10.1590/1414-431x20186839

Liu S, Liu X, Li J, Zhou H, Carr MJ, Zhang Z, Shi W (2019) Long noncoding RNAs: novel regulators of virus-host interactions. Rev Med Virol 29(4):e2046. https://doi.org/10.1002/rmv.2046

Liu Y, Ao X, Wang Y, Li X, Wang J (2022) Long non-coding RNA in gastric cancer: mechanisms and clinical implications for drug resistance. Front Oncol 12:841411. https://doi.org/10.3389/fonc.2022.841411

Livesey, G. (2014). Carbohydrate digestion, absorption, and fiber. Reference module in biomedical sciences Elsevier doi:https://doi.org/10.1016/b978-0-12-801238-3.00043-x

Lone SN, Nisar S, Masoodi T, Singh M, Rizwan A, Hashem S et al (2022) Liquid biopsy: a step closer to transform diagnosis, prognosis and future of cancer treatments. Mol Cancer 21(1):79. https://doi.org/10.1186/s12943-022-01543-7

Lou C, Zhao J, Gu Y, Li Q, Tang S, Wu Y, Tang J, Zhang C, Li Z, Zhang Y (2019) LINC01559 accelerates pancreatic cancer cell proliferation and migration through YAP-mediated pathway. J Cell Physiol 235(4):3928–3938. https://doi.org/10.1002/jcp.29288

Lu Q, Liang Y, Meng X, Zhao Y, Fan H, Hou S (2023) The role of long noncoding RNAs in intestinal health and diseases: a focus on the intestinal barrier. Biomolecules 13(11):1674. https://doi.org/10.3390/biom13111674

Ma L, Guo H, Zhao Y, Liu Z, Wang C, Bu J et al (2024) Liquid biopsy in cancer current: status, challenges and future prospects. Signal Transduct Target Ther 9(1):336. https://doi.org/10.1038/s41392-024-02021-w

Majeed Shah I, Ahmad Dar M, Ahmad Bhat K, Ahmad Dar T, Ahmad F, Mudasir Ahmad S (2022) Long non-coding RNAs: biogenesis, mechanism of action and role in different biological and pathological processes. In: Recent advances in non-coding RNAs. https://doi.org/10.5772/intechopen.104861

Mattick JS, Amaral PP, Carninci P, Carpenter S, Chang HY, Chen L-L, Chen R, Dean C, Dinger ME, Fitzgerald KA, Gingeras TR, Guttman M, Hirose T, Huarte M, Johnson R, Kanduri C, Kapranov P, Lawrence JB, Lee JT, Mendell JT (2023) Long non-coding RNAs: definitions, functions, challenges and recommendations. Nat Rev Mol Cell Biol 24(6):430–447. https://doi.org/10.1038/s41580-022-00566-8

Meng X, Bai X, Ke A, Li K, Lei Y, Ding S, Dai D (2024) Long non-coding RNAs in drug resistance of gastric cancer: complex mechanisms and potential clinical applications. Biomolecules 14(6):608. https://doi.org/10.3390/biom14060608

Miyamoto A, Weinmaster G (2009) Notch signal transduction: molecular and cellular mechanisms. Elsevier EBooks, pp 1259–1268. https://doi.org/10.1016/b978-008045046-9.01026-3

Ortega Moreno L, Chaparro M, Gisbert JP (2024) Long non-coding RNAs and their potential role as biomarkers in inflammatory bowel disease. Int J Mol Sci 25(16):8808. https://doi.org/10.3390/ijms25168808

Peng C, Li X, Yu Y, Chen J (2019) LncRNA GASL1 inhibits tumor growth in gastric carcinoma by inactivating the Wnt/β-catenin signaling pathway. Exp Ther Med 17(5):4039–4045. https://doi.org/10.3892/etm.2019.7409

Potemkin N, Clarkson AN (2023) Non-coding RNAs in stroke pathology, diagnostics, and therapeutics. Neurochem Int 162:105467–105467. https://doi.org/10.1016/j.neuint.2022.105467

Pourmehran Y, Sadri F, Hosseini SF, Mohammadi Y, Rezaei Z (2024) Exploring the influence of non-coding RNAs on NF-κB signaling pathway regulation in ulcerative colitis. Biomed Pharmacother 179:117390. https://doi.org/10.1016/j.biopha.2024.117390

Ramya Devi KT, Karthik D, Mahendran T, Jaganathan MK, Hemdev SP (2021) Long noncoding RNAs: role and contribution in pancreatic cancer. Transcription 12(1):12–27. https://doi.org/10.1080/21541264.2021.1922071

Ranjbar R, Ghasemian M, Maniati M, Khatami SH, Jamali N, Taheri-Anganeh M (2022) Gastrointestinal disorder biomarkers. Clin Chim Acta 530:13–26. https://doi.org/10.1016/j.cca.2022.02.013

Riquelme I, Pérez-Moreno P, Mora-Lagos B, Ili C, Brebi P, Roa JC (2023) Long non-coding RNAs (lncRNAs) as regulators of the PI3K/AKT/mTOR pathway in gastric carcinoma. Int J Mol Sci 24(7):6294. https://doi.org/10.3390/ijms24076294

Saini C, Vats P, Maharshi S, Baweja B, Nema R (2025) Involvement of lncRNA in cancer diagnosis and prognosis and clinical implications. Rep Pract Oncol Radiother 30(3):439–450. https://doi.org/10.5603/rpor.106487

Shakhpazyan NK, Mikhaleva LM, Bedzhanyan AL, Sadykhov NK, Midiber K, Konyukova AK, Kontorschikov AS, Maslenkina K, Orekhov AN (2023) Long non-coding RNAs in colorectal cancer: navigating the intersections of immunity, intercellular communication, and therapeutic potential. Biomedicine 11(9):2411–2411. https://doi.org/10.3390/biomedicines11092411

Shi Y, Sun H (2020) Down-regulation of lncRNA LINC00152 suppresses gastric cancer cell migration and invasion through inhibition of the ERK/MAPK signaling pathway. Onco Targets Ther 13:2115–2124. https://doi.org/10.2147/ott.s217452

Shin J-J, Park J, Shin H-S, Arab I, Suk K, Lee W-H (2024) Roles of lncRNAs in NF-κB-mediated macrophage inflammation and their implications in the pathogenesis of human diseases. Int J Mol Sci 25(5):2670–2670. https://doi.org/10.3390/ijms25052670

Siddiqui H, Al-Ghafari A, Choudhry H, Al Doghaither H (2019) Roles of long non-coding RNAs in colorectal cancer tumorigenesis: a review. Mol Clin Oncol 11(2):167–172. https://doi.org/10.3892/mco.2019.1872

Silva JMC, Teixeira EB, Matheus R, Ferraz RS, Moreira FC, Paulo, Calcagno DQ (2023) The landscape of lncRNAs in gastric cancer: from molecular mechanisms to potential clinical applications. Front Pharmacol 14:1237723. https://doi.org/10.3389/fphar.2023.1237723

Statello L, Guo C-J, Chen L-L, Huarte M (2021) Gene regulation by long non-coding RNAs and its biological functions. Nat Rev Mol Cell Biol 22(2):96–118. https://doi.org/10.1038/s41580-020-00315-9

Su M, Xiao Y-H, Ma J, Cao D, Zhou Y, Wang H, Liao Q, Wang W (2018) Long non-coding RNAs in esophageal cancer: molecular mechanisms, functions, and potential applications. J Hematol Oncol 11(1):118. https://doi.org/10.1186/s13045-018-0663-8

Sun Y, Sun F, Jin J, Xu W, Qian H (2023) Exosomal LncRNAs in gastrointestinal cancer: biological functions and emerging clinical applications. Cancer 15(3):959–959. https://doi.org/10.3390/cancers15030959

Sung WJ, Hong J (2023) Targeting lncRNAs of colorectal cancers with natural products. Front Pharmacol 13:1050032. https://doi.org/10.3389/fphar.2022.1050032

Tang PC-T, Zhang Y-Y, Li JS-F, Chan MK-K, Chen J, Tang Y, Zhou Y, Zhang D, Leung K-T, To K-F, Tang SC-W, Lan H-Y, Tang PM-K (2022) LncRNA-dependent mechanisms of transforming growth factor-β: from tissue fibrosis to cancer progression. Non-Coding RNA 8(3):36. https://doi.org/10.3390/ncrna8030036

Tellez-Gabriel M, Heymann D (2019) Exosomal lncRNAs: the newest promising liquid biopsy. Cancer Drug Resist 2(4):1002–1017. https://doi.org/10.20517/cdr.2019.69

Tian S-Q, Shen J-J, Sun D-P, Chen W-M (2025) lncRNA CHAF1B-2 contributes to the tumorigenesis of gastric cancer by activating the Wnt/β-catenin pathway. Sci Rep 15(1):7620. https://doi.org/10.1038/s41598-024-84344-w

Wang Y, Zhou L, Lu J, Jiang B, Liu C, Guo J, Xiao GG (2020) Research progress on long non-coding RNAs and their roles as potential biomarkers for diagnosis and prognosis in pancreatic cancer. Cancer Cell Int 20(1):457. https://doi.org/10.1186/s12935-020-01550-y

Wang B, Yuan C, Qie Y, Dang S (2023) Long non-coding RNAs and pancreatic cancer: a multifaceted view. Biomed Pharmacother 167:115601. https://doi.org/10.1016/j.biopha.2023.115601

Wazir H, Abid M, Essani B, Saeed H, Khan MA, Nasrullah FNU, Qadeer U, Khalid A, Varrassi G, Muzammil MA, Maryam A, Syed ARS, Shah AA, Kinger S, Ullah F, Wazir H, Abid M, Essani B, Saeed H, Khan MA (2023) Diagnosis and treatment of liver disease: current trends and future directions. Cureus 15(12):e49920. https://doi.org/10.7759/cureus.49920

Wei L, Sun J, Zhang N, Zheng Y, Wang X, Lv L, Liu J, Xu Y, Shen Y, Yang M (2020) Noncoding RNAs in gastric cancer: implications for drug resistance. Mol Cancer 19(1):62. https://doi.org/10.1186/s12943-020-01185-7

Weng Y, Ma J, Zhang J, Wang J (2018) Long non-coding RNA LINC01133 silencing exerts antioncogenic effect in pancreatic cancer through the methylation of DKK1 promoter and the activation of Wnt signaling pathway. Cancer Biol Ther 20(3):368–380. https://doi.org/10.1080/15384047.2018.1529110

Xin Y, Li X, Zhang M, Shang Z, Luo Z, Wang Y et al (2023) Fusobacterium nucleatum -induced exosomal HOTTIP promotes gastric cancer progression through the microRNA –885-3p/EphB2 axis. Cancer Sci 114(6):2360–2374. https://doi.org/10.1111/cas.15781

Yan Q, Wong W, Gong L, Yang J, Liang D, Chin K-Y, Dai S, Wang J (2024) Roles of long non-coding RNAs in esophageal cell squamous carcinoma (review). Int J Mol Med 54(2):72. https://doi.org/10.3892/ijmm.2024.5396

Yang C, Chen K (2022) Long non-coding RNA in esophageal cancer: a review of research Progress. Pathol Oncol Res 28:1610140. https://doi.org/10.3389/pore.2022.1610140

Yang X, Duan B, Zhou X (2017) Long non-coding RNA FOXD2-AS1 functions AS a tumor promoter in colorectal cancer by regulating EMT and notch signaling pathway. Eur Rev Med Pharmacol Sci 21(16):3586–3591. https://pubmed.ncbi.nlm.nih.gov/28925486/

Yoon J-H, Abdelmohsen K, Gorospe M (2013) Posttranscriptional gene regulation by long non-coding RNA. J Mol Biol 425(19):3723–3730. https://doi.org/10.1016/j.jmb.2012.11.024

You L, Wu Q, Xin Z, Zhong H, Zhou J, Jiao L, Song X, Ying B (2021) The long non-coding RNA HOXA11-AS activates ITGB3 expression to promote the migration and invasion of gastric cancer by sponging miR-124-3p. Cancer Cell Int 21(1):576. https://doi.org/10.1186/s12935-021-02255-6

Yuan L, Xu Z-Y, Ruan S-M, Mo S, Qin J-J, Cheng X-D (2020) Long non-coding RNAs towards precision medicine in gastric cancer: early diagnosis, treatment, and drug resistance. Mol Cancer 19(1):96. https://doi.org/10.1186/s12943-020-01219-0

Zeng L, Liao Q, Zou Z, Wen Y, Wang J, Liu C et al (2018) Long non-coding RNA XLOC_006753 promotes the development of multidrug resistance in gastric cancer cells through the PI3K/AKT/mTOR signaling pathway. Cell Physiol Biochem 51(3):1221–1236. https://doi.org/10.1159/000495499

Zhang J, Yin M, Peng G, Zhao Y (2018) CRNDE: an important oncogenic long non-coding RNA in human cancers. Cell Prolif 51(3):e12440. https://doi.org/10.1111/cpr.12440

Chapter 9
Long Non-coding RNAs in Kidney Diseases

Fabio Sallustio, Francesca Montenegro, Angela Picerno, Antonella Cicirelli, Giorgio Ladisa, Alessio Nigro, and Loreto Gesualdo

Abstract Long non-coding RNAs (lncRNAs) are increasingly recognized as critical regulators of gene expression in health and disease, with emerging roles in kidney pathophysiology. This chapter provides a comprehensive overview of the involvement of lncRNAs in major renal disorders, including diabetic nephropathy, IgA nephropathy, membranous nephropathy, lupus nephritis, and acute kidney injury. LncRNAs modulate key cellular processes such as inflammation, fibrosis, oxidative stress, apoptosis, and immune responses across different renal compartments and cell types. Specific lncRNAs have also been identified as non-invasive biomarkers in biofluids and potential therapeutic targets. Furthermore, we explore the regulatory functions of lncRNAs in adult renal stem/progenitor cells (ARPCs), emphasizing their role in tissue regeneration, cellular senescence, and repair following injury. The interplay between lncRNAs and signaling pathways in ARPCs suggests novel mechanisms of endogenous kidney repair and therapeutic avenues. Although preclinical evidence is compelling, clinical translation remains limited by species-specific differences and the complexity of lncRNA networks. Continued mechanistic studies and validation in human models are essential to fully harness the diagnostic and therapeutic potential of lncRNAs in nephrology.

Keywords Long non-coding RNAs (lncRNAs) · Kidney disease · Diabetic nephropathy · IgA nephropathy · Lupus nephritis · Acute kidney injury · Membranous nephropathy · Renal progenitor cells · Epigenetic regulation · RNA-based therapeutics

---

F. Sallustio (✉) · F. Montenegro · A. Picerno · L. Gesualdo
Department of Precision and Regenerative Medicine and Ionian Area, University of Bari Aldo Moro, Bari, Italy
e-mail: fabio.sallustio@uniba.it

A. Cicirelli · G. Ladisa
Department of Interdisciplinary Medicine, University of Bari Aldo Moro, Bari, Italy

A. Nigro
PersonGene, University of Bari Spinoff, Bari, Italy

L. M. Vaschetto (ed.), *Long Non-Coding RNAs, Human Health, and Diseases*, Epigenetics and Human Health 16, https://doi.org/10.1007/978-3-032-19668-2_9

Abbreviations

AKI	Acute kidney injury
ARPCs	Adult renal progenitor cells
CKD	Chronic kidney disease
DN	Diabetic nephropathy
ESKD	End-stage kidney disease
Gd-IgA1	Galactose-deficient immunoglobulin A1
IgAN	IgA nephropathy
LN	Lupus nephritis
lncRNA	Long non-coding RNA
MN	Membranous nephropathy
NEAT1	Nuclear-enriched abundant transcript 1
PBMCs	Peripheral blood mononuclear cells
RPTECs	Renal proximal tubular epithelial cells
TUG1	Taurine upregulated gene 1
XIST	X-inactive specific transcript

9.1 Introduction

Long non-coding RNAs (LncRNAs) have emerged as significant players in the pathophysiology of kidney diseases, offering potential avenues for novel diagnostic and therapeutic strategies.

Understanding the intricate roles of lncRNAs in kidney disease can enhance our ability to develop targeted treatments and improve patient outcomes. Ongoing research into lncRNAs may reveal further insights into their regulatory mechanisms, potentially leading to innovative treatment approaches for various renal conditions.

The exploration of lncRNAs in kidney diseases is crucial, as their dysregulation is linked to conditions such as acute kidney injury and chronic kidney disease and may provide new targets for therapeutic intervention in these prevalent disorders. The identification of specific lncRNAs associated with these conditions can change strategies in renal disease management. This may significantly improve treatment efficacy and minimize adverse effects by tailoring therapies to individual genetic profiles in kidney disease patients. This shift toward personalized medicine, focusing on the unique genetic and epigenetic profiles of patients, aligns with the emerging understanding of lncRNAs in kidney disease.

Moreover, the integration of lncRNA profiles into diagnostic frameworks may enable earlier detection and intervention in chronic kidney disease, ultimately reducing morbidity and mortality associated with this condition and improving overall patient quality of life. Additionally, leveraging lncRNA signatures could facilitate the development of novel therapeutic strategies tailored to individual patient needs.

Here we will analyze some findings on lncRNA associated with several kidney diseases, such as diabetic nephropathy, IgA nephropathy, acute kidney injury, membranous nephropathy, and lupus nephritis, considering the mechanisms by which they act and how they can influence these pathologies. In the last part of this chapter, we will consider the lncRNA function in human adult stem and progenitor cells, which are essential for tissue regeneration and play a critical role in tissue development and homeostasis.

9.2 LncRNAs in Diabetic Nephropathy (DN)

Diabetic nephropathy (DN) is a major microvascular complication of both type 1 and type 2 diabetes mellitus, and it represents the leading cause of end-stage kidney disease (ESKD) in Western countries (Navarro-González et al. 2011; Selby et al. 2020). DN is clinically characterized by persistent albuminuria, a progressive decline in renal function, and typical glomerular lesions. It is commonly associated with arterial hypertension and markedly increases the risk of cardiovascular (CV) morbidity and mortality, with estimates suggesting a 20- to 40-fold elevated risk for CV death among diabetic patients (Remuzzi et al. 2006; Selby et al. 2020).

The pathogenesis of DN is multifactorial and involves a complex interplay of mechanisms, including metabolic disturbances induced by chronic hyperglycemia, oxidative stress, inflammation, fibrosis, and endothelial dysfunction (Ke et al. 2025).

Given that environmental factors, such as nutritional status, strongly influence the development and progression of diabetes mellitus and diabetic nephropathy through epigenetic modifications, it is clear that in the complex scenario of DN, both traditional signaling pathways and epigenetic mechanisms, including chromatin modifications, DNA methylation, and ncRNAs, are involved (Kato and Natarajan 2019).

The role of lncRNAs in this context has recently attracted attention, given growing evidence that their dysregulated expression and primary sequence mutations are strongly associated with various human diseases, including cancer, Alzheimer's disease, and psoriasis (Alvarez and DiStefano 2013).

Moreover, the progression of renal damage in diabetic patients involves diverse cellular subtypes and multiple molecular pathways. As a result, numerous studies have focused on different lncRNAs and their effects on various renal cell types, including mesangial cells, podocytes, renal tubular epithelial cells, and glomerular endothelial cells.

Among these, podocytes have been extensively studied for their interaction with lncRNAs. For example, Long et al. demonstrated that the long non-coding RNA Tug1 plays a critical role in regulating peroxisome proliferator–activated receptor γ coactivator α (PGC-1α) transcription and mitochondrial energy metabolism in diabetic nephropathy. In diabetic mice, Tug1 expression is markedly reduced in podocytes, and its podocyte-specific overexpression restores PGC-1α levels, enhances mitochondrial bioenergetics, and alleviates key pathological features of DN (Jianyin et al. 2016).

MALAT1 is an lncRNA upregulated in the kidneys of diabetic mice and in mouse podocytes exposed to high glucose. Its dysregulation contributes to podocyte injury and proteinuria, potentially through a feedback loop with β-catenin, which, if hyper-activated, can contribute to kidney fibrosis (Hu et al. 2017).

With respect to mesangial cells (MCs), Hong et al. discovered that the lincRNA Gm4419 was upregulated in murine MCs under high-glucose conditions, mimicking the DN environment. Gm4419 promotes the expression of proinflammatory cytokines through the NF-κB (p50)/NLRP3 inflammasome inflammatory signaling pathway (Yi et al. 2017).

The lincRNA 1700020I14Rik acts as a protective factor in diabetic nephropathy by regulating the miR-34a-5p/Sirt1/HIF-1α signaling pathway. Its expression is downregulated in high-glucose conditions in db/db DN mice and murine MCs, while overexpression reduces mesangial cell proliferation and fibrosis, suggesting a potential renoprotective effect in DN (Li et al. 2018).

The expression of the lncRNA plasmacytoma variant translocation 1 (PVT1) increases under hyperglycemic conditions in primary human mesangial cells. In these cells, PVT1 promotes extracellular matrix deposition and may contribute to one of the major pathological features of diabetic nephropathy (Alvarez and DiStefano 2011).

Other lncRNAs involved in DN have been studied in renal tubular epithelial and glomerular cells, showing how these molecules may act across different kidney compartments.

The lncRNA MIAT, whose expression is reduced in diabetic rats, correlates with impaired kidney function, as indicated by lower serum creatinine and blood urea nitrogen levels. This reduction in MIAT expression is also observed in high-glucose-treated human renal tubular epithelial cells (HK-2) and is linked to the nuclear factor erythroid 2-related factor 2 (Nrf2) antioxidant response pathway, suggesting a protective role for MIAT in regulating Nrf2 signaling (Zhou et al. 2015).

Sun et al. found that the lncRNA Erbb4-IR promotes TGF-β/Smad3-dependent renal fibrosis in type 2 diabetic nephropathy (T2DN) by suppressing miR-29b in db/db mice. Targeted silencing of Erbb4-IR led to marked improvements in key disease indicators, including reduced microalbuminuria, lower serum creatinine levels, and diminished renal fibrosis, highlighting its potential as a therapeutic target for T2DN (Sun et al. 2018).

Kato et al. demonstrated that a megacluster of approximately 40 microRNAs and their host long non-coding RNA (lnc-MGC) are upregulated in the glomeruli of diabetic nephropathy models both in vivo and in vitro. This lncRNA is regulated by CHOP, a transcription factor linked to endoplasmic reticulum stress. Inhibiting lnc-MGC suppressed the cluster's miRNAs, extracellular matrix accumulation, and glomerular hypertrophy in diabetic mice (Kato et al. 2016).

However, dysregulation of lncRNA expression does not necessarily imply a direct contribution of these molecules to disease pathogenesis. For example, although the lncRNA HOX transcript antisense intergenic RNA (HOTAIR) is upregulated in the kidneys of humans and mice with diabetes, its inactivation in normal mice or in DN mice models does not significantly affect the renal phenotype in a biologically meaningful way (Majumder et al. 2019). While several studies have highlighted potential roles of lncRNAs in the pathogenesis of chronic diseases in vitro, caution is required when assigning pathological functions to individual lncRNAs without robust in vivo validation (Majumder et al. 2019).

Another key challenge in understanding lncRNAs' roles is their low conservation across species, making it difficult to apply murine model findings to humans. However, their functional and structural conservation may be higher than sequence conservation (Lorenzen and Thum 2016).

Lastly, the study of circulating lncRNAs could be promising for the prognosis of patients with diabetic nephropathy. For instance, the expression of lncRNA NR_033515 was found to be significantly elevated in the serum of DN patients, correlating with the different stages of DN. This lncRNA is also positively associated with diagnostic markers of kidney injury, such as KIM-1 and NGAL, and promotes the expression of proteins involved in cell proliferation and fibrosis, accelerating epithelial-mesenchymal transition (EMT) in DN (Gao et al. 2018a).

In summary, the growing body of evidence supporting the role of lncRNAs in the pathogenesis of DN will drive the development of new therapeutic strategies targeting these molecules, which were once considered "junk" RNA produced by aberrant DNA transcription. RNA-based therapies may offer a promising approach for personalized and precision treatment of DN, enabling the manipulation of specific protective and pathogenic lncRNAs based on the patient's genetic profile. However, several challenges remain, with the most pressing one being the need to translate preclinical findings into clinical applications.

9.3 LncRNAs in IgA Nephropathy (IgAN)

IgA nephropathy (IgAN) is the most common form of primary glomerulonephritis worldwide, with the highest prevalence in Asian-Pacific populations, and frequently progresses to end-stage kidney disease (ESKD) (Filippone et al. 2024).

The pathogenesis of IgAN is initiated by the production of aberrantly glycosylated galactose-deficient IgA1 (Gd-IgA1), which leads to the formation of immune complexes with anti-glycan IgG autoantibodies. These pathogenic IgA1-IgG complexes deposit in the glomerular mesangium, promoting inflammation, mesangial cell proliferation, and extracellular matrix expansion (Knoppova et al. 2021). Diagnosis requires renal biopsy and reveals a dominant or codominant IgA (mostly IgA1) immunodeposition pattern in the mesangium.

Long non-coding RNAs (lncRNAs) are increasingly studied in IgAN due to their regulatory functions in immunity, inflammation, and fibrosis—hallmarks of IgAN pathophysiology. It has been shown that 3D chromatin structure enables certain lncRNAs, termed "immune gene-priming lncRNAs," to be positioned near immune genes before activation (Fanucchi et al. 2019). Additionally, lncRNAs modulate transcription factors and fibrotic pathways, and some are released in exosomes or detected in blood, making them ideal candidates as biomarkers.

Microarray analysis on PBMCs from IgAN patients identified 149 differentially expressed lncRNAs interacting with seven mRNAs implicated in innate immunity (KEGG pathway analysis) (Zuo et al. 2017). Notably, lncRNA IL5RA was found to interact with LILRA5—an immunoreceptor involved in leukocyte activation—as well as with C1GALT1, a key enzyme in IgA1 glycosylation, previously associated with IgAN susceptibility (Wang et al. 2013). These findings highlight the role of lncRNAs in modulating innate immunity and IgA1 structure, potentially contributing to disease onset.

In plasma-derived exosomes from IgAN patients and healthy relatives, RNA sequencing revealed downregulation of lncRNA-G21551 in affected individuals (Guo et al. 2020). The nearby coding gene, FCGR3B, encodes a low-affinity IgG receptor involved in immune complex clearance. Polymorphisms in FCGR3B influence IgAN severity (Xu et al. 2007), suggesting that lncRNA-G21551 might impact the clearance of IgA-IgG immune complexes.

LncRNA PTTG3P, overexpressed in renal biopsies and urine of IgAN patients, showed correlation between tissue and urinary levels. Its upregulation in B cells led to increased proliferation, elevated cyclin D1 and Ki-67 expression, and proinflammatory cytokine production (IL-1β, IL-8) (Bi et al. 2021; Kaneko et al. 2019). This indicates that PTTG3P may enhance B-cell activation and inflammation, both of which are central to IgAN.

Fibrosis is a key driver of chronic kidney disease in IgAN. The ICAM-1-related lncRNA (ICR) was associated with fibrotic processes in IgAN biopsies. Inhibition of ICR in TGF-β1-treated HK2 cells reduced fibrosis via suppression of the Akt/mTOR pathway (Wen et al. 2022). Similarly, lnc-TSI was found elevated in IgAN kidney tissue, inversely correlating with fibrosis severity. It may act as a negative regulator of the TGF-β1/Smad pathway (Wang et al. 2018).

In pediatric IgAN patients, lncRNA FGD5-AS1 was shown to act via the FGD5-AS1/miR-196b-5p/PTEN axis. FGD5-AS1 and PTEN were downregulated in serum, while miR-196b-5p was upregulated. Restoration of FGD5-AS1 in vivo reduced glomerular IgA deposition and tissue injury (Li et al. 2021), suggesting therapeutic potential.

A recent GWAS in Chinese patients identified a novel IgAN susceptibility locus at 4p14, involving the immune gene-priming lncRNA LOC101060498 (Li et al. 2024). This lncRNA interacts with the RHOH gene, important for T-cell development and signaling (Ahmad Mokhtar et al. 2021; He et al. 2014). RHOH inhibits Th17 differentiation in autoimmune diseases like psoriasis (Tamehiro et al. 2019), which is relevant since IgAN patients exhibit Th17/Treg imbalance (Lin et al. 2012).

Furthermore, seven transcription factors were shown to be cis-regulated by six differentially expressed lncRNAs in IgAN patients, affecting cell proliferation and apoptosis pathways (Zhai et al. 2024).

These findings demonstrate the multifaceted roles of lncRNAs in IgAN: influencing IgA glycosylation, B-cell activation, T cell differentiation, fibrosis, and inflammation. Their presence in biological fluids such as blood and urine makes them accessible targets for longitudinal monitoring and potential diagnostic or prognostic applications.

For instance, lower serum H19 levels correlated with worse renal outcomes, highlighting its prognostic value (He et al. 2022). Additionally, HOTAIR emerged as a potential biomarker and contributor to IgAN pathogenesis in a multi-layered network analysis integrating gene, TF, miRNA, and lncRNA expression (Gholaminejad et al. 2021).

LncRNAs can also be affected by genetic variants. SNPs in MIR31HG and MIR17HG were associated with increased IgAN risk in Chinese cohorts (Yuan et al. 2020), underlining the genetic regulation of lncRNA function in disease.

Considering all these findings together, there is strong evidence that lncRNAs play critical regulatory roles in IgAN pathogenesis. However, many studies remain observational, and larger, mechanistically focused investigations are needed. Expanding our understanding of lncRNA function may yield novel diagnostic tools and therapeutic strategies for IgAN.

9.4 Long Non-coding RNAs in Membranous Nephropathy

Membranous nephropathy (MN) is an autoimmune glomerular disease characterized by the thickening of glomerular capillary walls due to immune complex deposition. In approximately 80% of cases, no secondary cause is identified, classifying the disease as predominantly primary. The remaining cases are associated with chronic kidney conditions, drug exposure, or infections (Claudio 2023). The identification of the phospholipase A2 receptor (PLA2R) as the first major antigen in MN has paved the way for discovering additional antigens; however, the etiopathogenesis of primary MN remains largely elusive (Ronco et al. 2021).

Recent research has focused on identifying novel diagnostic biomarkers, including microRNAs (miRNAs) and long non-coding RNAs (lncRNAs). Among these, the lncRNA X-inactive specific transcript (XIST) has emerged as a promising candidate (Huang et al. 2014; Jin et al. 2019). Physiologically, XIST is consistently expressed in females, where it coats the X chromosome during the initiation of X chromosome inactivation (Lee and Bartolomei 2013). However, aberrant expression of XIST has also been implicated in pathological conditions, including MN. In a study by Yen-Sung Huang et al., a podocyte model of MN exposed to lipopolysaccharides (LPS) demonstrated significantly increased levels of XIST. Similar upregulation was observed in a corresponding mouse model, and XIST was readily detectable in the urine of both MN mice and human patients. Moreover, these mice exhibited reduced levels of the histone mark H3K27me3, suggesting that epigenetic dysregulation may contribute to the elevated expression of XIST.

Another study by Jin et al. investigated the role of XIST in podocyte injury and found that it interacts with miR-217, a miRNA downregulated in MN. XIST contains a binding site for miR-217 and, upon upregulation, potentially induced by angiotensin II, can sequester miR-217, impairing its function. Since miR-217 acts as a negative regulator of Toll-like receptor 4 (TLR4), its inhibition by XIST leads to increased TLR4 expression, thereby promoting disease progression (Jin et al. 2019).

Another lncRNA, nuclear-enriched abundant transcript 1 (NEAT1), has also been implicated in MN. Pei Pi et al. examined the effects of albumin stimulation in vitro on podocytes and tubular epithelial cells and replicated MN in vivo using a cationic bovine serum albumin model. Unlike in podocytes, NEAT1 was significantly upregulated in tubular cells, where its overexpression was associated with increased apoptosis and cell cycle arrest. The study suggests that NEAT1 may enhance expression of the pro-apoptotic BH3-only protein NOXA (PMAIP1), which in turn suppresses anti-apoptotic proteins, leading to cellular damage (Pi et al. 2021).

These observations underscore the distinct cellular responses in MN between glomerular and tubular compartments. Although emerging data reveal a differential epigenetic landscape between these nephron segments, a comprehensive understanding of the epigenetic networks underlying MN pathogenesis remains incomplete.

From a systems biology perspective, bioinformatics studies comparing gene expression profiles between MN patients and healthy controls offer valuable insight. For example, Guangyu Zhou et al. employed miRNA and lncRNA datasets to identify differentially expressed genes and pathways. Their integrative analysis led to the construction of an mRNA-lncRNA co-expression network and a competing endogenous RNA (ceRNA) network. The study revealed a set of immune-related mRNAs and a ceRNA network comprising four mRNAs, three lncRNAs (LINC00052, LINC00641, N4BP2L2-IT2), and five miRNAs (Zhou et al. 2021). However, limitations such as small sample size and reliance on publicly available datasets highlight the need for experimental validation.

In conclusion, MN is a complex and often cryptogenic disease, particularly in its primary form. Among the lncRNAs emerging as potential biomarkers and therapeutic targets, XIST and NEAT1 stand out for their dual diagnostic and functional relevance. XIST appears to modulate inflammatory pathways, while NEAT1 may contribute to apoptotic processes in tubular cells. These findings underscore the potential of lncRNAs not only as biomarkers but also as novel targets in the therapeutic landscape of MN.

9.5 lncRNAs in Systemic Lupus Erythematosus and Lupus Nephritis

Systemic lupus erythematosus (SLE) is a chronic autoimmune disease characterized by aberrant activation of both B and T lymphocyte compartments. This dysregulated immune response leads to the excessive production of autoantibodies and the subsequent formation of immune complexes, which activate inflammatory cascades, including the complement system. These processes culminate in multisystem clinical manifestations, most notably affecting the kidneys, skin, nervous system, and cardiovascular system (Kiriakidou and Ching 2020; Nagy et al. 2005; Yu et al. 2021).

The global incidence of SLE ranges from approximately 1 to 10 per 100,000 individuals, while its prevalence typically varies between 20 and 70 per 100,000 individuals, depending on geographic and demographic factors (Pons-Estel et al. 2010). T cells in SLE exhibit aberrant expression of the CD3 complex, which is essential for antigen recognition and intracellular signaling (Crispín et al. 2010).

Concurrently, B-cell homeostasis is disrupted, with a reduction in naïve B cells and an expansion of antibody-secreting plasma cells in peripheral blood. This shift enhances autoantibody production and promotes immune complex-mediated tissue injury (Odendahl et al. 2000). Additionally, B cells contribute to disease propagation by presenting autoantigens to T cells, thereby sustaining immune activation (Chan et al. 1999).

Advances in high-throughput sequencing technologies have facilitated the identification of novel non-invasive biomarkers for autoimmune diseases. Among them, long non-coding RNAs (lncRNAs) have emerged as important regulatory molecules in SLE pathogenesis. Their stability in biofluids such as plasma, serum, and urine renders them particularly attractive as diagnostic and prognostic biomarkers (Wang et al. 2012; Zhou et al. 2019). Additionally, urinary miRNAs have also been proposed as sensitive indicators of renal involvement in SLE patients.

Several lncRNAs have been implicated in the immunopathogenesis of SLE, either through transcriptional regulation of inflammatory mediators or modulation of immune cell function. For example, four lncRNAs—lnc0949, lnc0591, lnc1992, and lnc3995—are differentially expressed following innate immune stimulation of THP-1 macrophages and regulate the expression of key proinflammatory cytokines, including TNF-α and IL-6 (Li et al. 2014; Linker-Israeli et al. 1991; McCarthy et al. 2014; Nagafuchi et al. 1993). Another well-characterized lncRNA, GAS5 (growth arrest-specific transcript 5), is typically downregulated in the plasma of SLE patients, and its levels inversely correlate with disease activity. Interestingly, GAS5 is paradoxically overexpressed in CD4+ T cells, suggesting a cell-specific regulatory function in autoimmunity (Li et al. 2017; Suo et al. 2017; Wu et al. 2017; Ye et al. 2019).

Nuclear Enriched Abundant Transcript 1 (NEAT1), originally described in the context of viral infection, modulates innate immunity via activation of the MAPK pathway (Saha et al. 2006; Zhang et al. 2013, 2016). In SLE, NEAT1 expression is upregulated in monocytes and correlates with elevated levels of inflammatory cytokines and chemokines, including IL-6, CCL2, CCL5, and CXCL10 (Li et al. 2010). Its overexpression in renal tissue promotes Th1 cell recruitment, contributing to glomerular injury in lupus nephritis (LN) (Masutani et al. 2001). Additionally, two intronic lncRNAs—AC007278.2 and AC007278.3—modulate the expression of the IL18RAP subunit and are upregulated in SLE patients and LN kidney biopsies (Ihim et al. 2022; Rasuli et al. 2024; Swain 2001; Umare et al. 2019; You et al. 2021).

HOX transcript antisense intergenic RNA (HOTAIR) is known for its epigenetic regulatory functions. In SLE, HOTAIR activates the NF-κB pathway by promoting

IκBα degradation, leading to the transcription of inflammatory cytokines such as TNF-α, IL-6, and IL-18 (Kong et al. 2022; Obaid et al. 2018; Paniri and Akhavan-Niaki 2020; Wang et al. 2021). Elevated HOTAIR expression has been reported in several autoimmune diseases and is positively associated with markers of inflammation, nephritis severity, and proteinuria in LN (Mahmoud et al. 2022; Medhat et al. 2022). However, the correlation between HOTAIR and TNF-α in LN remains inconclusive, and further studies are required to validate its use as a biomarker.

Metastasis-Associated Lung Adenocarcinoma Transcript 1 (MALAT-1), also known as NEAT2, is another lncRNA significantly upregulated in SLE patients, particularly in monocytes (Fang et al. 2016; Yang et al. 2017). It regulates the expression of IL-21, a cytokine involved in disease activity, and may serve as a biomarker to differentiate between SLE and LN subtypes (Shbeer and Ahmed Robadi 2024). MALAT-1 expression correlates positively with serum creatinine and negatively with anti-dsDNA levels, supporting its relevance as a marker of renal involvement (Saleh et al. 2020).

Taurine Upregulated Gene 1 (TUG1), initially linked to cancer biology, is downregulated in PBMCs of SLE patients, especially in those with LN. Monocytes exhibit the most significant reduction, and their expression correlates inversely with disease severity. These findings suggest that TUG1 may serve as a non-invasive biomarker capable of distinguishing LN from non-renal SLE manifestations (Cao et al. 2020; Ji et al. 2016; Li et al. 2016a, b; Song et al. 2015).

In conclusion, a growing body of evidence supports the role of lncRNAs in the pathogenesis of SLE and LN. These molecules influence key immunological processes, including cytokine production, immune cell activation, and renal inflammation. Due to their stability in biological fluids and disease-specific expression patterns, lncRNAs represent promising biomarkers for disease stratification and monitoring. However, further validation studies are necessary to confirm their clinical utility and to facilitate their integration into routine practice.

9.6 Long Non-coding RNAs in Acute Kidney Injury

Acute kidney injury (AKI) is a critical clinical condition characterized by a rapid decline in renal function, resulting in the accumulation of nitrogenous waste products such as urea and creatinine, disturbances in acid-base balance, and fluid and electrolyte dysregulation (Haase-Fielitz et al. 2023). Globally, AKI affects approximately 10 million individuals and poses a significant burden, particularly in hospitalized and critically ill patients. Its prevalence in intensive care units (ICUs) reaches up to 35%, with associated mortality rates ranging from 37% to 60% (Chawla et al. 2017; Coca et al. 2012).

The predominant causes of AKI include sepsis, septic shock, ischemia-reperfusion injury, and exposure to nephrotoxic agents. Sepsis-related AKI accounts for over 50% of all AKI cases, with mortality ranging between 20.9% and 56.8%, depending on the severity (Bilgili et al. 2014; Stasi et al. 2021). This increasing incidence has elevated AKI to a major public health concern. Prompt diagnosis and a deeper understanding of the underlying molecular mechanisms are essential for improving patient outcomes.

Currently, AKI diagnosis relies heavily on serum creatinine concentration and urine output, both of which are late markers of renal dysfunction. Consequently, significant efforts are underway to identify novel, early biomarkers. Approximately 5–6% of critically ill patients require renal replacement therapy (RRT), and many survivors face a heightened risk of progression to chronic kidney disease (CKD) or end-stage renal disease (ESRD) (Chawla et al. 2017; Coca et al. 2012).

Glomerulonephritis contributes to nearly 10% of adult AKI cases, often manifesting as rapidly progressive glomerulonephritis (RPGN), with etiologies including anti-glomerular basement membrane (anti-GBM) disease, small-vessel vasculitis, IgA nephropathy, thrombotic microangiopathies, and lupus nephritis. Additionally, glomerulonephritis-related AKI may arise from overlapping injuries such as acute tubular necrosis secondary to hypoperfusion, nephrotic syndrome, or drug-induced tubular toxicity (Pesce et al. 2021).

Emerging evidence indicates that long non-coding RNAs (lncRNAs) are significantly dysregulated in AKI and play crucial roles in its pathogenesis (Ma et al. 2021). A comprehensive microarray analysis revealed 5361 upregulated and 5928 downregulated lncRNAs in patients with septicemia-induced AKI (Giannuzzi et al. 2023), highlighting the widespread transcriptional alterations associated with this condition.

Among the most studied lncRNAs in AKI is HOX transcript antisense RNA (HOTAIR). Originally recognized for its role in chromatin remodeling, HOTAIR is also highly expressed in renal progenitor cells, where it regulates cell cycle dynamics, cellular senescence, and the secretion of anti-aging proteins such as α-Klotho (Giannuzzi et al. 2023; Picerno et al. 2022). Recent studies demonstrated its involvement in septic AKI: in rodent models, overexpression of HOTAIR significantly reduced proinflammatory cytokines (TNF-α and IL-1β), downregulated miR-34a, and upregulated the anti-apoptotic factor Bcl-2. This supports a protective function for HOTAIR in AKI via modulation of the miR-34a/Bcl-2 axis (Jiang et al. 2019b).

In addition to HOTAIR, several other lncRNAs have been implicated in AKI through the modulation of inflammation and apoptosis. Notably, Metastasis-Associated Lung Adenocarcinoma Transcript 1 (MALAT1) exacerbates inflammation and tissue damage in AKI, especially in lipopolysaccharide (LPS)-induced models. MALAT1 exerts its effects via the MALAT1/miR-146a/NF-κB axis, where its upregulation suppresses miR-146a, thereby enhancing NF-κB signaling and proinflammatory cytokine production (Ding et al. 2018). Silencing MALAT1 leads to increased miR-146a levels and reduced expression of proinflammatory mediators, highlighting its potential as a therapeutic target (Lu et al. 2021; Tian et al. 2018). Hypoxia-inducible factor 1-α (HIF-1α) triggered the increased expression of MALAT1, which in turn inhibited the expression of IL-6, TNF-α, and NF-kB (Tian et al. 2018).

Another key lncRNA, Maternally Expressed Gene 3 (MEG3), contributes to AKI pathogenesis by regulating TNF-α expression via the miR-181b-5p axis. In hypoxia-induced renal injury models, inhibition of MEG3 reduced inflammation and apoptosis, suggesting its potential as a therapeutic target (Lu et al. 2021).

In contrast to the deleterious roles of MALAT1 and MEG3, the lncRNA Taurine-Upregulated Gene 1 (TUG1) appears to confer protection in AKI. Overexpression of TUG1 in LPS-treated HK-2 cells mitigated inflammation and apoptosis through modulation of the NF-κB pathway and activation of the Nrf2 antioxidant response pathway (Duan et al. 2017; M'baya-Moutoula et al. 2018; Wang et al. 2020). These findings suggest that TUG1 may serve as a promising therapeutic candidate for sepsis-induced AKI.

The lncRNA HOXA cluster antisense RNA 2 (HOXA-AS2) also plays a protective role in sepsis-induced acute kidney injury (AKI). In both animal and cell models, it has been observed that the expression levels of HOXA-AS2 were reduced under septic conditions, while its overexpression attenuated lipopolysaccharide (LPS)-induced damage (Wu et al. 2020). Moreover, it has been found that HOXA-AS2 is negatively correlated with miR-106b-5p, which counteracts the protective effect of the lncRNA. Therefore, HOXA-AS2 exhibits a protective effect in sepsis-induced AKI by counteracting miR-106b-5p and hindering the Wnt/β-catenin and NF-κB signaling pathways (Wu et al. 2020).

Another lncRNA deeply implicated in AKI is Nuclear Enriched Abundant Transcript 1 (NEAT1). NEAT1 promotes kidney injury by activating the NLRP3 inflammasome and upregulating the expression of proinflammatory cytokines, such as IL-1β and IL-18 (Xue et al. 2024). Recent studies have shown that NEAT1 is significantly upregulated in both sepsis-induced AKI and ischemia-reperfusion injury (IRI) models. In these contexts, NEAT1 silencing improved cell viability, reduced apoptosis, and downregulated the inflammatory response.

Of particular interest is the NEAT1_2 variant, which was found to be upregulated in renal tubular cells following IRI. NEAT1_2 promotes apoptosis by sequestering miR-129-5p, preventing it from repressing key apoptotic genes like FADD, CASP-8, and CASP-3 (Ma et al. 2022). Furthermore, in LPS-induced AKI models, NEAT1 acts as a molecular sponge for miR-22-3p, and its inhibition reverses inflammatory damage by enhancing miR-22-3p activity. NEAT1 also represses miR-27a-3p, thereby exacerbating apoptosis in tubular epithelial cells (Jiang et al. 2019a; Zhou et al. 2022).

Growth Arrest Specific 5 (GAS5) is another lncRNA involved in IRI-induced AKI. Its expression is elevated following IRI and is suppressed by ischemic preconditioning. GAS5 acts by sponging miR-21, a microRNA that inhibits thrombospondin-1 (TSP-1), a mediator of tissue injury and fibrosis. This GAS5/miR-21/TSP-1 axis represents a novel mechanism by which GAS5 promotes apoptosis and inflammation in AKI (Geng et al. 2020).

Additional lncRNAs associated with AKI include TCONS_00016406 (also known as lncRNA6406), which reduces oxidative stress and apoptosis via the miR-687/PTEN axis in LPS-induced AKI (Liu et al. 2020; Yang et al. 2022), and

XLOC-032768, which attenuates cisplatin-induced AKI by decreasing apoptosis and TNF-mediated inflammation (Zhou et al. 2020).

Similarly, LRNA9884 is upregulated in AKI and promotes the transcriptional activation of proinflammatory cytokines through the NF-κB signaling pathway (Zhang et al. 2020). Another notable lncRNA, XIST, was found to enhance apoptosis and inflammation in IRI models by modulating the miR-124-3p/ITGB1 axis (Chen et al. 2020).

Psoriasis Susceptibility-Related RNA Gene Induced by Stress (PRINS) also contributes to IRI-induced AKI by upregulating the chemokine RANTES, which facilitates leukocyte infiltration and tissue damage, thereby aggravating renal dysfunction (Yu et al. 2016).

In conclusion, lncRNAs regulate a diverse array of molecular pathways implicated in inflammation, oxidative stress, fibrosis, and apoptosis in AKI. As such, they represent not only potential biomarkers for early diagnosis but also promising targets for novel therapeutic strategies. Although the field remains in its early stages, continued research into the functional roles of lncRNAs is crucial for translating these findings into clinical applications.

9.7 The Impact of LncRNAs on Stem Cells in Renal Diseases

Human adult stem and progenitor cells are essential for tissue regeneration and play a critical role in tissue development and homeostasis. Their function is regulated by a complex network of environmental signals, chemical mediators, and epigenetic mechanisms. Among these regulators, long non-coding RNAs (lncRNAs) have emerged as key modulators of stem cell activity.

LncRNAs not only contribute to the maintenance of stemness but also mediate the response of tissue-resident stem/progenitor cells to injury. For instance, using transcriptome sequencing and tiling arrays, the lncRNA ANCR was identified as a crucial regulator of gene expression required to maintain the undifferentiated state of epidermal progenitor cells.

In the context of osteogenesis, lncRNAs such as MALAT1, NEAT1, DANCR, SNHG1, MIR22HG, and LINC00314 influence stem cell differentiation into osteoblasts, with implications in osteoporosis. Their effects are primarily mediated through key signaling pathways, including PTEN/AKT, MAPK, and STAT (Cui et al. 2019b; Jin et al. 2020; Zheng et al. 2019). NEAT1 plays a pivotal role in neuronal differentiation, apoptosis, and progenitor cell migration by modulating the Wnt/β-catenin pathway through the activation of miR-124 (Cui et al. 2019a). Similarly, lncRNA AK137033 inhibits the osteogenic differentiation potential of adipose-derived stem cells in diabetic osteoporosis by epigenetically regulating the Wnt pathway via DNA methylation of the sFRP2 promoter (Zhang et al. 2018) (Fig. 9.1).

Stem cells' self-renewal and differentiation potential can, in certain contexts, be hijacked toward tumorigenesis. Regulatory networks involving transcription factors

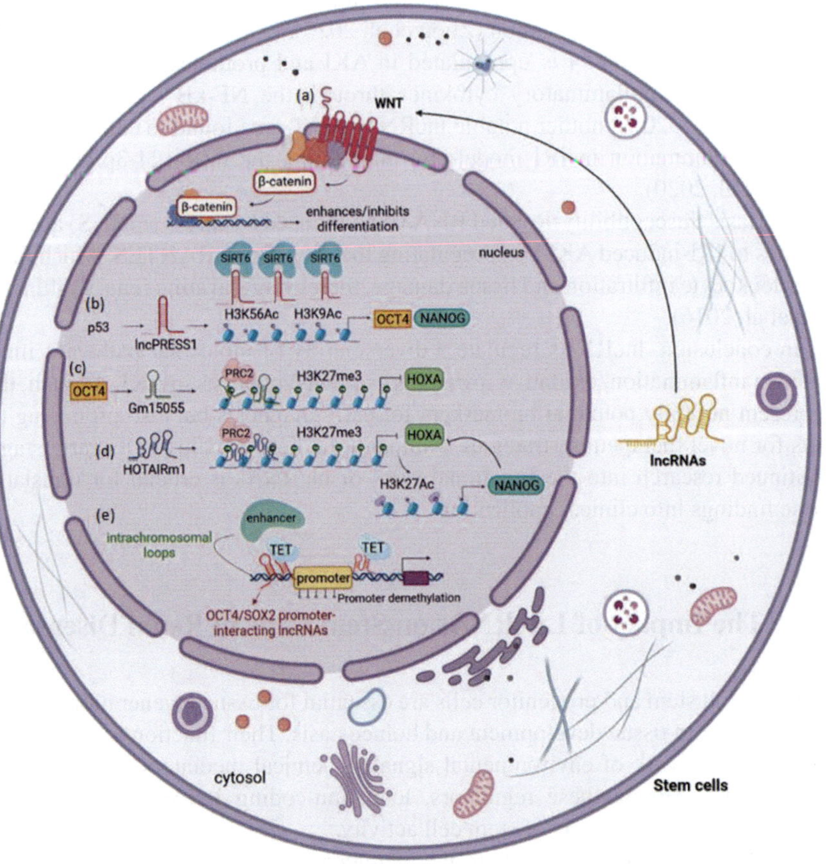

Fig. 9.1 Regulatory mechanisms of lncRNAs in stem cells. (**a**) Several lncRNAs interact in the canonical Wnt signaling cascade and influence the expression of genes by inducing or inhibiting cell differentiation. (**b**) The p53-regulated long non-coding PRESS1 physically interacts with SIRT6 and inhibits its attachment to chromatin, controls a gene network that promotes the pluripotency of hESCs (human embryonic stem cells) by maintaining high levels of histone H3K56 and H3K9 acetylation in the promoters of pluripotency genes such as OCT4 and NANOG. (**c**) The lncRNA Gm15055, whose expression is influenced by OCT4, represses Hoxa gene expression by recruiting PRC2 to the cluster and maintaining the H3k27me3 modification on HoxA promoters in mESCs (mouse embryonic stem cells); (**d**) The nuclear lncRNA HOTAIRm1 also regulates HoxA expres-sion leading to trimethylation of histone H3K27 and epigenetic silencing of the gene; furthermore, it promotes the acetylation of H3K27 in the enhancer site of the NANOG gene by upregulating its expression and in turn inhibiting HoxA, creating a reciprocal regulatory loop which increases the stemmating effect. (**e**) Some lncRNAs enhance epigenetic reprogramming by coordinating intrachromosomal looping and by recruiting the TET2 demethylase, promoting DNA demethylation at the OCT4 and SOX2 promoters. (From Giannuzzi et al. (2023) under Creative Commons Attribution (CC BY) license 4.0)

like SOX2 and KLF4, along with polycomb group proteins, are often dysregulated in cancer stem cells (CSCs) and involve feedback loops with specific lncRNAs (He et al. 2021; Herreros-Villanueva et al. 2013). For instance, HOTAIR is overexpressed in breast cancer and is a marker of metastatic potential (Gupta et al. 2010). Further research into lncRNAs in CSC biology may yield promising clinical biomarkers and therapeutic targets.

9.8 LncRNAs in Human Adult Renal Stem/Progenitor Cells

While the role of lncRNAs in renal pathophysiology is increasingly recognized, their function in the healthy kidney—particularly in adult renal stem/progenitor cells (ARPCs)—remains largely unexplored. ARPCs represent a highly promising population for developing novel treatments for acute and chronic kidney injury. These cells, isolated from both tubules and glomeruli, share morphological and transcriptional features, though they also exhibit distinct profiles (Angelotti et al. 2012; Bruno et al. 2009; Lazzeri et al. 2018; Sallustio et al. 2010, 2017). Due to their multipotent nature, ARPCs can differentiate into adipocytes, osteoblasts, and tubular epithelial-like cells. Notably, CD133 + CD24+ ARPCs have been shown to support tubular regeneration in models of glycerol-induced acute kidney injury, reducing damage and accelerating recovery (Sagrinati et al. 2006).

They respond to nephrotoxic insults such as cisplatin. Upon TLR2 activation, ARPCs release inhibin-A and decorin, both directly and via microvesicles, promoting regeneration of renal proximal tubular epithelial cells (RPTECs) (Franzin et al. 2023; Sallustio et al. 2017). Moreover, ARPCs modulate immune homeostasis by activating regulatory T cells (Tregs) and modifying double-negative T cells (T_DN), essential for immune tolerance (Curci et al. 2021). ARPCs also contribute to renal repair by modulating inflammation and immune responses. For instance, after lipopolysaccharide (LPS) exposure, ARPCs secrete antimicrobial molecules such as CXCL6, SAA4, and BPIFA2, which block LPS-induced endothelial-to-mesenchymal transition (Sallustio et al. 2019).

However, these regenerative properties may be compromised by renal senescence, which leads to progenitor dysfunction and impaired tissue repair (Picerno et al. 2021, 2022). Given the growing evidence that lncRNAs are central players in cellular reprogramming, understanding their role in ARPC biology is critical.

LncRNAs role in renal pathogenesis and stem cell biology is increasingly recognized, with specific lncRNAs like HOTAIR and TUG1 being highlighted for their impact on renal stem/progenitor cells and kidney injury. The lncRNA TUG1, carried by exosomes derived from human urine-derived stem cells, plausibly renal progenitor cells, has shown protective effects against renal ischemia/reperfusion injury. It regulates ferroptosis, a form of cell death, by interacting with the RNA-binding protein SRSF1 and stabilizing ACSL4 mRNA. This interaction helps ameliorate kidney injury and suggests a therapeutic potential for TUG1-rich exosomes in treating AKI (Sun et al. 2022).

A recent genome-wide lncRNA expression screen in ARPCs identified 611 differentially expressed lncRNAs distinguishing ARPCs from RPTECs. Pathway analysis revealed that several ARPC-specific lncRNAs regulate cell cycle progression. Among these, HOTAIR was found to be a central regulator. Using CRISPR/Cas9-mediated gene editing, researchers created HOTAIR knockout ARPC lines and demonstrated their essential role in preventing apoptosis and sustaining self-renewal. HOTAIR supports ARPC proliferation and longevity by modulating the expression of the CD133 stemness marker and promoting the secretion of α-Klotho, an anti-aging protein with paracrine effects on renal tissues (Picerno et al. 2022) (Fig. 9.2). Notably, α-Klotho deficiency in mice is associated with aging-related phenotypes, which may result from stem cell

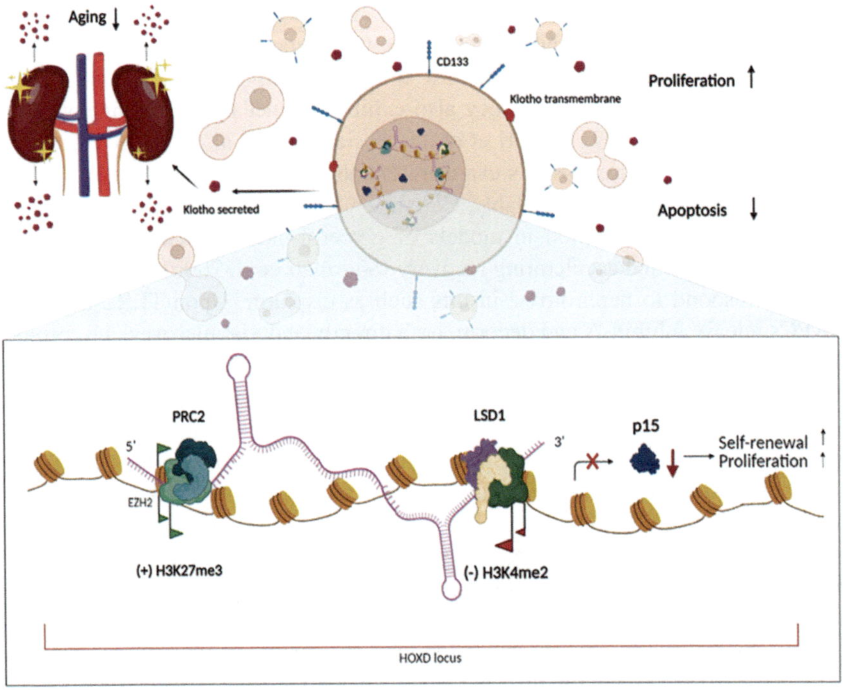

Fig. 9.2 HOTAIR regulates ARPC stemness properties. HOTAIR regulates the proliferation and self-renewal capacity of ARPCs, limits their apoptosis, and regulates the CD133 expression, an inverse marker of senescence and functional marker of stemness of ARPCs. HOTAIR also regulates the secretion of α-Klotho, which in turn can influence the surrounding tissues and therefore modulate tissue aging, playing an important role in preventing cells from becoming senescent in the short term. These mechanisms are regulated through the p15 epigenetic silencing by HOTAIR. This lncRNA acts as a molecular scaffold to link the PRC2 and LSD1 protein complex and coordinates the chromatin targeting of these proteins. The complex leads to histone H3K27 trimethylation and H3K4 demethylation in p15 gene promoter. Trimethylation of H3K27 leads to the silencing of the cyclin p15, and to keep the ARPC cell cycle active, sustaining the self-renewal and proliferation. (From Picerno et al. (2022: 11), by permission of Oxford University Press)

depletion or their differentiation toward fibrotic lineages (Bian et al. 2015). HOTAIR also represses the expression of the cell cycle inhibitor p15 by promoting trimethylation of histone H3K27 (H3K27me3) at the p15 promoter region. This epigenetic modification keeps ARPCs cycling and avoids premature senescence (Picerno et al. 2022).

Similar mechanisms are used by leukemia stem cells (Gao et al. 2018b) and by epidermal stem cells during re-epithelialization of injured tissue, where HOTAIR expression is upregulated to sustain their regenerative function (Shi et al. 2020). The interplay between lncRNAs and stem cells is detailed in Chap. 3 by Perez-Campos et al.

9.9 Conclusions

Long non-coding RNAs (lncRNAs) have emerged as pivotal regulators in the pathogenesis and progression of various kidney diseases, acting through diverse molecular pathways that govern inflammation, apoptosis, fibrosis, oxidative stress, immune modulation, and cellular metabolism. The dysregulation of specific lncRNAs has been consistently observed in conditions such as diabetic nephropathy, IgA nephropathy, membranous nephropathy, lupus nephritis, and acute kidney injury, underscoring their relevance as potential diagnostic and prognostic biomarkers, as well as therapeutic targets. Furthermore, lncRNAs exert significant control over the functional properties of human adult renal stem/progenitor cells (ARPCs), influencing their self-renewal, differentiation, and regenerative capacities—functions that are crucial for renal repair mechanisms following injury. Notably, exosome-derived lncRNAs have demonstrated promising roles in intercellular communication and renoprotection, particularly in the context of ischemia-reperfusion injury.

Despite compelling preclinical data, several challenges hinder the clinical translation of lncRNA-based strategies. These include limited conservation across species, the complexity of their regulatory networks, and the need for robust in vivo validation and mechanistic elucidation. Future research should prioritize the integration of high-throughput transcriptomic profiling with functional genomics and gene editing approaches to define causal relationships and therapeutic potential. Ultimately, the incorporation of lncRNA signatures into precision nephrology frameworks may revolutionize the early diagnosis, risk stratification, and treatment of renal diseases, paving the way for RNA-based personalized medicine.Compliance with Ethical Standards

Conflict of Interest The authors declare that they have no conflict of interest.

Ethical Approval This chapter is a review of previously published accounts; as such, no animal or human studies were performed.

References

Ahmad Mokhtar AM, Hashim IF, Mohd Zaini Makhtar M, Salikin NH, Amin-Nordin S (2021) The role of RhoH in TCR signalling and its involvement in diseases. Cells 10(4):950. https://doi.org/10.3390/cells10040950

Alvarez ML, DiStefano JK (2011) Functional characterization of the plasmacytoma variant translocation 1 gene (PVT1) in diabetic nephropathy. PLoS One 6(4):e18671. https://doi.org/10.1371/JOURNAL.PONE.0018671

Alvarez ML, DiStefano JK (2013) The role of non-coding RNAs in diabetic nephropathy: potential applications as biomarkers for disease development and progression. Diabetes Res Clin Pract 99(1):1–11. https://doi.org/10.1016/J.DIABRES.2012.10.010

Angelotti ML, Ronconi E, Ballerini L, Peired A, Mazzinghi B, Sagrinati C, Parente E, Gacci M, Carini M, Rotondi M, Fogo AB, Lazzeri E, Lasagni L, Romagnani P (2012) Characterization of renal progenitors committed toward tubular lineage and their regenerative potential in renal tubular injury. Stem Cells 30(8):1714–1725. https://doi.org/10.1002/stem.1130

Bi M, Shi J, Zhao Y, Li C (2021) LncRNA PTTG3P induced aberrant glycosylated IgA1 production and B cell growth in IgA nephropathy. Environ Sci Pollut Res 28(40):56606–56614. https://doi.org/10.1007/s11356-021-13335-5

Bian A, Neyra JA, Zhan M, Hu MC (2015) Klotho, stem cells, and aging. Clin Interv Aging 10:1233. https://doi.org/10.2147/CIA.S84978

Bilgili B, Haliloğlu M, Cinel İ (2014) Sepsis and acute kidney injury. Turk J Anaesthesiol Reanim 42(6):294. https://doi.org/10.5152/TJAR.2014.83436

Bruno S, Bussolati B, Grange C, Collino F, Di Cantogno LV, Herrera MB, Biancone L, Tetta C, Segoloni G, Camussi G (2009) Isolation and characterization of resident mesenchymal stem cells in human glomeruli. Stem Cells Dev 18(6):867–880. https://doi.org/10.1089/scd.2008.0320

Cao HY, Li D, Wang YP, Lu HX, Sun J, Li HB (2020) Clinical significance of reduced expression of lncRNA TUG1 in the peripheral blood of systemic lupus erythematosus patients. Int J Rheum Dis 23(3):428–434. https://doi.org/10.1111/1756-185X.13786

Chan OTM, Hannum LG, Haberman AM, Madaio MP, Shlomchik MJ (1999) A novel mouse with B cells but lacking serum antibody reveals an antibody-independent role for B cells in murine lupus. J Exp Med 189(10):1639–1647. https://doi.org/10.1084/JEM.189.10.1639

Chawla LS, Bellomo R, Bihorac A, Goldstein SL, Siew ED, Bagshaw SM, Bittleman D, Cruz D, Endre Z, Fitzgerald RL, Forni L, Kane-Gill SL, Hoste E, Koyner J, Liu KD, MacEdo E, Mehta R, Murray P, Nadim M et al (2017) Acute kidney disease and renal recovery: consensus report of the acute disease quality initiative (ADQI) 16 workgroup. Nat Rev Nephrol 13(4):241–257. https://doi.org/10.1038/nrneph.2017.2

Chen F, Hu Y, Xie Y, Zhao Z, Ma L, Li Z, Tan W (2020) Total glucosides of paeony alleviate cell apoptosis and inflammation by targeting the long noncoding RNA XIST/MicroRNA-124-3p/ITGB1 Axis in renal ischemia/reperfusion injury. Mediat Inflamm 2020(1):8869511. https://doi.org/10.1155/2020/8869511

Claudio P (2023) Primary membranous nephropathy: an endless story. J Nephrol 36(2):563–574. https://doi.org/10.1007/S40620-022-01461-3/TABLES/1

Coca SG, Singanamala S, Parikh CR (2012) Chronic kidney disease after acute kidney injury: a systematic review and meta-analysis. Kidney Int 81(5):442–448. https://doi.org/10.1038/KI.2011.379

Crispín JC, Liossis S-NC, Kis-Toth K, Lieberman LA, Kyttaris VC, Juang Y-T, Tsokos GC (2010) Pathogenesis of human systemic lupus erythematosus: recent advances. Trends Mol Med 16(2):47–57. https://doi.org/10.1016/j.molmed.2009.12.005

Cui Y et al (2019a) LncRNA Neat1 regulates neuronal differentiation via the Wnt/β-catenin pathway. Neural Regen Res 14(3):430–436. https://doi.org/10.4103/1673-5374.245466

Cui Y, Yin Y, Xiao Z, Zhao Y, Chen B, Yang B, Xu B, Song H, Zou Y, Ma X, Dai J (2019b) LncRNA Neat1 mediates miR-124-induced activation of Wnt/β-catenin signaling in spinal cord neural progenitor cells. Stem Cell Res Ther 10(1):400. https://doi.org/10.1186/S13287-019-1487-3

Curci C, Picerno A, Chaoul N, Stasi A, De Palma G, Franzin R, Pontrelli P, Castellano G, Pertosa GB, Macchia L, Di Lorenzo VF, Sabbà C, Gallone A, Gesualdo L, Sallustio F (2021) Adult renal stem/progenitor cells can modulate t regulatory cells and double negative t cells. Int J Mol Sci 22(1):1–16. https://doi.org/10.3390/ijms22010274

Ding Y, Guo F, Zhu T, Li J, Gu D, Jiang W, Lu Y, Zhou D (2018) Mechanism of long non-coding RNA MALAT1 in lipopolysaccharide-induced acute kidney injury is mediated by the miR-146a/NF-κB signaling pathway. Int J Mol Med 41(1):446–454. https://doi.org/10.3892/IJMM.2017.3232

Duan LJ, Ding M, Hou LJ, Cui YT, Li CJ, Yu DM (2017) Long noncoding RNA TUG1 alleviates extracellular matrix accumulation via mediating microRNA-377 targeting of PPARγ in diabetic nephropathy. Biochem Biophys Res Commun 484(3):598–604. https://doi.org/10.1016/J.BBRC.2017.01.145

Fang Z, Zhang S, Wang Y, Shen S, Wang F, Hao Y, Li Y, Zhang B, Zhou Y, Yang H (2016) Long non-coding RNA MALAT-1 modulates metastatic potential of tongue squamous cell carcinomas partially through the regulation of small proline rich proteins. BMC Cancer 16(1):706. https://doi.org/10.1186/S12885-016-2735-X

Fanucchi S, Fok ET, Dalla E, Shibayama Y, Börner K, Chang EY, Stoychev S, Imakaev M, Grimm D, Wang KC, Li G, Sung W-K, Mhlanga MM (2019) Immune genes are primed for robust transcription by proximal long noncoding RNAs located in nuclear compartments. Nat Genet 51(1):138–150. https://doi.org/10.1038/s41588-018-0298-2

Filippone EJ, Gulati R, Farber JL (2024) Contemporary review of IgA nephropathy. Front Immunol 15:1436923. https://doi.org/10.3389/fimmu.2024.1436923

Franzin R, Stasi A, De Palma G, Picerno A, Curci C, Sebastiano S, Campioni M, Cicirelli A, Rizzo A, Di Lorenzo VF, Pontrelli P, Pertosa GB, Castellano G, Gesualdo L, Sallustio F (2023) Human adult renal progenitor cells prevent cisplatin-nephrotoxicity by inducing CYP1B1 overexpression and miR-27b-3p Down-regulation through extracellular vesicles. Cells 12(12):1655. https://doi.org/10.3390/cells12121655

Gao J, Wang W, Wang F, Guo C (2018a) LncRNA-NR_033515 promotes proliferation, fibrogenesis and epithelial-to-mesenchymal transition by targeting miR-743b-5p in diabetic nephropathy. Biomed Pharmacother 106:543–552. https://doi.org/10.1016/J.BIOPHA.2018.06.104

Gao S, Zhou B, Li H, Huang X, Wu Y, Xing C, Yu X, Ji Y (2018b) Long noncoding RNA HOTAIR promotes the self-renewal of leukemia stem cells through epigenetic silencing of p15. Exp Hematol 67:32–40.e3. https://doi.org/10.1016/J.EXPHEM.2018.08.005

Geng X, Song N, Zhao S, Xu J, Liu Y, Fang Y, Liang M, Xu X, Ding X (2020) LncRNA GAS5 promotes apoptosis as a competing endogenous RNA for miR-21 via thrombospondin 1 in ischemic AKI. Cell Death Discov 6(1):19. https://doi.org/10.1038/S41420-020-0253-8

Gholaminejad A, Gheisari Y, Jalali S, Roointan A (2021) Comprehensive analysis of IgA nephropathy expression profiles: identification of potential biomarkers and therapeutic agents. BMC Nephrol 22(1):137. https://doi.org/10.1186/s12882-021-02356-4

Giannuzzi F, Maiullari S, Gesualdo L, Sallustio F (2023) The Mission of long non-coding RNAs in human adult renal stem/progenitor cells and renal diseases. Cells 12(8):1115. https://doi.org/10.3390/CELLS12081115

Guo N, Zhou Q, Huang X, Yu J, Han Q, Nong B, Xiong Y, Liang P, Li J, Feng M, Lv J, Yang Q (2020) Identification of differentially expressed circulating exosomal lncRNAs in IgA nephropathy patients. BMC Immunol 21(1):16. https://doi.org/10.1186/s12865-020-00344-1

Gupta RA, Shah N, Wang KC, Kim J, Horlings HM, Wong DJ, Tsai MC, Hung T, Argani P, Rinn JL, Wang Y, Brzoska P, Kong B, Li R, West RB, Van De Vijver MJ, Sukumar S, Chang HY (2010) Long non-coding RNA HOTAIR reprograms chromatin state to promote cancer metastasis. Nature 464(7291):1071–1076. https://doi.org/10.1038/nature08975

Haase-Fielitz A, Ann MacMillan-Crow L, Turgut F, Awad AS, Abdel-Rahman EM (2023) Acute kidney injury: medical causes and pathogenesis. J Clin Med 12(1):375. https://doi.org/10.3390/JCM12010375

He B, Chen C, Teng L, Tan K (2014) Global view of enhancer–promoter interactome in human cells. Proc Natl Acad Sci 111(21):E2191–E2199. https://doi.org/10.1073/pnas.1320308111

He Y, Jiang X, Duan L, Xiong Q, Yuan Y, Liu P, Jiang L, Shen Q, Zhao S, Yang C, Chen Y (2021) LncRNA PKMYT1AR promotes cancer stem cell maintenance in non-small cell lung cancer via activating Wnt signaling pathway. Mol Cancer 20(1):156. https://doi.org/10.1186/S12943-021-01469-6

He L, Wang H, He P, Jiang Y, Ma F, Wang J, Hu J (2022) Serum long noncoding RNA H19 and CKD progression in IgA nephropathy. J Nephrol 36(2):397–406. https://doi.org/10.1007/s40620-022-01536-1

Herreros-Villanueva M, Zhang JS, Koenig A, Abel EV, Smyrk TC, Bamlet WR, De Narvajas AAM, Gomez TS, Simeone DM, Bujanda L, Billadeau DD (2013) SOX2 promotes dedifferentiation and imparts stem cell-like features to pancreatic cancer cells. Oncogene 2(8):e61. https://doi.org/10.1038/oncsis.2013.23

Hu M, Wang R, Li X, Fan M, Lin J, Zhen J, Chen L, Lv Z (2017) LncRNA MALAT1 is dysregulated in diabetic nephropathy and involved in high glucose-induced podocyte injury via its interplay with β-catenin. J Cell Mol Med 21(11):2732–2747. https://doi.org/10.1111/JCMM.13189

Huang YS, Hsieh HY, Shih HM, Sytwu HK, Wu CC (2014) Urinary Xist is a potential biomarker for membranous nephropathy. Biochem Biophys Res Commun 452(3):415–421. https://doi.org/10.1016/J.BBRC.2014.08.077

Ihim SA, Abubakar SD, Zian Z, Sasaki T, Saffarioun M, Maleknia S, Azizi G (2022) Interleukin-18 cytokine in immunity, inflammation, and autoimmunity: biological role in induction, regulation, and treatment. Front Immunol 13:919973. https://doi.org/10.3389/FIMMU.2022.919973

Ji TT, Huang X, Jin J, Pan SH, Zhuge XJ (2016) Inhibition of long non-coding RNA TUG1 on gastric cancer cell transference and invasion through regulating and controlling the expression of miR-144/c-Met axis. Asian Pac J Trop Med 9(5):508–512. https://doi.org/10.1016/J.APJTM.2016.03.026

Jiang X, Li D, Shen W, Shen X, Liu Y (2019a) LncRNA NEAT1 promotes hypoxia-induced renal tubular epithelial apoptosis through downregulating miR-27a-3p. J Cell Biochem 120(9):16273–16282. https://doi.org/10.1002/JCB.28909

Jiang ZJ, Zhang MY, Fan ZW, Sun WL, Tang Y (2019b) Influence of lncRNA HOTAIR on acute kidney injury in sepsis rats through regulating miR-34a/Bcl-2 pathway. Eur Rev Med Pharmacol Sci 23(8):3512–3519. https://doi.org/10.26355/EURREV_201904_17717

Jianyin L, Badal SS, Zengchun Y, Yin W, Ayanga BA, Galvan DL, Green NH, Chang BH, Overbeek PA, Danesh FR (2016) Long noncoding RNA Tug1 regulates mitochondrial bioenergetics in diabetic nephropathy. J Clin Invest 126(11):4205. https://doi.org/10.1172/JCI87927

Jin L, Pan M, Ye H, Zheng Y, Chen Y, Huang W, Xu X, Zheng S (2019) Down-regulation of the long non-coding RNA XIST ameliorates podocyte apoptosis in membranous nephropathy via the miR-217-TLR4 pathway. Exp Physiol 104(2):220–230. https://doi.org/10.1113/EP087190

Jin C, Jia L, Tang Z, Zheng Y (2020) Long non-coding RNA MIR22HG promotes osteogenic differentiation of bone marrow mesenchymal stem cells via PTEN/AKT pathway. Cell Death Dis 11(7):601. https://doi.org/10.1038/S41419-020-02813-2

Kaneko N, Kurata M, Yamamoto T, Morikawa S, Masumoto J (2019) The role of interleukin-1 in general pathology. Inflamm Regen 39(1):12. https://doi.org/10.1186/s41232-019-0101-5

Kato M, Natarajan R (2019) Epigenetics and epigenomics in diabetic kidney disease and metabolic memory. Nat Rev Nephrol 15(6):327. https://doi.org/10.1038/S41581-019-0135-6

Kato M, Wang M, Chen Z, Bhatt K, Oh HJ, Lanting L, Deshpande S, Jia Y, Lai JYC, O'Connor CL, Wu Y, Hodgin JB, Nelson RG, Bitzer M, Natarajan R (2016) An endoplasmic reticulum stress-regulated lncRNA hosting a microRNA megacluster induces early features of diabetic nephropathy. Nat Commun 7(1):12864. https://doi.org/10.1038/ncomms12864

Ke J, Fan Y, Zhang S (2025) Effects of PARP1 inhibitor PJ-34 on TGFα, IL-6, and IL-1β levels in diabetic nephropathy. J Immunol 214(2):304–315. https://doi.org/10.1093/JIMMUN/VKAE024

Kiriakidou M, Ching CL (2020) Systemic lupus erythematosus. Ann Intern Med 172(11):ITC82–ITC96. https://doi.org/10.7326/AITC202006020

Knoppova B, Reily C, King RG, Julian BA, Novak J, Green TJ (2021) Pathogenesis of IgA nephropathy: current understanding and implications for development of disease-specific treatment. J Clin Med 10(19):4501. https://doi.org/10.3390/jcm10194501

Kong W, Yin G, Zheng S, Liu X, Zhu A, Yu P, Zhang J, Shan Y, Ying R, Jin H (2022) Long noncoding RNA (lncRNA) HOTAIR: pathogenic roles and therapeutic opportunities in gastric cancer. Genes Dis 9(5):1269–1280. https://doi.org/10.1016/J.GENDIS.2021.07.006

Lazzeri E, Angelotti ML, Peired A, Conte C, Marschner JA, Maggi L, Mazzinghi B, Lombardi D, Melica ME, Nardi S, Ronconi E, Sisti A, Antonelli G, Becherucci F, De Chiara L, Guevara RR, Burger A, Schaefer B, Annunziato F et al (2018) Endocycle-related tubular cell hypertrophy and progenitor proliferation recover renal function after acute kidney injury. Nat Commun 9(1):1344. https://doi.org/10.1038/s41467-018-03753-4

Lee JT, Bartolomei MS (2013) X-inactivation, imprinting, and long noncoding RNAs in health and disease. Cell 152(6):1308–1323. https://doi.org/10.1016/J.CELL.2013.02.016

Li Y, Lee PY, Reeves WH (2010) Monocyte and macrophage abnormalities in systemic lupus erythematosus. Arch Immunol Ther Exp 58(5):355–364. https://doi.org/10.1007/S00005-010-0093-Y/FIGURES/2

Li Z, Chao TC, Chang KY, Lin N, Patil VS, Shimizu C, Head SR, Burns JC, Rana TM (2014) The long noncoding RNA THRIL regulates TNFα expression through its interaction with hnRNPL. Proc Natl Acad Sci USA 111(3):1002–1007. https://doi.org/10.1073/pnas.1313768111

Li Z, Shen J, Chan MTV, Wu WKK (2016a) TUG1: a pivotal oncogenic long non-coding RNA of human cancers. Cell Prolif 49(4):471–475. https://doi.org/10.1111/CPR.12269

Li J, Zhang M, An G, Ma Q (2016b) LncRNA TUG1 acts as a tumor suppressor in human glioma by promoting cell apoptosis. Exp Biol Med 241(6):644. https://doi.org/10.1177/1535370215622708

Li J, Wu GC, Zhang TP, Yang XK, Chen SS, Li LJ, Xu SZ, Lv TT, Leng RX, Pan HF, Ye DQ (2017) Association of long noncoding RNAs expression levels and their gene polymorphisms with systemic lupus erythematosus. Sci Rep 7(1):1–8. https://doi.org/10.1038/s41598-017-15156-4

Li A, Peng R, Sun Y, Liu H, Peng H, Zhang Z (2018) LincRNA 1700020I14Rik alleviates cell proliferation and fibrosis in diabetic nephropathy via miR-34a-5p/Sirt1/HIF-1α signaling. Cell Death Dis 9(5):461. https://doi.org/10.1038/S41419-018-0527-8

Li Y, Xia M, Peng L, Liu H, Chen G, Wang C, Yuan D, Liu Y, Liu H (2021) Downregulation of miR-214-3p attenuates mesangial hypercellularity by targeting PTEN-mediated JNK/c-Jun signaling in IgA nephropathy. Int J Biol Sci 17(13):3343–3355. https://doi.org/10.7150/ijbs.61274

Li M, Hao X, Shi D, Cheng S, Zhong Z, Cai L, Jiang M, Ding L, Ding L, Wang C, Yu X (2024) Identification of susceptibility loci and relevant cell type for IgA nephropathy in Han Chinese by integrative genome-wide analysis. Front Med 18(5):862–877. https://doi.org/10.1007/s11684-024-1086-2

Lin F-J, Jiang G-R, Shan J-P, Zhu C, Zou J, Wu X-R (2012) Imbalance of regulatory T cells to Th17 cells in IgA nephropathy. Scand J Clin Lab Invest 72(3):221–229. https://doi.org/10.3109/00365513.2011.652158

Linker-Israeli M, Deans R, Wallace D, Prehn J, Ozeri-Chen T, Klinenberg J (1991) Elevated levels of endogenous IL-6 in systemic lupus erythematosus. A putative role in pathogenesis. J Immunol 147(1):117–123. https://doi.org/10.4049/JIMMUNOL.147.1.117

Liu X, Zhu N, Zhang B, Xu SB (2020) Long noncoding RNA TCONS_00016406 attenuates lipopolysaccharide-induced acute kidney injury by regulating the miR-687/PTEN pathway. Front Physiol 11:533358. https://doi.org/10.3389/FPHYS.2020.00622/BIBTEX

Lorenzen JM, Thum T (2016) Long noncoding RNAs in kidney and cardiovascular diseases. Nat Rev Nephrol 12(6):360–373. https://doi.org/10.1038/nrneph.2016.51

Lu HY, Wang GY, Zhao JW, Jiang HT (2021) Knockdown of lncRNA MALAT1 ameliorates acute kidney injury by mediating the miR-204/APOL1 pathway. J Clin Lab Anal 35(8):e23881. https://doi.org/10.1002/JCLA.23881

M'baya-Moutoula E, Louvet L, Molinié R, Guerrera IC, Cerutti C, Fourdinier O, Nourry V, Gutierrez L, Morlière P, Mesnard F, Massy ZA, Metzinger-Le Meuth V, Metzinger L (2018) A multi-omics analysis of the regulatory changes induced by miR-223 in a monocyte/macrophage cell line. Biochim Biophys Acta Mol basis Dis 1864(8):2664–2678. https://doi.org/10.1016/J.BBADIS.2018.05.010

Ma T, Jia H, Ji P, He Y, Chen L (2021) Identification of the candidate lncRNA biomarkers for acute kidney injury: a systematic review and meta-analysis. Expert Rev Mol Diagn 21(1):77–89. https://doi.org/10.1080/14737159.2021.1873131

Ma T, Li H, Liu H, Peng Y, Lin T, Deng Z, Jia N, Chen Z, Wang P (2022) Neat1 promotes acute kidney injury to chronic kidney disease by facilitating tubular epithelial cells apoptosis via sequestering miR-129-5p. Mol Ther 30(10):3313–3332. https://doi.org/10.1016/j.ymthe.2022.05.019

Mahmoud RH, Fouad NA, Hefzy EM, Shaker OG, Ahmed TI, Hussein HA, Nasr MH, Zaki OM, Abdelghaffar NK, Abdelaleem OO (2022) The potential role of serum expression profile of long non coding RNAs, Cox2 and HOTAIR as novel diagnostic biomarkers in systemic lupus erythematosus. PLoS One 17(8):e0268176. https://doi.org/10.1371/JOURNAL.PONE.0268176

Majumder S, Hadden MJ, Thieme K, Batchu SN, Niveditha D, Chowdhury S, Yerra VG, Advani SL, Bowskill BB, Liu Y, Vakili H, Alghamdi TA, White KE, Geldenhuys L, Siddiqi FS, Advani A (2019) Dysregulated expression but redundant function of the long non-coding RNA HOTAIR in diabetic kidney disease. Diabetologia 62(11):2129–2142. https://doi.org/10.1007/S00125-019-4967-1/FIGURES/7

Masutani K, Akahoshi M, Tsuruya K, Tokumoto M, Ninomiya T, Kohsaka T, Fukuda K, Kanai H, Nakashima H, Otsuka T, Hirakata H (2001) Predominance of Th1 immune response in diffuse proliferative lupus nephritis. Arthritis Rheum 44(9):2097–2106. https://doi.org/10.1002/1529-0131(200109)44:9<2097::AID-ART360>3.0.CO;2-6

McCarthy EM, Smith S, Lee RZ, Cunnane G, Doran MF, Donnelly S, Howard D, O'Connell P, Kearns G, Gabhann JN, Jefferies CA (2014) The association of cytokines with disease activity and damage scores in systemic lupus erythematosus patients. Rheumatology 53(9):1586–1594. https://doi.org/10.1093/RHEUMATOLOGY/KET428

Medhat E, Ayeldeen G, Ahmed HH, Shaker O, Gheita T, Ashour SS (2022) HOTAIR and THRIL long non coding RNAs and their target genes in rheumatoid arthritis patients. Rep Biochem Mol Biol 10(4):614–621. https://doi.org/10.52547/rbmb.10.4.697

Nagafuchi H, Suzuki N, Mizushima Y, Sakane T (1993) Constitutive expression of IL-6 receptors and their role in the excessive B cell function in patients with systemic lupus erythematosus. J Immunol 151(11):6525–6534. https://doi.org/10.4049/JIMMUNOL.151.11.6525

Nagy G, Koncz A, Perl A (2005) T- and B-cell abnormalities in systemic lupus erythematosus. Crit Rev Immunol 25(2):123–140. https://doi.org/10.1615/CRITREVIMMUNOL.V25.I2.30

Navarro-González JF, Mora-Fernández C, De Fuentes MM, García-Pérez J (2011) Inflammatory molecules and pathways in the pathogenesis of diabetic nephropathy. Nat Rev Nephrol 7(6):327–340. https://doi.org/10.1038/nrneph.2011.51

Obaid M, Udden SMN, Deb P, Shihabeddin N, Zaki MH, Mandal SS (2018) LncRNA HOTAIR regulates lipopolysaccharide-induced cytokine expression and inflammatory response in macrophages. Sci Rep 8(1):1–18. https://doi.org/10.1038/s41598-018-33722-2

Odendahl M, Jacobi A, Hansen A, Feist E, Hiepe F, Burmester GR, Lipsky PE, Radbruch A, Dörner T (2000) Disturbed peripheral B lymphocyte homeostasis in systemic lupus erythematosus. J Immunol 165(10):5970–5979. https://doi.org/10.4049/JIMMUNOL.165.10.5970

Paniri A, Akhavan-Niaki H (2020) Emerging role of IL-6 and NLRP3 inflammasome as potential therapeutic targets to combat COVID-19: role of lncRNAs in cytokine storm modulation. Life Sci 257:118114. https://doi.org/10.1016/J.LFS.2020.118114

Pesce F, Stea ED, Rossini M, Fiorentino M, Piancone F, Infante B, Stallone G, Castellano G, Gesualdo L (2021) Glomerulonephritis in AKI: from pathogenesis to therapeutic intervention. Front Med 7:582272. https://doi.org/10.3389/FMED.2020.582272/PDF

Pi P, Yin Q, Xiao L, Luo D (2021) Long non-coding RNA Neat1 triggers renal tubular epithelial cell apoptosis via activating BH3-only protein in membranous nephropathy. Autoimmunity 54(8):539–546. https://doi.org/10.1080/08916934.2021.1972289

Picerno A, Stasi A, Franzin R, Curci C, di Bari I, Gesualdo L, Sallustio F (2021) Why stem/progenitor cells lose their regenerative potential. World J Stem Cells 13(11):1714–1732. https://doi.org/10.4252/wjsc.v13.i11.1714

Picerno A, Giannuzzi F, Curci C, De Palma G, Di Chiano M, Simone S, Franzin R, Gallone A, Di Lorenzo VF, Stasi A, Pertosa GB, Sabbà C, Gesualdo L, Sallustio F (2022) The long non-coding RNA HOTAIR controls the self-renewal, cell senescence, and secretion of anti-aging protein α-klotho in human adult renal progenitor cells. Stem Cells 40(10):963–975. https://doi.org/10.1093/STMCLS/SXAC054

Pons-Estel GJ, Alarcón GS, Scofield L, Reinlib L, Cooper GS (2010) Understanding the epidemiology and progression of systemic lupus erythematosus NIH public access. Semin Arthritis Rheum 39(4):257. https://doi.org/10.1016/j.semarthrit.2008.10.007

Rasuli E, Javidi-Aghdam K, Akbarzadeh-Khiavi M, Abdshah A, Gadakchi L, Jafarpour M, Khabbazi A, Farajnia S, Safary A, Shaykh-Baygloo N (2024) Immunoregulatory role of AC007278.3 and HOTAIR long non-coding RNAs in lupus nephritis: potential biomarkers and therapeutic targets. Mol Biol Rep 51(1):1075. https://doi.org/10.1007/S11033-024-10019-4/FIGURES/2

Remuzzi G, Macia M, Ruggenenti P (2006) Prevention and treatment of diabetic renal disease in type 2 diabetes: the BENEDICT study. J Am Soc Nephrol 17(Suppl. 2):S90–S97. https://doi.org/10.1681/ASN.2005121324

Ronco P, Beck L, Debiec H, Fervenza FC, Hou FF, Jha V, Sethi S, Tong A, Vivarelli M, Wetzels J (2021) Membranous nephropathy. Nat Rev Dis Primers 7(1):1–23. https://doi.org/10.1038/s41572-021-00303-z

Sagrinati C, Netti GS, Mazzinghi B, Lazzeri E, Liotta F, Frosali F, Ronconi E, Meini C, Gacci M, Squecco R, Carini M, Gesualdo L, Francini F, Maggi E, Annunziato F, Lasagni L, Serio M, Romagnani S, Romagnani P (2006) Isolation and characterization of multipotent progenitor cells from the Bowman's capsule of adult human kidneys. J Am Soc Nephrol 17(9):2443–2456. https://doi.org/10.1681/ASN.2006010089

Saha S, Murthy S, Rangarajan PN (2006) Identification and characterization of a virus-inducible non-coding RNA in mouse brain. J Gen Virol 87(7):1991–1995. https://doi.org/10.1099/VIR.0.81768-0/CITE/REFWORKS

Saleh AA, Kasem HE, Zahran E, El-Hefnawy SM (2020) Dysregulation of cell-free long non-coding RNAs (NEAT2, CTC-471J1.2 and lnc-DC) in Egyptian systemic lupus and lupus nephritis patients. Meta Gene 24:100665. https://doi.org/10.1016/J.MGENE.2020.100665

Sallustio F, De Benedictis L, Castellano G, Zaza G, Loverre A, Costantino V, Grandaliano G, Schena FP (2010) TLR2 plays a role in the activation of human resident renal stem/progenitor cells. FASEB J 24(2):514–525. https://doi.org/10.1096/fj.09-136481

Sallustio F, Curci C, Aloisi A, Toma CC, Marulli E, Serino G, Cox SN, De Palma G, Stasi A, Divella C, Rinaldi R, Schena FP (2017) Inhibin-A and Decorin secreted by human adult renal stem/progenitor cells through the TLR2 engagement induce renal tubular cell regeneration. Sci Rep 7(1):8225. https://doi.org/10.1038/s41598-017-08474-0

Sallustio F, Stasi A, Curci C, Divella C, Picerno A, Franzin R, de Palma G, Rutigliano M, Lucarelli G, Battaglia M, Staffieri F, Crovace A, Pertosa GB, Castellano G, Gallone A, Gesualdo L (2019) Renal progenitor cells revert LPS-induced endothelial-to-mesenchymal transition by secreting CXCL6, SAA4, and BPIFA2 antiseptic peptides. FASEB J 33(10):10753–10766. https://doi.org/10.1096/fj.201900351R

Selby NM, Bmbs B, Dm M, Taal MW, Mmed B (2020) An updated overview of diabetic nephropathy: diagnosis, prognosis, treatment goals and latest guidelines. Diabetes Obes Metab 22(S1):3–15. https://doi.org/10.1111/DOM.14007

Shbeer AM, Ahmed Robadi I (2024) The role of Interleukin-21 in autoimmune diseases: mechanisms, therapeutic implications, and future directions. Cytokine 173:156437. https://doi.org/10.1016/J.CYTO.2023.156437

Shi Y, Yang R, Tu L, Liu D (2020) Long non-coding RNA HOTAIR promotes burn wound healing by regulating epidermal stem cells. Mol Med Rep 22(3):1811–1820. https://doi.org/10.3892/MMR.2020.11268/HTML

Song J, Kim D, Han J, Kim Y, Lee M, Jin EJ (2015) PBMC and exosome-derived Hotair is a critical regulator and potent marker for rheumatoid arthritis. Clin Exp Med 15(1):121–126. https://doi.org/10.1007/S10238-013-0271-4

Stasi A, Franzin R, Divella C, Sallustio F, Curci C, Picerno A, Pontrelli P, Staffieri F, Lacitignola L, Crovace A, Cantaluppi V, Medica D, Ronco C, de Cal M, Lorenzin A, Zanella M, Pertosa GB, Stallone G, Gesualdo L, Castellano G (2021) PMMA-based continuous hemofiltration modulated complement activation and renal dysfunction in LPS-induced acute kidney injury. Front Immunol 12:605212. https://doi.org/10.3389/FIMMU.2021.605212/BIBTEX

Sun SF, Tang PMK, Feng M, Xiao J, Huang XR, Li P, Ma RCW, Lan HY (2018) Novel lncRNA Erbb4-IR promotes diabetic kidney injury in db/db mice by targeting miR-29b. Diabetes 67(4):731–744. https://doi.org/10.2337/DB17-0816

Sun Z, Wu J, Bi Q, Wang W (2022) Exosomal lncRNA TUG1 derived from human urine-derived stem cells attenuates renal ischemia/reperfusion injury by interacting with SRSF1 to regulate ASCL4-mediated ferroptosis. Stem Cell Res Ther 13(1):1–16. https://doi.org/10.1186/S13287-022-02986-X/FIGURES/9

Suo QF, Sheng J, Qiang FY, Tang ZS, Yang YY (2017) Association of long non-coding RNA GAS5 and miR-21 levels in CD4+ T cells with clinical features of systemic lupus erythematosus. Exp Ther Med 15(1):345. https://doi.org/10.3892/ETM.2017.5429

Swain SL (2001) Interleukin 18: tipping the balance towards a T helper cell 1 response. J Exp Med 194(3):f11. https://doi.org/10.1084/JEM.194.3.F11

Tamehiro N, Nishida K, Sugita Y, Hayakawa K, Oda H, Nitta T, Nakano M, Nishioka A, Yanobu-Takanashi R, Goto M, Okamura T, Adachi R, Kondo K, Morita A, Suzuki H (2019) Ras homolog gene family H (RhoH) deficiency induces psoriasis-like chronic dermatitis by promoting TH17 cell polarization. J Allergy Clin Immunol 143(5):1878–1891. https://doi.org/10.1016/j.jaci.2018.09.032

Tian H, Wu M, Zhou P, Huang C, Ye C, Wang L (2018) The long non-coding RNA MALAT1 is increased in renal ischemia-reperfusion injury and inhibits hypoxia-induced inflammation. Ren Fail 40(1):527–533. https://doi.org/10.1080/0886022X.2018.1487863

Umare V, Pradhan V, Nath S, Rajadhyaksha A, Ghosh K, Nadkarni AH (2019) Impact of functional IL-18 polymorphisms on genetic predisposition and diverse clinical manifestations of the disease in Indian SLE patients. Lupus 28(4):545–554. https://doi.org/10.1177/0961203319834677

Wang G, Tam LS, Kwan BCH, Li EKM, Chow KM, Luk CCW, Li PKT, Szeto CC (2012) Expression of miR-146a and miR-155 in the urinary sediment of systemic lupus erythematosus. Clin Rheumatol 31(3):435–440. https://doi.org/10.1007/S10067-011-1857-4

Wang W, Sun Y, Fu Y, Yu X, Li M (2013) Interaction of C1GALT1–IL5RA on the susceptibility to IgA nephropathy in southern Han Chinese. J Hum Genet 58(1):40–46. https://doi.org/10.1038/jhg.2012.133

Wang P, Luo M-L, Song E, Zhou Z, Ma T, Wang J, Jia N, Wang G, Nie S, Liu Y, Hou F (2018) Long noncoding RNA *lnc-TSI* inhibits renal fibrogenesis by negatively regulating the TGF-β/Smad3 pathway. Sci Transl Med 10(462):eaat2039. https://doi.org/10.1126/scitranslmed.aat2039

Wang W, Zhang S, Yang F, Xie J, Chen J, Li Z (2020) Diosmetin alleviates acute kidney injury by promoting the TUG1/Nrf2/HO-1 pathway in sepsis rats. Int Immunopharmacol 88:106965. https://doi.org/10.1016/J.INTIMP.2020.106965

Wang Y, Yi K, Liu X, Tan Y, Jin W, Li Y, Zhou J, Wang H, Kang C (2021) HOTAIR up-regulation activates NF-κB to induce immunoescape in gliomas. Front Immunol 12:785463. https://doi.org/10.3389/FIMMU.2021.785463/FULL

Wen L, Zhao Z, Li F, Ji F, Wen J (2022) ICAM-1 related long noncoding RNA is associated with progression of IgA nephropathy and fibrotic changes in proximal tubular cells. Sci Rep 12(1):9645. https://doi.org/10.1038/s41598-022-13521-6

Wu GC, Li J, Leng RX, Li XP, Li XM, Wang DG, Pan HF, Ye DQ (2017) Identification of long non-coding RNAs GAS5, linc0597 and lnc-DC in plasma as novel biomarkers for systemic lupus erythematosus. Oncotarget 8(14):23650. https://doi.org/10.18632/ONCOTARGET.15569

Wu H, Wang J, Ma Z (2020) Long noncoding RNA HOXA-AS2 mediates microRNA-106b-5p to repress sepsis-engendered acute kidney injury. J Biochem Mol Toxicol 34(4):e22453. https://doi.org/10.1002/JBT.22453

Xu G, He Q, Shou Z, Wang H, Zhang X, Wang Y, Chen Y, Chen J (2007) NA1/NA2 heterozygote of Fcgr3b is a risk factor for progression of IgA nephropathy in Chinese. J Clin Lab Anal 21(5):298–302. https://doi.org/10.1002/jcla.20189

Xue R, Yiu WH, Chan KW, Lok SWY, Zou Y, Ma J, Li H, Chan LYY, Huang XR, Lai KN, Lan HY, Tang SCW (2024) Long non-coding RNA NEAT1, NOD-like receptor family protein 3 inflammasome, and acute kidney injury. J Am Soc Nephrol 35(8):998–1015. https://doi.org/10.1681/ASN.0000000000000362

Yang H, Liang N, Wang M, Fei Y, Sun J, Li Z, Xu Y, Guo C, Cao Z, Li S, Jiao Y (2017) Long noncoding RNA MALAT-1 is a novel inflammatory regulator in human systemic lupus erythematosus. Oncotarget 8(44):77400–77406

Yang L, Wang B, Ma L, Fu P (2022) An update of long-noncoding RNAs in acute kidney injury. Front Physiol 13:849403. https://doi.org/10.3389/FPHYS.2022.849403/PDF

Ye DQ, Wu GC, Hu Y, Guan SY, Pan HF (2019) Differential plasma expression profiles of long non-coding RNAs reveal potential biomarkers for systemic lupus erythematosus. Biomolecules 9(6):206. https://doi.org/10.3390/BIOM9060206

Yi H, Peng R, Zhang LY, Sun Y, Peng HM, Liu HD, Yu LJ, Li AL, Zhang YJ, Jiang WH, Zhang Z (2017) LincRNA-Gm4419 knockdown ameliorates NF-κB/NLRP3 inflammasome-mediated inflammation in diabetic nephropathy. Cell Death Dis 8(2):e2583. https://doi.org/10.1038/cddis.2016.451

You Y, Zhao X, Wu Y, Mao J, Ge L, Guo J, Zhao C, Chen D, Song Z (2021) Integrated transcriptome profiling revealed that elevated long non-coding RNA-AC007278.2 expression repressed CCR7 transcription in systemic lupus erythematosus. Front Immunol 12:615859. https://doi.org/10.3389/FIMMU.2021.615859/BIBTEX

Yu TM, Palanisamy K, Sun KT, Day YJ, Shu KH, Wang IK, Shyu WC, Chen P, Chen YL, Li CY (2016) RANTES mediates kidney ischemia reperfusion injury through a possible role of HIF-1α and LncRNA PRINS. Sci Rep 6:18424. https://doi.org/10.1038/SREP18424

Yu H, Nagafuchi Y, Fujio K (2021) Clinical and immunological biomarkers for systemic lupus erythematosus. Biomolecules 11(7):928. https://doi.org/10.3390/BIOM11070928

Yuan H, Li S, Wang L, Zhao X, Xue L, Lei X, Fu R (2020) Genetic variants of the MIR31HG gene are related to a risk of IgA nephropathy. Int Immunopharmacol 84:106533. https://doi.org/10.1016/j.intimp.2020.106533

Zhai Y, Tian H, Zhang W, Sun S, Zhao Z (2024) Genome-wide analysis of long noncoding RNAs as cis-acting regulators of transcription factor-encoding genes in IgA nephropathy. PLoS One 19(5):e0304301. https://doi.org/10.1371/journal.pone.0304301

Zhang Q, Chen CY, Yedavalli VSRK, Jeang KT (2013) NEAT1 long noncoding RNA and paraspeckle bodies modulate HIV-1 posttranscriptional expression. MBio 4(1):e00596–e00512. https://doi.org/10.1128/mBio.00596-12

Zhang F, Wu L, Qian J, Qu B, Xia S, La T, Wu Y, Ma J, Zeng J, Guo Q, Cui Y, Yang W, Huang J, Zhu W, Yao Y, Shen N, Tang Y (2016) Identification of the long noncoding RNA NEAT1 as a novel inflammatory regulator acting through MAPK pathway in human lupus. J Autoimmun 75:96–104. https://doi.org/10.1016/J.JAUT.2016.07.012

Zhang Y et al (2018) LncRNA AK137033 suppresses osteogenesis via sFRP2 promoter methylation in diabetic osteoporosis. Bone 112:90–101. https://doi.org/10.1016/j.bone.2018.04.013

Zhang Y, Tang PMK, Niu Y, García Córdoba CA, Huang XR, Yu C, Lan HY (2020) Long non-coding RNA LRNA9884 promotes acute kidney injury via regulating NF-kB-mediated transcriptional activation of MIF. Front Physiol 11:590027. https://doi.org/10.3389/FPHYS.2020.590027/BIBTEX

Zheng S, Wang YB, Yang YL, Chen BP, Wang CX, Li RH, Huang D (2019) LncRNA MALAT1 inhibits osteogenic differentiation of mesenchymal stem cells in osteoporosis rats through MAPK signaling pathway. Eur Rev Med Pharmacol Sci 23(11):4609–4617. https://doi.org/10.26355/EURREV_201906_18038

Zhou L, Xu DY, Sha WG, Shen L, Lu GY, Yin X (2015) Long non-coding MIAT mediates high glucose-induced renal tubular epithelial injury. Biochem Biophys Res Commun 468(4):726–732. https://doi.org/10.1016/J.BBRC.2015.11.023

Zhou Z, Sun B, Huang S, Zhao L (2019) Roles of circular RNAs in immune regulation and autoimmune diseases. Cell Death Dis 10(7):1–13. https://doi.org/10.1038/s41419-019-1744-5

Zhou X, Jiang K, Luo H, Wu C, Yu W, Cheng F (2020) Novel lncRNA XLOC_032768 alleviates cisplatin-induced apoptosis and inflammatory response of renal tubular epithelial cells through TNF-α. Int Immunopharmacol 83:106472. https://doi.org/10.1016/J.INTIMP.2020.106472

Zhou G, Jiang N, Zhang W, Guo S, Xin G (2021) Biomarker identification in membranous nephropathy using a long non-coding RNA-mediated competitive endogenous RNA network. Interdiscip Sci 13(4):615–623. https://doi.org/10.1007/S12539-021-00466-Z

Zhou Y, Wang Y, Li Q, Dong KE, Chen C, Mao E, Jiang W (2022) Downregulation of lncRNA NEAT1 alleviates sepsis-induced acute kidney injury. Cent Eur J Immunol 47(1):8–19. https://doi.org/10.5114/CEJI.2022.115628

Zuo N, Li Y, Liu N, Wang L (2017) Differentially expressed long non-coding RNAs and mRNAs in patients with IgA nephropathy. Mol Med Rep 16(5):7724–7730. https://doi.org/10.3892/mmr.2017.7542

Chapter 10
Long Non-coding RNAs in Bone Diseases

Iyyappan Saranya, Dilipkumar Preetha, Sasi Nivruthi, Nivetha Suresh, and Nagarajan Selvamurugan

Abstract Long non-coding RNAs (lncRNAs) play a key role in several biological processes. They regulate gene expression at both the transcriptional and post-transcriptional levels by interacting with microRNAs (miRNAs) and RNA-binding proteins. In bone biology, lncRNAs play a major role in the regulation of bone remodeling via osteoblasts and osteoclasts. Dysregulation of lncRNAs has been implicated in bone diseases, such as osteoporosis, osteoarthritis, and bone cancer. Differential expression patterns of lncRNAs highlight their potential as biomarkers for the diagnosis and prediction of bone disorders. Furthermore, the regulation of signaling pathways by lncRNAs could be used to exploit their role as therapeutic targets for bone disorders. This chapter explores the mechanism of action through which lncRNAs regulate gene expression in bone health and disease, focusing on their emerging roles as biomarkers and therapeutics for skeletal disorders.

Keywords Osteoporosis · Fracture healing · Osteosarcoma · LncRNAs · Drug resistance

Abbreviations

AS	Ankylosing spondylitis
BMD	Bone mineral density
BMP2	Bone morphogenetic protein 2
BMSCs	Bone marrow-derived mesenchymal stem cells
ceRNAs	Competing endogenous RNAs
DFH	Delayed fracture healing

I. Saranya · D. Preetha · S. Nivruthi · N. Suresh · N. Selvamurugan (✉)
Department of Biotechnology, School of Bioengineering, SRM Institute of Science and Technology, Kattankulathur, Tamil Nadu, India
e-mail: si5013@srmist.edu.in; pd3269@srmist.edu.in; ns7876@srmist.edu.in; ns9518@srmist.edu.in; selvamun@srmist.edu.in

L. M. Vaschetto (ed.), *Long Non-Coding RNAs, Human Health, and Diseases*, Epigenetics and Human Health 16, https://doi.org/10.1007/978-3-032-19668-2_10

lncRNA	Long non-coding RNAs
miRNA	Micro RNA
ncRNA	Non-coding RNAs
OA	Osteoarthritis
OP	Osteoporosis
OS	Osteosarcoma
OVX	Ovariectomized
SNHG	Small Nucleolar RNA Host Gene
TGF-β	Transforming growth factor-β
TNF-α	Tumor necrosis factor-α
XIST	X-Inactive-Specific Transcript

10.1 Introduction

Approximately 98% of the transcribed genome comprises non-coding RNAs (ncRNAs), which are crucial regulators of various physiological and pathological processes, including skeletal disorders and tumorigenesis (Nemeth et al. 2024). Based on their functional characteristics, ncRNAs are broadly classified into structural (housekeeping) and regulatory ncRNAs. Structural ncRNAs, including transfer RNAs (tRNAs) and ribosomal RNAs (rRNAs), play essential roles in fundamental cellular processes such as protein synthesis (Tamblin-Hopper et al. 2024). In contrast, regulatory ncRNAs influence gene expression through diverse mechanisms and are further categorized into small ncRNAs (<200 nucleotides) and long ncRNAs (lncRNAs) (>200 nucleotides) based on their length. Small ncRNAs include microRNAs (miRNAs), PIWI-interacting RNAs (piRNAs), small interfering RNAs (siRNAs), and small nuclear RNAs (snRNAs) (Márton et al. 2025). The class of ncRNAs, known as lncRNAs, has a variety of structural patterns and modes of action. The lncRNAs are transcribed using RNA polymerase II. Based on their genomic origin, biogenesis, and regulatory functions, lncRNAs are further classified into various subtypes, including intronic, enhancer-associated, promoter-associated, sense, antisense, intergenic, and bidirectional lncRNAs, as well as linear and circular RNAs. According to their distinct gene regulation mechanisms, lncRNAs are functionally separated into three categories: cis-acting lncRNAs, trans-acting lncRNAs, and competing endogenous RNAs (ceRNAs) (Basu et al. 2025). Through intricate molecular processes, lncRNAs regulate gene expression at both transcriptional and post-transcriptional levels (Usman et al. 2024). They act as ceRNAs and interact with other ncRNAs by binding to miRNAs and decreasing the miRNA-mediated repression of target mRNAs (Mir 2024). Moreover, lncRNAs regulate chromatin remodeling and alter transcription accessibility by interacting with transcription factors and chromatin-modifying complexes. These intricate processes demonstrate the mechanism by which lncRNAs regulate gene expression and maintain cellular homeostasis (Xing and Rodeo 2024).

Given their crucial roles in regulating gene expression and maintaining cellular homeostasis, lncRNAs have been increasingly implicated in the pathogenesis of various skeletal disorders, including osteoporosis (OP), osteoarthritis (OA), ankylosing spondylitis (AS), osteosarcoma (OS), and bone fractures (Esawy et al. 2023; Jankowski et al. 2023). Understanding how ncRNAs contribute to these conditions provides valuable insights into potential diagnostic markers and therapeutic targets. Furthermore, lncRNAs control the osteogenic differentiation of stem cells, suggesting that they may play a crucial role as molecular regulators in the onset of bone diseases and in treatment development. LncRNAs can stimulate osteogenic signaling pathways, which in turn promote osteogenic differentiation and bone formation. For instance, lncRNA MALAT1 promotes osteoblast differentiation of bone marrow-derived mesenchymal stem cells (BMSCs) by sponging miR-129-5p (Yin et al. 2022). Similarly, lncRNA USP2-AS1 promotes osteogenic differentiation in human BMSCs by interacting with lysine demethylase 3A (KDM3A) and ETS1 (proto-oncogene), which leads to the upregulation of ubiquitin carboxyl-terminal hydrolase 2 (USP2) expression and activation of the Wnt/β-catenin signaling pathway (Luo et al. 2024). Some lncRNAs suppress the formation of new bone, mainly by increasing the activity of osteoclasts or preventing the differentiation of osteoblasts. For example, the lncRNA AW011738 suppresses osteogenesis in MC3T3-E1 cells via the lncRNA AW011738/miR-24-2-5p/TREM1 axis (Liu et al. 2024a). Similarly, in MC3T3-E1 cells, MIR17HG has been associated with tumor necrosis factor-α (TNF-α)-induced inhibition of osteogenic differentiation and PANoptosis (apoptosis, pyroptosis, and necroptosis) (Liu et al. 2024a).

The role of lncRNAs in bone remodeling and repair further underscores their importance as they influence the activity of bone cells and extracellular matrix (ECM) formation to maintain bone strength and integrity. In this chapter, we summarize the roles of lncRNAs in various bone diseases. In particular, novel insights into the molecular mechanisms of lncRNAs in diseases such as OP, OS, OA, and fractures are discussed. We also focused on the role of lncRNAs in drug resistance and their potential as biomarkers of bone disease, injury, and regeneration.

10.2 LncRNAs in Bone Diseases

Bone is a dynamic and mineralized connective tissue that plays several physiological roles in the body, constantly undergoing remodeling to maintain structural integrity and homeostasis. This remodeling is controlled by two bone cell types: osteoclasts and osteoblasts. Osteoclast-osteoblast interactions are tightly regulated by various factors to achieve a balance between bone formation and resorption (Li et al. 2024c). Disruption of homeostasis by various factors compromises the structure, strength, and function of bones, resulting in various bone diseases. These include nutritional deficiencies and environmental and genetic factors. Genetic factors are important in diseases such as osteogenesis imperfecta, which are caused by mutations in collagen-producing genes and result in brittle bones (Arya et al. 2024).

Deficiencies in nutrients such as calcium and vitamin D can cause bone weakening and lead to diseases such as OP (Forlino and Marini 2016). Lifestyle factors such as lack of physical activity, smoking, and alcohol abuse can also have an adverse effect on bones and lead to bone disorders. Trauma and infections, such as OA and fractures, directly affect bone tissue and result in inflammation and bone injury (Sheng et al. 2021).

Among the molecular regulators involved, lncRNAs have gained attention for their ability to modulate the activity of bone cells through diverse mechanisms. For example, at the transcriptional level, MALAT1 promotes latent transforming growth factor-β binding protein 3 (LTBP3) transcription in mesenchymal stem cells (MSCs) by recruiting the transcription factor Sp1 to the LTBP3 promoter, which in turn, increases transforming growth factor-β (TGF-β) secretion and its bioavailability in bone. Post-transcriptionally, MALAT1 influences the distribution of serine/arginine (SR) splicing factors within nuclear speckle domains, thereby directing alternative splicing (Li et al. 2014; Yoon et al. 2013). These regulatory mechanisms ensure a fine balance between bone resorption and formation, thereby playing crucial roles in bone metabolism and health (Fig. 10.1) (Li et al. 2021). The following section discusses the roles of lncRNAs in various bone disorders and their regulatory functions in osteogenesis and bone resorption.

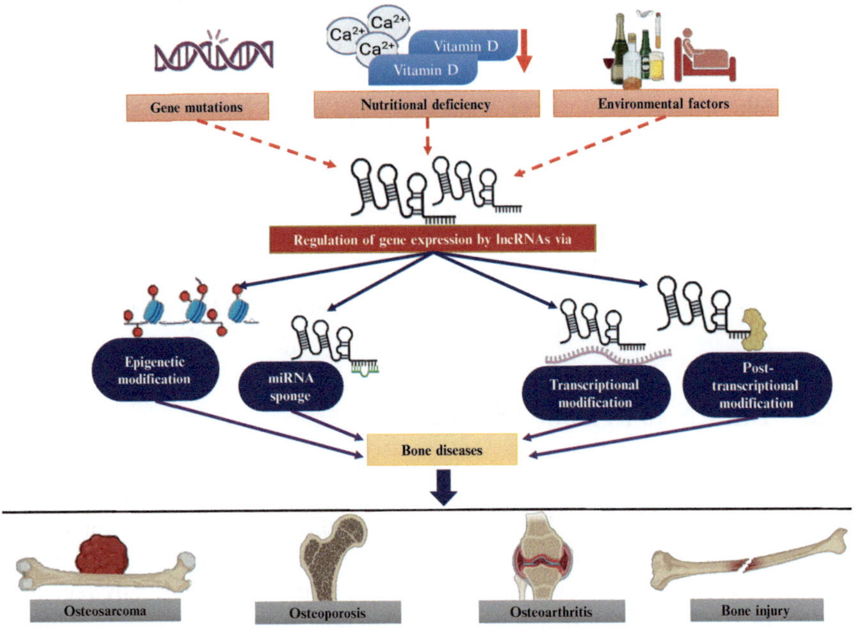

Fig. 10.1 Expression of lncRNAs is moderated by various factors, such as genetic and environmental factors. LncRNAs regulate bone diseases through various mechanisms

10.2.1 Regulatory Role of lncRNAs in OP

OP is a bone disease caused by low bone mineral density (BMD) and structural damage, resulting in elevated fragility and fracture risks. With increasing life expectancy, the incidence of OP continues to increase, especially among the elderly (Sozen et al. 2017). LncRNAs play key roles in OP by regulating the activity of osteoblasts and osteoclasts. LncRNA KCNQ1OT1 has been shown to promote BMSC osteogenesis by sponging miR-214, leading to the upregulation of bone morphogenetic protein 2 (BMP2) (Wang et al. 2019a). Similarly, lncRNA-OG enhances osteogenesis through the activation of heterogeneous nuclear ribonucleoprotein K-mediated BMP signaling (Tang et al. 2019). In contrast, lncRNAs such as nuclear paraspeckle assembly transcript 1 (NEAT1), AK077216, and X-inactive specific transcript (XIST) have been reported to stimulate osteoclastogenesis, highlighting the dual role of lncRNAs in regulating bone remodeling (Zhang et al. 2020b, 2022a; Liu et al. 2019). Thus, emerging evidence suggests that lncRNAs play pivotal roles in the pathogenesis and therapeutic management of OP by regulating osteoblastogenesis, osteoclastogenesis, and bone remodeling pathways (Fig. 10.2). MALAT1 is a highly conserved nuclear lncRNA that is involved in osteoblastogenesis and osteoclastogenesis. MALAT1 deficiency leads to OP because of its role in binding to TEAD3, which prevents the activation of NFATc1,

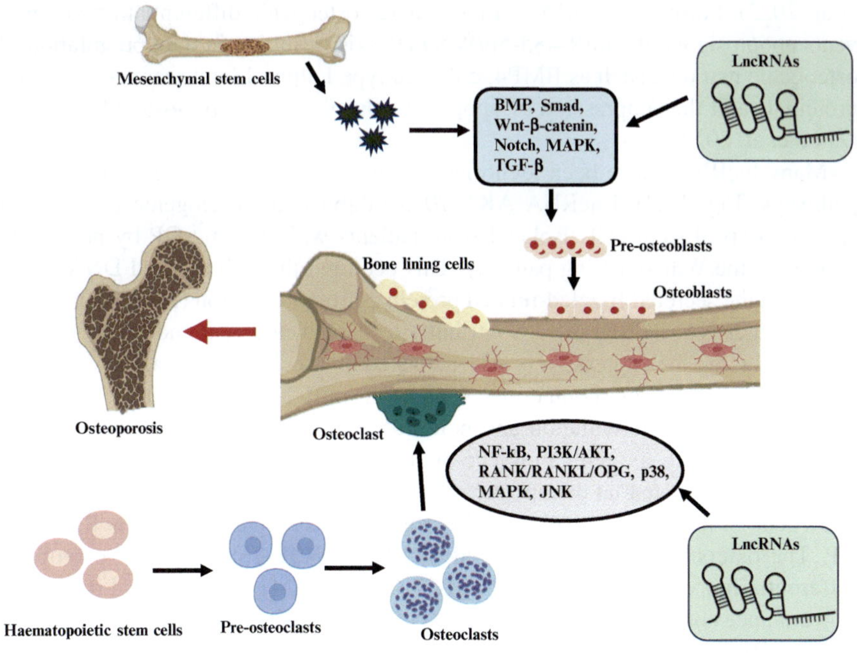

Fig. 10.2 LncRNAs regulate osteoblast and osteoclast differentiation via various signaling pathways to participate in the development of osteoporosis

a key regulator of osteoclast differentiation (Zhao et al. 2022). The lncRNA HAGLR, which is downregulated in postmenopausal OP, regulates the osteogenic differentiation of BMSCs via miR-182-5p and HOXA10 expression, suggesting its potential as a therapeutic target (Huang et al. 2023d). In OP, LINC00205 functions as a ceRNA by sponging miR-26b-5p, leading to lysine (K)-specific methyltransferase 2C upregulation and the modulation of osteogenic differentiation, contributing to OP progression (Wang et al. 2023). RAD51-AS1 influenced OP by activating the TGF-β pathway through interactions with Y-box binding protein 1 (RNA/DNA binding protein), suppressing SMAD7 and SMURF2 translation (Li et al. 2023a). XIST promotes osteoclastogenesis by stabilizing sphingosine kinase 1 (SPHK1), which activates the extracellular signal-regulated kinase-1/−2 (ERK-1/−2) pathway via sphingosine-1-phosphate signaling. Silencing of XIST or SPHK1 disrupts this pathway (Zhang et al. 2022a).

Several lncRNAs contribute to bone homeostasis by regulating the osteoblast function. A study by Qian et al. found that lncRNA AWPPH maintains bone microarchitecture by modulating the type I collagen α1/α2 ratio, which is essential for osteoblast activity (Qian et al. 2022). Another study demonstrated that the plant flavanol quercetin enhances osteogenesis by activating Wnt/β-catenin and suppressing nuclear factor-kappa B (NF-κB) signaling, counteracting TNF-α-induced BMSCs impairment. The effects of quercetin were also shown to be mediated by MALAT1, and its osteogenic effect was restored by treatment in an ovariectomized (OVX)-induced OP model, highlighting its potential as a therapeutic agent (Feng et al. 2023). Furthermore, MALAT1 facilitates osteogenic differentiation and prevents apoptosis via the miR-485-5p/WNT7B axis, resulting in the upregulation of osteogenic markers, such as BMP4, collagen type I alpha 1, and secreted phosphoprotein 1, and the suppression of apoptosis-related genes in pre-osteoblastic cells (Zhou et al. 2023).

Many lncRNAs have been identified as key regulators of OP-related signaling pathways (Fig. 10.2). LncRNA AK137033 inhibited the osteogenic potential of adipose-derived stem cells isolated from patients with diabetic OP by negatively regulating the Wnt signaling pathway. This occurred through reduced DNA methylation in the secreted frizzled-related protein 2 promoter region (Peng et al. 2022). Another lncRNA, PCBP1-AS1, contributes to the development and progression of OP by inducing the expression of p21-activated kinase 2 (PAK2, a negative regulator of osteogenesis) via competitive binding to miR-126-5p (Li 2023). H19 promotes osteoclast differentiation and bone resorption by upregulating cathepsin K expression through sponging miR-29c-3p. Notably, H19 knockout enhanced BMD, underscoring its potential therapeutic relevance (Li et al. 2025).

Recent studies have explored the roles of exosome-derived lncRNAs in OP. The osteoclast-derived exosomal lncRNA AW011738 represses osteoblast differentiation via the AW011738/miR-24-2-5p/TREM1 pathway, and its inhibition enhances osteoblast differentiation and bone density in OVX mice (Liu et al. 2024a). Bioactive glass nanoparticles and BMSC-derived extracellular vesicles loaded with lncRNA NRON inhibited osteoclastogenesis by preventing NFATc1 nuclear translocation, revealing a novel therapeutic strategy for OP (Yang et al.

2022). This evidence underscores the essential role of lncRNAs in OP and provides insights into their potential as diagnostic markers and therapeutic targets. Further exploration of the lncRNA-mediated mechanisms may pave the way for innovative treatment strategies to improve bone health and mitigate the progression of OP.

10.2.2 Role of lncRNAs in OA

OA is a degenerative joint disease characterized by the loss of articular cartilage, formation of bone spurs (osteophytes), periarticular cysts, and sclerotic bone, which manifests clinically as pain, stiffness, and an altered gait. OA is a complex multifactorial disorder with many risk factors, including advanced age, obesity, joint overuse, and family history. LncRNAs are involved in the development of degenerative bone disorders by regulating osteogenic differentiation, bone resorption, and matrix remodeling. LncRNAs significantly regulate signaling pathways like BMP, Wnt, and TGF-β, affecting osteoblast and osteoclast function. Elucidation of the pathways regulated by lncRNAs may offer new therapeutic targets for OA. Zheng et al. revealed that small nucleolar RNA host gene 14 (SNHG14) acts as a sponge to suppress miR-137, thereby promoting chondrocyte apoptosis, inflammation, and ECM degradation. The knockdown of SNHG14 significantly alleviated lipopolysaccharide (LPS)-induced chondrocyte damage, highlighting its pathogenic role in OA (Zheng et al. 2023). The lncRNA MINCR exerts a protective effect by modulating the miR-146a-5p/BMPR2 axis, promoting chondrocyte proliferation and suppressing apoptosis and ECM degeneration (Li et al. 2022a). Zhou et al. investigated the mechanosensitive role of lncRNA H19 in OA and demonstrated that its downregulation suppressed autophagy. LncRNA H19 targets miR-148a and activates autophagy in chondrocytes, which in turn delays chondrocyte degeneration and alleviates the development of OA (Zhou et al. 2025b).

Exosomal lncRNAs play critical roles in the pathogenesis of OA. Wang et al. identified a ceRNA network involving the LYRM4-AS1/GRPR/miR-6515-5p axis that regulates chondrocyte viability and inflammation. Exosome therapy reversed interleukin-1 (IL-1)-induced apoptosis and inflammatory responses, underscoring its therapeutic potential (Wang et al. 2021b). Another study identified WDR11-AS1 as a key lncRNA that enhanced ECM production and SOX9 mRNA stabilization by interacting with the RNA-binding protein PABPC1. The downregulation of WDR11-AS1 in OA cartilage promotes ECM degradation, making it a promising therapeutic target (Huang et al 2023b). Shen et al. reported that lncRNA TUC339 was downregulated in patients with OA with increased expression of inflammatory cytokines, indicating its potential as a biomarker. BMSC-derived exosomes alleviate OA by upregulating TUC339, promoting M2 macrophage polarization, reducing inflammation, and enhancing chondrocyte viability (Shen et al. 2023b). Zhang et al. demonstrated that lncRNA GAS-5 is low in patients with OA, and the expression of this lncRNA is associated with higher pro-inflammatory cytokines and lower anti-inflammatory cytokines, implicating its role in immune modulation and therapeutic potential (Zhang et al. 2021c).

Several reports have indicated the involvement of lncRNAs in OA via key signaling pathways. A recent study highlighted the protective role of the lncRNA CRNDE (colorectal neoplasia differentially expressed) in injury-induced OA by enhancing dapper antagonist of catenin-1 expression through epigenetic modifications. This restrained the activation of Wnt/β-catenin signaling to impede the progression of OA. CRNDE overexpression alleviated cartilage injury, synovitis, ECM degradation, inflammation, and apoptosis (Zhang et al. 2022d). Other lncRNAs such as HOTAIR, NEAT1, and SNHG14 also contribute to the progression of OA. HOTAIR is overexpressed in OA chondrocytes and induces inflammation through the phosphatase and TENsin homolog/phosphatidylinositol 3-kinase/protein kinase B (PTEN/PI3K/AKT) pathways. Baicalin treatment ameliorates OA by inhibiting HOTAIR and restoring PTEN signaling (Chen et al. 2023). Huang et al. demonstrated that NEAT1 downregulation in OA enhances miR-374b-5p expression while reducing post-glycosylphosphatidylinositol attachment to protein 1 (PGAP1) levels, leading to increased apoptosis and inflammation in chondrocytes. Targeting the NEAT1/miR-374b-5p/PGAP1 axis is a potential therapeutic strategy for OA (Huang et al. 2023a). BMSC-derived exosomal SNHG7 mitigated inflammation and ferroptosis in IL-1β-induced chondrocytes by modulating the miR-485-5p/FSP1 axis (Wang et al. 2024b). The lncRNA AC006064.4-201 is downregulated in senescent chondrocytes and OA cartilage and can protect against chondrocyte senescence by interacting with polypyrimidine tract-binding protein 1 (Shen et al. 2023a). Wang et al. demonstrated that SNHG14 is overexpressed in OA, promoting inflammation via miR-124-3p sponging and activation of FSTL-1/NLRP3 and TLR4/NF-κB signaling. SNHG14 downregulation suppressed inflammation and OA progression, highlighting its potential as a therapeutic target (Wang et al. 2021a). The function of SNHG1 in OA has previously been identified. These results demonstrate its protective role in chondrocytes. SNHG1 overexpression enhanced cell viability and inhibited apoptosis in IL-1β-treated chondrocytes by activating the PI3K/AKT pathway and inhibiting autophagy in vitro. Under in vivo conditions, SNHG1 overexpression prevented chondrocyte loss in OA-treated mice, supporting its therapeutic potential (Wang et al. 2024b). The role of lncRNA NKILA in IL-1β-induced inflammation in chondrocytes has been reported. IL-1β downregulates NKILA, resulting in increased NF-κB activity, reduced chondrocyte proliferation, and enhanced apoptosis, thereby accelerating cartilage degeneration, whereas NKILA overexpression alleviates OA symptoms by inhibiting NF-κB signaling, making it a promising drug candidate (Wang et al. 2024a). These findings highlight the multifunctional roles of lncRNAs in OA pathogenesis and their potential as biomarkers and therapeutic targets, thereby offering new directions for future research and clinical applications.

10.2.3 Role of lncRNAs in OS

OS is the most common and aggressive type of malignant bone tumor and predominantly affects adolescents and young adults. Despite advancements in treatment strategies, including surgery, chemotherapy, and radiation therapy, the prognosis remains poor owing to high rates of metastasis and recurrence. Accumulating evidence suggests that lncRNAs contribute to OS progression through multiple mechanisms, including modulation of gene expression, interaction with miRNAs, and regulation of signaling pathways such as Wnt/β-catenin, PI3K/AKT/mTOR, NF-κB, Notch, Hippo, and Hypoxia-inducible factor 1-alpha.

LncRNAs are implicated in OS tumorigenesis through the overexpression of oncogenic lncRNAs and the inhibition of tumor-suppressive lncRNAs. In recent years, a significant number of oncogenic lncRNAs, such as 91H, HULC, FGFR3-AS1, MALAT1, BCAR4, HIF2PUT, TUG1, UCA1, HOTTIP, and HOTAIR, have been identified and implicated in cell growth and OS proliferation. Through intricate interactions with miRNAs, lncRNAs can induce DNA methylation and damage and regulate gene expression. Multiple lncRNAs are associated with OS development by acting as ceRNAs and modulating important oncogenic pathways. LncRNA ODRUL promotes OS progression by interacting with miR-3182, leading to increased MMP2 expression, which is associated with tumor growth and invasion (Zhu et al. 2017). PCAT6 sponging miR-185-5p activated the TGF-β pathway, whereas SNHG10 and MRPL23-AS1 sponging miR-182-5p and miR-30b, respectively, promoted OS cell invasion and proliferation via the Wnt/β-catenin signaling (Zhu et al. 2020a, b; Zhang et al. 2021b). Similarly, lncRNAs AWPPH and LINC00665 facilitated OS progression by sponging miRNAs, which resulted in the activation of the Wnt/β-catenin pathway (Li et al. 2019; Bai et al. 2023). AP003352.1 modulated OS malignancy by driving lysophosphatidic acid receptor 3 (LPAR3) expression through the miR-141-3p axis, while lncRNA XLOC_005950 regulates OS progression via miR-542-3p-mediated modulation of the muscle isoform of the phosphofructokinase gene, PFKM (phosphofructokinase muscle) expression (Yu et al. 2023; Jia et al. 2021). MELTF-AS1 increases OS metastasis by regulating MMP14 via miR-485-5p interaction. CBR3-AS1 promotes OS by sponging miR-140-5p and activating DEAD-box helicase 54 to upregulate nuclear casein and cyclin-dependent kinase substrate 1, thereby modulating the mTOR signaling pathway (Yao et al. 2022; Ding et al. 2021). The lncRNA SNHG15 promotes OS progression by regulating the miR-346/TRAF4 axis (Chen and Xu 2020).

Although most members of the lncRNA family promote the occurrence and development of OS, several lncRNAs have been shown to actively inhibit OS development. The tumor-suppressive roles of GAS5 and ZNF281 have been reported previously. They sponge oncogenic miRNAs and inhibit the pathways associated with OS proliferation and invasion. GAS5 functions as a tumor suppressor by sponging miR-663a, which regulates Rho GTPase and RHOB expression, thereby suppressing OS cell proliferation and invasion (Yao et al. 2020). ZNF281 downregulates the protein kinase ROCK1 by upregulating miR-144 expression, thereby inhibiting OS cell invasion and migration (Shao et al. 2020). The Hippo signaling pathway is activated by LINC01089 to prevent the progression of OS, while LINC00588 acts as a ceRNA for miRNA-1972 to inhibit tumor growth in OS (Zhang et al. 2022c; Zhou et al. 2020). Some lncRNAs directly inhibit proliferation and migration of OS cells. LINC00665 encodes LINC00665_18aa, a short peptide that inhibits tumor growth by suppressing OS cell migration and proliferation through CREB1/RPS6KA3 interaction (Pan et al. 2023). LncRNA-p21 inhibits OS proliferation by upregulating PTEN through the miR-130b/PTEN/AKT axis, efficiently inhibiting the AKT signaling pathway (Han and Liu 2018). The lncRNA FENDRR enhances the sensitivity of OS cells to doxorubicin by downregulating genes associated with drug resistance, such as ABCB1 and ABCC1, thereby inhibiting tumor growth (Kun-Peng et al. 2017). Furthermore, by regulating miR-375, lncRNA TUSC7 inhibits OS migration and proliferation, whereas LINC01278 decreases OS proliferation and induces apoptosis via the miR-134-5p/KRAS axis (Fig. 10.3) (Wang et al. 2020b; Zhang et al. 2021a).

Several lncRNA-based prognostic models have been developed, including the age-associated survival lncRNA (AASR) signature, which comprises five lncRNAs (GAS5, AC093627.22, UNC5B-AS1, AC083900.1, and AP000802.1), to predict overall survival in patients with OS. Another model based on endoplasmic reticulum stress lncRNAs included five lncRNAs (AC023157.3, AL031673.1,

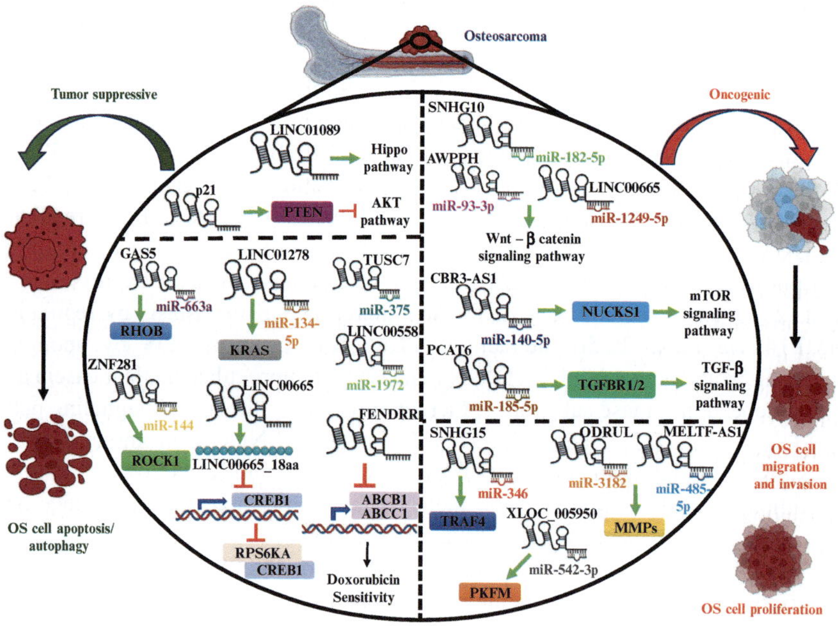

Fig. 10.3 Mechanisms of lncRNA in osteosarcoma progression. LncRNAs are involved in oncogenic and tumor-suppressive roles by acting as molecular sponges for microRNAs, regulating the expression of genes involved in various signaling pathways leading to osteosarcoma cell proliferation, invasion and migration, apoptosis, and autophagy

LINC02298, LINC02328, and SNHG26) and served as a prognostic signature for OS (He et al. 2024; Xu et al. 2024). AP003352.1, in the miR-141-3p/LPAR3 axis, is a prospective biomarker (Yu et al. 2023). Zhao et al. identified six cellular senescence-related lncRNAs with independent prognostic significance, where high expression of lncRNAs AL512330.1, SENCR, LINC02710, and AL022328.1 was associated with poor outcomes, while expression of lncRNAs AC090559.1 and AC091271.1 was correlated with better survival in juvenile patients with OS (Zhao et al. 2024b). Furthermore, lncRNAs, including HOXA-AS2 and FAL1, are correlated with worse prognosis, and HNF1A-AS1 can distinguish patients with OS from healthy persons (Xiao et al. 2024; Wang et al. 2018; Cai et al. 2017). Various upregulated lncRNAs, including ER3-IT1, SND1-IT1, and AGAP1-IT1, and downregulated lncRNAs, such as NRG1-IT1 and FGF14-IT1, further underscore their relevance to OS pathology (Rothzerg et al. 2022). Other lncRNAs that are significant prognostic indicators include BC050642 and LINC01614 (Yang et al. 2019; Cai et al. 2021). In addition, various lncRNAs, including ZEB-AS1 and DICER1-AS1, play important roles in the advancement of OS, highlighting their role as efficient biomarkers of the disease (Liu and Lin 2016; Hu et al. 2018). Thus, several lncRNAs significantly influence OS progression by regulating key oncogenic pathways, acting as oncogenes or tumor suppressors, and impacting the metastasis, proliferation, and apoptosis of OS cells (Table 10.1). Their dysregulation emphasizes their potential as diagnostic biomarkers and therapeutic targets, with emerging prognostic models assisting in patient outcome prediction.

10.2.4 Role of lncRNAs in Osteonecrosis

Femoral head osteonecrosis (ONFH) is a condition characterized by disrupted bone homeostasis and is frequently linked to dysregulated lncRNAs that affect the differentiation, survival, and apoptosis of BMSCs. The lncRNA GAS5 was markedly downregulated in patients with ONFH and rat models, resulting in decreased bone volume and trabecular thickness. The reduced expression impairs the osteogenic differentiation of BMSCs, indicating that GAS5 can be used as an early diagnostic biomarker and a potential molecular target for ONFH treatment (Liu et al. 2022). Likewise, lncRNA FGD5-AS1 promotes cell proliferation and inhibits apoptosis in steroid-induced ONFH (SONFH). It acts as a ceRNA of miR-296-5p, preventing the inhibition of STAT3, a central regulator of cell survival and differentiation. This mechanism identified FGD5-AS1 as a potential drug target for SONFH (Wu et al. 2022). Furthermore, lncRNA TMEM235 (transmembrane protein 235) augmented the therapeutic effect of BMSCs transplantation in early SONFH by alleviating hypoxia-induced apoptosis. TMEM235 competes with miR-34a-3p to suppress BIRC5 inhibition, an apoptosis inhibitor essential for the survival of BMSCs in a hypoxic environment. This interaction increases cell viability and osteogenic differentiation, making TMEM235 a potential target for inducing bone regeneration in SONFH (Zhang et al. 2022b). The lncRNA NORAD acts as a central regulator in

Table 10.1 List of lncRNAs and their mechanism of action in osteosarcoma

Sl. No	LncRNA	Expression in OS	Regulatory role in OS	Mechanism of action	References
1.	LINC00689	Upregulated	Oncogenic	Promoted OS cell growth and invasion by targeting the miR-129-5p/NUSAP1 axis.	Zhang et al. (2025a)
2.	PHF3-3	Upregulated	Oncogenic	Lnc-PHF3-3 contributed to doxorubicin resistance in OS by sponging miR-142-3p and enhancing HMGB1 expression.	Zhou et al. (2025a)
3.	HOXA-AS3	Upregulated	Oncogenic	Interacted with miR-218-5p and facilitated OS cell proliferation and invasion via the miR-218-5p/FOXP1 axis.	Li et al. (2024d)
4.	LINC01137	Upregulated	Oncogenic	Enhanced OS cell proliferation, migration, and invasion, and modulated macrophage polarization within the tumor microenvironment.	Tang et al. (2024)
5.	PVT1	Upregulated	Oncogenic	Promoted OS progression by inhibiting ferroptosis through activating the STAT3/GPX4 axis.	Li et al. (2024a)
6.	LAMTOR5-AS1	Upregulated	Oncogenic	Regulated chemotherapy-induced oxidative stress via the miR-34a-3p/SIRT1/HNF4A axis in OS cells.	Zhao et al. (2024c)
7.	GClnc1	Upregulated	Oncogenic	Promoted OS cell progression and migration by negatively regulating RNA-binding protein NONO ubiquitination.	Zhang et al. (2024)
8.	MALAT1	Upregulated	Oncogenic	Suppressed cell apoptosis and induced cell proliferation, migration, invasion, and epithelial-mesenchymal transition (EMT) by inhibiting miR-590-3p in OS.	Zhao et al. (2022)
9.	SELPLG-2:1	Upregulated	Oncogenic	Compete for binding with miR-10a-5p, promoting OS cell proliferation and invasion, and inhibiting apoptosis.	Li et al. (2022b)

(continued)

Table 10.1 (continued)

Sl. No	LncRNA	Expression in OS	Regulatory role in OS	Mechanism of action	References
10.	LINC01419	Upregulated	Oncogenic	Promoted OS cell proliferation, motility, and EMT by targeting the miR-519a-3p/PDRG1 axis.	Gu et al. (2020)
11.	TUG1	Upregulated	Oncogenic	Functioned as a ceRNA and sponged miR-144-3p to promote EZH2 expression, resulting in OS cell proliferation.	Cao et al. (2017)
12.	FBXL19-AS1	Upregulated	Oncogenic	Negatively regulated miR-346 and promoted OS cell proliferation, migration, and invasion.	Pan et al. (2018)
13.	NEAT1	Upregulated	Oncogenic	Interacted with miR-194 and negatively modulated its expression, enhancing the proliferation and invasion of OS cells.	Wang et al. (2017a)
14.	DHRS4-AS1	Downregulated	Tumor suppressive	Inhibited OS cell proliferation and promoted apoptosis via targeting the miR-362-5p.	Tang et al. (2025)
15.	C1QTNF1-AS1	Downregulated	Tumor suppressive	Suppressed OS progression by inhibiting miR-34a-5p, leading to the upregulation of LDHA and PDK3 expression in OS cells.	Zhang et al. (2025b)

SONFH by enhancing BMSCs proliferation and osteogenic differentiation and suppressing apoptosis. Mechanistically, NORAD inhibits miR-26a-5p, which targets osteoclast-related signaling pathways. Through sequestration of miR-26a-5p, NORAD suppresses the inhibition of critical osteogenic factors, thus promoting bone formation and inhibiting bone resorption. This indicates that the NORAD/miR-26a-5p axis could serve as a potential target for SONFH treatment (Fu et al. 2021). Taken together, these results highlight the important functions of lncRNAs and miRNAs in ONFH pathogenesis and suggest novel therapeutic approaches, especially for SONFH. Regulation of these ncRNAs may promote BMSCs survival, facilitate osteogenesis, and delay or prevent disease progression.

10.2.5 Role of lncRNAs in Fracture Healing

A fracture is a medical condition in which a bone is broken owing to high-force impact, stress, or pathological conditions such as OP and OS. Usually, bone structure and function are fully restored after healing. However, abnormal bone healing affects approximately 10% of patients with fractures (Kang et al. 2024). Dysregulation of osteoblasts is considered the direct cause of delayed fracture healing (DFH) because effective bone formation depends on the number and activity of osteoblasts (Mi et al. 2022). LncRNAs function as critical regulators, capable of both enhancing and suppressing osteogenic repair processes through various mechanisms of action like regulating the expression of genes involved in bone formation and remodeling, interacting with miRNAs, and mediating signaling pathways (Aurilia et al. 2021). Elevated levels of the lncRNA CASC11 have been associated with DFH. CASC11 interacts with miR-150-3p and suppresses the expression of osteogenic marker genes, such as runt-related transcription factor 2 (Runx2) and osteocalcin, which are crucial for bone healing (Wu et al. 2024). LINC00339 negatively regulates miR-16-5p, which leads to the inhibition of osteoblast proliferation and promotion of apoptosis, thereby impairing the healing process (Li et al. 2024e). Moreover, decreased levels of the lncRNA MAGI2-AS3 have been observed in patients with DFH. MAGI2-AS3 promotes fracture healing by downregulating miR-223–3p, resulting in enhanced osteoblast proliferation and reduced apoptosis (Dong et al. 2024).

LncRNAs often enhance the expression of osteogenic genes by acting as ceRNAs. Studies have shown that lncRNA ENST00000563492 promotes osteogenesis by acting as a ceRNA for miR-205-5p. By sequestering miR-205-5p, ENST00000563492 allows the expression of osteogenic genes, thereby facilitating bone formation and healing (Ouyang et al. 2020). In the case of GAS5, its downregulation and miR-29a-3p regulation promoted carpal navicular fracture healing and inhibited apoptosis (Liu et al. 2024b). HOXA11-AS inhibits cell proliferation and promotes osteoblast apoptosis by sponging miR-124-3p, thereby promoting DFH (Wang et al. 2017b).

LncRNAs have also been found to mediate various signaling pathways that are crucial for bone healing. As we mentioned earlier, these pathways include the Wnt/β-catenin, MAPK, TGF-β/BMP, and Notch. By regulating these pathways, lncRNAs influence bone cell proliferation, differentiation, and apoptosis. For example, the lncRNA TRPM2-AS plays a role in osteoblast proliferation and differentiation via the miR-545-3p/BMP2 pathway (Kang et al. 2024). LncRNA KCNQ1OT1 plays a vital role in DFH by inducing cell proliferation and inhibiting cell apoptosis through the activation of the Wnt/β-catenin signaling pathway (Gu et al. 2019). These regulatory functions of lncRNAs ensure that bone regeneration processes are well coordinated and efficient.

10.2.6 Role of lncRNAs in Other Bone Disorders

Intervertebral disc degeneration (IVDD) is a musculoskeletal disorder that causes severe pain and disabilities. Proliferation and apoptosis of nucleus pulposus (NP) cells, along with the uncontrolled expression of ECM components, are the major causal factors of IVDD. This dysregulation is modulated by lncRNA expression. For instance, HOTAIR inhibits apoptosis of NP cells by acting as an miR-34a sponge, which leads to the upregulation of Bcl-2 expression. However, HOTAIR expression was downregulated during IVDD progression (Yu et al. 2018). Alternatively, the upregulation of lncRNAs such as GAS5, TUG1, and RMRP leads to the progression of IVDD. GAS5 sponged miR-155 and RMRP sponged miR-206 to regulate the apoptosis and proliferation of NP cells, respectively, while TUG1 activated the Wnt/beta-catenin signaling pathway to promote IVDD (Jiang et al. 2021). Multiple myeloma (MM) is a cancer that originates from plasma cells in the bone marrow. It is often associated with bone pain, lesions, limb weakness, and anemia. Several lncRNAs have been shown to promote MM through mechanisms such as sponging miRNAs, increasing drug resistance, and inhibiting apoptosis. Upregulation of MALAT1 regulates the expression of SOX13 by sponging miR-1271-5p, thereby promoting MM cell proliferation, migration, and invasion (Liu et al. 2020). In contrast, OIP5-AS1 is a downregulated lncRNA that inhibits MM cell proliferation via the OIP5-AS1-miR-410-KLF10/PTEN/AKT signaling axis (Yang et al. 2017). LncRNAs act as both oncogenes and tumor suppressors in acute myeloid leukemia (AML). For instance, H19, HOTAIR, and UCA1 are upregulated and sponge miR-19a-3p, miR-193a, and miR-126, respectively, to promote AML progression. The lncRNAs NEAT1 and MEG3 are downregulated in AML. Overexpression of NEAT1 leads to miR-23a-3p downregulation, thereby inhibiting AML progression, whereas overexpression of MEG3 upregulates p53 (tumor suppressor gene) expression, leading to AML suppression (Gourvest et al. 2019).

Ankylosing spondylitis (AS) is a chronic inflammatory disease characterized by severe pain and joint stiffness. LncRNAs may serve as potential biomarkers in AS. These lncRNAs are involved in immune and inflammatory response pathways, including protein ubiquitination regulation, major histocompatibility complex class I-mediated antigen processing and presentation, activation of MAPK, and IL-17 signaling pathways. Elevated H19 expression was found to be an indicator of AS and promotes AS via the H19/miR-675-5p/VDR/IL17A axis by mediating the inflammatory response (Esawy et al. 2023). These studies indicate that lncRNAs regulate bone disorders through various mechanisms, such as controlling gene expression, affecting cell behavior, and modulating immune responses. Further studies on these molecular mechanisms may highlight the potential of lncRNAs as diagnostic markers and therapeutic agents for bone disorders.

10.3 Potential of lncRNAs as Biomarkers in Bone Diseases

LncRNAs can serve as biomarkers for various bone diseases owing to their differential expression in bone tissues. For example, specific lncRNAs are upregulated or downregulated in conditions such as OP, OA, and bone cancer. For example, LINC00941 is upregulated and inhibits osteoblast proliferation and stimulates apoptosis via the miR-335-5p/KAT7 axis, thereby promoting DFH (Liu et al. 2025). In contrast, Chen et al. reported the downregulation of the lncRNA HAGLR in DFH. However, when it was overexpressed, it sponged miR-214-3p and promoted BMP2 expression, resulting in improved fracture healing (Chen and Yang 2021). In OS, LINC00689 was associated with disease progression via the miR-129-5p/NUSAP1 axis, whereas XLOC_006786 was found to suppress OS proliferation, invasion, and metastasis by sponging miR-491-5p (Zhang et al. 2025a; Xu et al. 2023). Similarly, in OP, LPS-induced TMC3-AS1 expression is upregulated and promotes disease progression by targeting miR-708. GAS5 was found to be downregulated, and overexpression of GAS5 led to the suppression of OP by promoting the osteogenesis of BMSCs through the miR-135a-5p/FOXO1 axis (Chen and Dai 2022; Wang et al. 2019b). Chen et al. revealed that lncRNA AWPPH is downregulated in osteonecrosis. Its expression upregulates Runx2 expression to promote osteoblast differentiation, thereby inhibiting osteonecrosis (Chen et al. 2020). Elevated levels of lncRNA RMRP upregulate retinol-binding protein 4 and activate the JNK signaling pathway by binding to FOXC1, thereby accelerating LPS-induced apoptosis and inflammation in OA chondrocytes (Li et al. 2024b). Thus, the upregulation or downregulation of lncRNAs makes them valuable indicators for the early diagnosis and monitoring of disease progression. Research on lncRNAs in bone diseases is still in its early stages, but its potential is vast. Ongoing studies aim to uncover the precise mechanisms by which lncRNAs influence bone health and to develop lncRNA-based biomarkers (Fig. 10.4) (Kamal et al. 2024).

10.4 LncRNAs Mediate Drug Resistance in Bone Diseases

The management of bone diseases, particularly bone malignancies, generally involves neoadjuvant chemotherapy, surgical resection, and subsequent adjuvant chemotherapy. However, the efficacy of these conventional treatment methods is frequently hindered by primary or acquired drug resistance. Several studies have identified lncRNAs that play roles in chemoresistance in bone cancers, including OS and AML. For instance, Linc00467 promotes adriamycin resistance in AML, but the downregulation of another lncRNA, XIST, decreases MYC expression, increases drug sensitivity, and induces apoptosis in AML cells (Rao et al. 2020; Wang et al. 2020a). AML-derived exosomal lncRNA SNHG5 promotes a cancer-associated fibroblast-like phenotype and autophagy in AML-MSCs through its interaction with PTBP1, which stabilizes autophagy-related 5 mRNA, resulting in

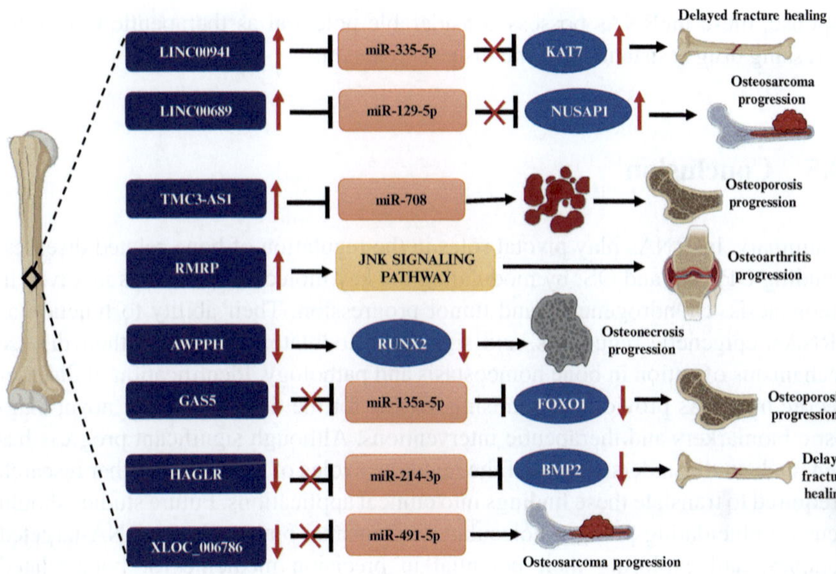

Fig. 10.4 LncRNAs are differentially expressed in various bone diseases, which make them potential biomarkers or therapeutic targets

enhanced chemoresistance (Song et al. 2025). Several lncRNAs, including SNHG14, PVT1, and ANRIL, have been implicated in the promotion of drug resistance in OS. SNHG14 enhances OS resistance to the drug nutlin 3a via the miR-206/SLC7A11 axis, whereas lncRNA PVT1 overexpression augments gemcitabine resistance by activating the DNMT1/PI3K/AKT signaling pathway (Li et al. 2023b; Sun et al. 2019). In contrast, ANRIL functions as a chemosensitive and prognostic biomarker, and its knockdown significantly increases OS cell sensitivity to both cisplatin and doxorubicin (Lee et al. 2021).

Several specific lncRNAs have been highlighted as possible therapeutic targets owing to their tendency to enhance chemosensitivity in OS. For example, MSC-AS1 downregulation inhibits OS progression and increases cisplatin sensitivity by binding to miR-142 (Zhang et al. 2020a). Overexpression of lncRNA CTA suppresses autophagy and promotes chemotherapy-induced apoptosis, thereby enhancing doxorubicin sensitivity (Wang et al. 2017c). FENDRR, an lncRNA downregulated in doxorubicin-resistant OS cells, has been shown to reverse drug resistance by suppressing the expression of ATP-binding cassette transporters ABCB1 and ABCC1 (Kun-Peng et al. 2017). MEG3 enhances OS chemosensitivity by regulating antitumor immunity via the miR-21-5p/p53 pathway and autophagy (Huang et al. 2023c). Other lncRNAs, including SARCC, NORAD, and ODRUL, regulated OS chemoresistance through various mechanisms involving miRNAs and key signaling pathways (Zhu et al. 2017; Wen et al. 2020; Xie et al. 2020). LINC00161 has been shown to mitigate OS drug resistance by targeting the miR-645/IFIT2 signaling axis (Wang et al. 2016). Because of their regulatory functions in the chemotherapy response, these lncRNAs possess considerable potential as therapeutic targets for addressing drug resistance in bone disease treatment.

10.5 Conclusion

In summary, lncRNAs play pivotal roles in the regulation of bone-related diseases, including OP, OA, and OS, by modulating the key molecular pathways involved in osteogenesis, chondrogenesis, and tumor progression. Their ability to function as ceRNAs, epigenetic regulators, and signaling modulators highlights their diverse mechanisms of action in bone homeostasis and pathology. Identification of disease-specific lncRNAs provides a promising avenue for the development of novel diagnostic biomarkers and therapeutic interventions. Although significant progress has been made in the understanding of the regulatory roles of lncRNAs, further research is required to translate these findings into clinical applications. Future studies should focus on elucidating precise molecular mechanisms, optimizing lncRNA-targeted therapies, and exploring their potential in precision medicine for bone-related disorders.

Acknowledgments This research received financial assistance from the Department of Science and Technology, Science and Engineering Research Board (DST-SERB) (Anusandhan National Research Foundation, CRG/2023/000212, awarded to N. S.), and the Department of Science and Technology (DST), India (INSPIRE Fellowship: 2021/IF210073, granted to I. S.).

Author Contributions I. Saranya, D. Preetha, S. Nivruthi, and S. Nivetha wrote the manuscript. N. Selvamurugan reviewed and edited the manuscript and secured funding for the project.

Compliance with Ethical Standards

Conflict of Interest The authors declare that they do not have any conflicts of interest.

Ethical Approval This chapter does not contain any studies with human participants performed by any of the authors.

References

Arya PN, Saranya I, Selvamurugan N (2024) RUNX2 regulation in osteoblast differentiation: a possible therapeutic function of the lncRNA and miRNA-mediated network. Differentiation 140:100803

Aurilia C, Donati S, Palmini G, Miglietta F, Iantomasi T, Brandi ML (2021) The involvement of long non-coding RNAs in bone. Int J Mol Sci 22(8):3909

Bai J, Zhang X, Jiang F, Shan H, Gao X, Bo L, Zhang Y (2023) A feedback loop of LINC00665 and the Wnt signaling pathway expedites osteosarcoma cell proliferation, invasion, and epithelial-mesenchymal transition. Orthop Surg 15(1):286–300

Basu S, Nadhan R, Dhanasekaran DN (2025) Long non-coding RNAs in ovarian cancer: mechanistic insights and clinical applications. Cancer 17(3):472

Cai L, Lv J, Zhang Y, Li J, Wang Y, Yang H (2017) The lnc RNA HNF 1A-AS 1 is a negative prognostic factor and promotes tumorigenesis in osteosarcoma. J Cell Mol Med 21(11):2654–2662

Cai Q, Zhao X, Wang Y, Li S, Wang J, Xin Z, Li F (2021) LINC01614 promotes osteosarcoma progression via miR-520a-3p/SNX3 axis. Cell Signal 83:109985

Cao J, Han X, Qi X, Jin X, Li X (2017) TUG1 promotes osteosarcoma tumorigenesis by upregulating EZH2 expression via miR-144-3p. Int J Oncol 51(4):1115–1123

Chen S, Dai M (2022) Lipopolysaccharide-induced lncRNA TMC3-AS1 is highly expressed in osteoporosis and promotes osteoblast apoptosis by suppressing the formation of mature miR-708. Int J Gen Med Volume 15:3345–3352

Chen X, Xu H (2020) LncRNA SNHG15 regulates osteosarcoma progression in vitro and in vivo via sponging miR-346 and regulating TRAF4 expression. Open Life Sci 15(1):423–436

Chen J, Yang Y (2021) LncRNA HAGLR absorbing miR-214-3p promotes BMP2 expression and improves tibial fractures. Am J Transl Res 13(10):11065

Chen X, Li J, Liang D, Zhang L, Wang Q (2020) LncRNA AWPPH participates in the development of non-traumatic osteonecrosis of femoral head by upregulating Runx2. Exp Ther Med 19(1):153–159

Chen X, Liu J, Sun Y, Wen J, Zhou Q, Ding X, Zhang X (2023) Correlation analysis of differentially expressed long non-coding RNA HOTAIR with PTEN/PI3K/AKT pathway and inflammation in patients with osteoarthritis and the effect of baicalin intervention. J Orthop Surg Res 18(1):34

Ding L, Liu T, Qu Y, Kang Z, Guo L, Zhang H, Jiang J, Qu F, Ge W, Zhang S (2021) lncRNA MELTF-AS1 facilitates osteosarcoma metastasis by modulating MMP14 expression. Mol Ther Nucleic Acids 26:787–797

Dong Z, Hu B, Wang S, Wang M, Sun S, Liu X, Li D, Wu D (2024) LncRNA MAGI2-AS3 promotes fracture healing through downregulation of miR-223-3p. J Orthop Surg Res 19(1):370

Esawy MM, Ebaid AM, Abd-Elhameed A, Thagfan FA, Mubaraki MA, Alazzouni AS, Dkhil MA, Shabana MA (2023) Assessment of circulating lncRNA H19 in ankylosing spondylitis patients and its correlation with disease activity. J Pers Med 13(6):914

Feng L, Yang Z, Hou N, Wang M, Lu X, Li Y, Wang H, Wang Y, Bai S, Zhang X, Lin Y (2023) Long non-coding RNA Malat1 increases the rescuing effect of quercetin on tnfα-impaired bone marrow stem cell osteogenesis and ovariectomy-induced osteoporosis. Int J Mol Sci 24(6):5965

Forlino A, Marini JC (2016) Osteogenesis imperfecta. Lancet 387(10028):1657–1671

Fu D, Yang S, Lu J, Lian H, Qin K (2021) LncRNA NORAD promotes bone marrow stem cell differentiation and proliferation by targeting miR-26a-5p in steroid-induced osteonecrosis of the femoral head. Stem Cell Res Ther 12:1–10

Gourvest M, Brousset P, Bousquet M (2019) Long noncoding RNAs in acute myeloid leukemia: functional characterization and clinical relevance. Cancer 11(11):1638

Gu H, Li Z, Lv XF, Zhao AB, Zhu MY, Zhang Y (2019) LncRNA KCNQ1OT1 delayed fracture healing through the Wnt/β-catenin pathway. European Review for Medical & Pharmacological Sciences 23(11):4575

Gu Z, Wu S, Wang J, Zhao S (2020) Long non-coding RNA LINC01419 mediates miR-519a-3p/PDRG1 axis to promote cell progression in osteosarcoma. Cancer Cell Int 20(1):147

Han W, Liu J (2018) LncRNA-p21 inhibited the proliferation of osteosarcoma cells via the miR-130b/PTEN/AKT signaling pathway. Biomed Pharmacother 97:911–918

He Y, Huang X, Ma Y, Yang G, Cui Y, Lv X, Zhao R, Jin H, Tong Y, Zhang X, Li J (2024) A novel aging-associated lncRNA signature for predicting prognosis in osteosarcoma. Sci Rep 14(1):1386

Hu XH, Dai J, Shang HL, Zhao ZX, Hao YD (2018) High levels of long non-coding RNA DICER1-AS1 are associated with poor clinical prognosis in patients with osteosarcoma. Eur Rev Med Pharmacol Sci 22(22):7640–7645

Huang F, Su Z, Yang J, Zhao X, Xu Y (2023a) Downregulation of lncRNA NEAT1 interacts with miR-374b-5p/PGAP1 axis to aggravate the development of osteoarthritis. J Orthop Surg Res 18(1):670

Huang H, Yan J, Lan X, Guo Y, Sun M, Zhao Y, Zhang F, Sun J, Lu S (2023b) LncRNA WDR11-AS1 promotes extracellular matrix synthesis in osteoarthritis by directly interacting with RNA-binding protein PABPC1 to stabilize SOX9 expression. Int J Mol Sci 24(1):817

Huang X, Zhang W, Pu F, Zhang Z (2023c) LncRNA MEG3 promotes chemosensitivity of osteosarcoma by regulating antitumor immunity via miR-21-5p/p53 pathway and autophagy. Genes Dis 10(2):531–541

Huang Y, Tao M, Yan S, He X (2023d) Long non-coding RNA Homeobox D gene cluster antisense growth-associated long noncoding RNA/microRNA-182-5p/Homeobox protein A10 alleviates postmenopausal osteoporosis via accelerating osteoblast differentiation of bone marrow mesenchymal stem cells. J Orthop Surg Res 18(1):726

Jankowski M, Farzaneh M, Ghaedrahmati F, Shirvaliloo M, Moalemnia A, Kulus M, Ziemak H, Chwarzyński M, Dzięgiel P, Zabel M, Piotrowska-Kempisty H (2023) Unveiling mesenchymal stem cells' regenerative potential in clinical applications: insights in miRNA and lncRNA implications. Cells 12(21):2559

Jia Z, Wang Y, Sun X, Zhao X, Zhang Y, Xu S, Wang Y, Li Y (2021) Effect of lncRNA XLOC_005950 knockout by CRISPR/Cas9 gene editing on energy metabolism and proliferation in osteosarcoma MG63 cells mediated by hsa-miR-542-3p. Oncol Lett 22(3):669

Jiang J, Sun Y, Xu G, Wang H, Wang L (2021) The role of miRNA, lncRNA and circRNA in the development of intervertebral disk degeneration. Exp Ther Med 21(6):555

Kamal A, Swellam M, Shalaby NM, Darwish MK, El-Nahrery EM (2024) Silent players, loud impact: unveiling the therapeutic potentials of LncRNAs. J Transl Genetics Genomics 8(2):162–185

Kang R, Huang L, Zeng T, Ma J, Jin D (2024) Long non-coding TRPM2-AS regulates fracture healing by targeting miR-545-3p/Bmp2. J Orthop Surg Res 19(1):466

Kun-Peng Z, Xiao-Long M, Chun-Lin Z (2017) LncRNA FENDRR sensitizes doxorubicin-resistance of osteosarcoma cells through down-regulating ABCB1 and ABCC1. Oncotarget 8(42):71881

Lee AM, Ferdjallah A, Moore E, Kim DC, Nath A, Greengard E, Huang RS (2021) Long non-coding RNA ANRIL as a potential biomarker of chemosensitivity and clinical outcomes in osteosarcoma. Int J Mol Sci 22(20):11168

Li Z (2023) LncRNA PCBP1-AS1 induces osteoporosis by sponging miR-126-5p/PAK2 axis. Bone Joint Res 12(6):375–386

Li B, Chen P, Qu J, Shi L, Zhuang W, Fu J, Li J, Zhang X, Sun Y, Zhuang W (2014) Activation of LTBP3 gene by a long noncoding RNA (lncRNA) MALAT1 transcript in mesenchymal stem cells from multiple myeloma. J Biol Chem 289(42):29365–29375

Li C, Wang F, Wei B, Wang L, Kong D (2019) LncRNA AWPPH promotes osteosarcoma progression via activation of Wnt/β-catenin pathway through modulating miR-93-3p/FZD7 axis. Biochem Biophys Res Commun 514(3):1017–1022

Li D, Yang C, Tian Y, Chen Z, Qian A, Yin C (2021) Roles and mechanism of long noncoding RNAs in bone diseases. In: Noncoding RNAs and bone. Springer, pp 95–128

Li D, Wang X, Yi T, Zhang L, Feng L, Zhang M, He Y, Gang S (2022a) LncRNA MINCR attenuates osteoarthritis progression via sponging miR-146a-5p to promote BMPR2 expression. Cell Cycle 21(22):2417–2432

Li S, Zeng M, Yang L, Tan J, Yang J, Guan H, Kuang M, Li J (2022b) Lnc-SELPLG-2: 1 enhanced osteosarcoma oncogenesis via hsa-miR-10a-5p and the BTRC cascade. BMC Cancer 22(1):1044

Li B, Wang J, Xu F, Wang Q, Liu Q, Wang G, Miao D, Sun Q (2023a) LncRNA RAD51-AS1 regulates human bone marrow mesenchymal stem cells via interaction with YBX1 to ameliorate osteoporosis. Stem Cell Rev Rep 19(1):170–187

Li L, Zhang Y, Gao Y, Hu Y, Wang R, Wang S, Li Y, He Y, Yuan C (2023b) LncSNHG14 promotes nutlin3a resistance by inhibiting ferroptosis via the miR-206/SLC7A11 axis in osteosarcoma cells. Cancer Gene Ther 30(5):704–715

Li G, Feng J, Huang S, Li Q (2024a) LncRNA-PVT1 inhibits Ferroptosis through activating STAT3/GPX4 Axis to promote osteosarcoma progression. Front Biosci Landmark 29(6):207

Li J, Zhou G, Chen T, Lin Q, Yang Q (2024b) LncRNA RMRP promotes chondrocyte injury by regulating the FOXC1/RBP4 axis. Cent Eur J Immunol 49(1):366–382

Li JX, Qu YD, Xia CL, Zhang W, Wang SS, Ou SJ, Yang Y, Qi Y, Xu CP (2024c) Analysis of PANoptosis-related ceRNA network reveals lncRNA MIR17HG involved in osteogenic differentiation inhibition impaired by tumor necrosis factor-α. Mol Biol Rep 51(1):909

Li R, Chen P, Zhou Y, Lang Y, Zhou C, Ren J, Maimaitiyimin A, Chen Z, Liu C, Mainike A, Ding L (2024d) LncRNA HOXA-AS3 promotes cell proliferation and invasion via targeting miR-218-5p/FOXP1 axis in osteosarcoma. Sci Rep 14(1):16581

Li Y, Sun Y, Ma K, Wang S, Wang Z, Huang L (2024e) Functional mechanism and clinical implications of LINC00339 in delayed fracture healing. J Orthop Surg Res 19(1):511

Li H, Zheng F, Tao A, Wu T, Zhan X, Tang H, Cui X, Ma Z, Li C, Jiang J, Wang Y (2025) LncRNA H19 promotes osteoclast differentiation by sponging miR-29c-3p to increase expression of cathepsin K. Bone 192:117340

Liu C, Lin J (2016) Long noncoding RNA ZEB1-AS1 acts as an oncogene in osteosarcoma by epigenetically activating ZEB1. Am J Transl Res 8(10):4095

Liu C, Cao Z, Bai Y, Dou C, Gong X, Liang M, Dong R, Quan H, Li J, Dai J, Kang F (2019) LncRNA AK077216 promotes RANKL-induced osteoclastogenesis and bone resorption via NFATc1 by inhibition of NIP45. J Cell Physiol 234(2):1606–1617

Liu N, Feng S, Li H, Chen X, Bai S, Liu Y (2020) Long non-coding RNA MALAT1 facilitates the tumorigenesis, invasion and glycolysis of multiple myeloma via miR-1271-5p/SOX13 axis. J Cancer Res Clin Oncol 146:367–379

Liu G, Luo S, Lei Y, Jiao M, Cao R, Guan H, Tian R, Wang K, Yang P (2022) Osteogenesis-related long noncoding RNA GAS5 as a novel biomarker for osteonecrosis of femoral head. Front Cell Dev Biol 10:857612

Liu J, Wang B, Chen H, Yu X, Cao X, Zhang H (2024a) Osteoclast-derived exosomes influence osteoblast differentiation in osteoporosis progression via the lncRNA AW011738/miR-24-2-5p/TREM1 axis. Biomed Pharmacother 178:117231

Liu X, Tian L, Deng Z, Guo Y, Zhang S (2024b) Zoledronic acid accelerates bone healing in carpal navicular fracture via silencing long non-coding RNA growth arrest specificity 5 to modulate MicroRNA-29a-3p expression. Mol Biotechnol 66(11):3238–3251

Liu L, Yang Y, Sun P (2025) LINC00941 affects the proliferation, apoptosis and differentiation of osteoblasts by regulating the miR-335-5p/KAT7 axis. J Orthop Surg Res 20(1):75

Luo W, Zhang N, Wang Z, Chen H, Sun J, Yao C, Zhang Y (2024) LncRNA USP2-AS1 facilitates the osteogenic differentiation of bone marrow mesenchymal stem cells by targeting KDM3A/ETS1/USP2 to activate the Wnt/β-catenin signaling pathway. RNA Biol 21(1):89–101

Márton É, Varga A, Domoszlai D, Buglyó G, Balázs A, Penyige A, Balogh I, Nagy B, Szilágyi M (2025) Non-coding RNAs in cancer: structure, function, and clinical application. Cancer 17(4):579

Mi B, Chen L, Xiong Y, Yang Y, Panayi AC, Xue H, Hu Y, Yan C, Hu L, Xie X, Lin Z (2022) Osteoblast/osteoclast and immune cocktail therapy of an exosome/drug delivery multifunctional hydrogel accelerates fracture repair. ACS Nano 16(1):771–782

Mir MA (2024) Illuminating the pathogenic role of SARS-CoV-2: insights into competing endogenous RNAs (ceRNAs) regulatory networks. Infect Genet Evol 122:105613

Nemeth K, Bayraktar R, Ferracin M, Calin GA (2024) Non-coding RNAs in disease: from mechanisms to therapeutics. Nat Rev Genet 25(3):211–232

Ouyang Z, Tan T, Zhang X, Wan J, Zhou Y, Jiang G, Yang D, Liu T (2020) LncRNA ENST00000563492 promoting the osteogenesis–angiogenesis coupling process in bone

mesenchymal stem cells (BMSCs) by functions as a ceRNA for miR-205-5p. Cell Death Dis 11(6):486
Pan R, He Z, Ruan W, Li S, Chen H, Chen Z, Liu F, Tian X, Nie Y (2018) lncRNA FBXL19-AS1 regulates osteosarcoma cell proliferation, migration and invasion by sponging miR-346. Onco Targets Ther 11:8409–8420
Pan J, Liu M, Duan X, Wang D (2023) A short peptide LINC00665_18aa encoded by lncRNA LINC00665 suppresses the proliferation and migration of osteosarcoma cells through the regulation of the CREB1/RPS6KA3 interaction. PLoS One 18(6):e0286422
Peng S, Gao Y, Shi S, Zhao D, Cao H, Fu T, Cai X, Xiao J (2022) LncRNA-AK137033 inhibits the osteogenic potential of adipose-derived stem cells in diabetic osteoporosis by regulating Wnt signaling pathway via DNA methylation. Cell Prolif 55(1):e13174. https://doi.org/10.1111/cpr.13174
Qian G, Yu Y, Dong Y, Hong Y, Wang M (2022) LncRNA AWPPH is downregulated in osteoporosis and regulates type I collagen α1 and α2 ratio. Arch Physiol Biochem 128(5):1297–1301
Rao CB, Luo D, Lin ZT, Xie MY, Hu Y, Peng Q, Jiang H, Zhang ZH, Lu XM (2020) Expression of long non-coding RNA linc00467 in childhood acute myeloid leukemia and its role in drug resistance. Zhongguo Dang dai er ke za zhi= Chinese J Contemp Pediatr 22(7):734–738
Rothzerg E, Xu J, Wood D (2022) Identification of differentially expressed intronic transcripts in osteosarcoma. Noncoding RNA 8(6):73
Shao Y, Tong Z, Wei J, Yang T (2020) LncRNA-zinc finger protein 281 downregulates rho-associated coiled-coil containing protein kinase 1 by upregulating miR-144 in osteosarcoma. Oncol Lett 20(4):1–1
Shen P, Gao J, Huang S, You C, Wang H, Chen P, Yao T, Gao T, Zhou B, Shen S, Zhao X (2023a) LncRNA AC006064. 4–201 serves as a novel molecular marker in alleviating cartilage senescence and protecting against osteoarthritis by destabilizing CDKN1B mRNA via interacting with PTBP1. Biomark Res 11(1):39
Shen X, Qin J, Wei Z, Liu F (2023b) Bone marrow mesenchymal stem cell exosome-derived lncRNA TUC339 influences the progression of osteoarthritis by regulating synovial macrophage polarization and chondrocyte apoptosis. Biomed Pharmacother 167:115488
Sheng B, Li X, Nussler AK, Zhu S (2021) The relationship between healthy lifestyles and bone health: a narrative review. Medicine 100(8):e24684
Song Y, Hu L, Cheng J, Li Z, Zheng J (2025) LncRNA SNHG5 induces CAFs-like phenotype and autophagy of AML-MSCs via PTBP1/ATG5 axis to confer chemoresistance of AML cells. Cell Signal 128:111625
Sozen T, Özişik L, Çalik Başaran NURSEL (2017) An overview and management of osteoporosis. Eur J Rheumatol 4(1)
Sun ZY, Jian YK, Zhu HY, Li B (2019) lncRNAPVT1 targets miR-152 to enhance chemoresistance of osteosarcoma to gemcitabine through activating c-MET/PI3K/AKT pathway. Pathol Res Pract 215(3):555–563
Tamblin-Hopper P, Kiss-Toth E, Sudbery I, Young D, Wilkinson JM (2024) The potential therapeutic applications of long non-coding RNAs. J Transl Genetics Genomics 8(3):225–243
Tang SA, Xie Z, Wang P, Li J, Wang S, Liu W, Li M, Wu X, Su H, Cen S, Ye G (2019) LncRNA-OG promotes the osteogenic differentiation of bone marrow-derived mesenchymal stem cells under the regulation of hnRNPK. Stem Cells 37(2):270–283
Tang N, Chen Y, Su Y, Zhang S, Huang T (2024) The role of disulfidptosis-associated LncRNA-LINC01137 in osteosarcoma biology and its regulatory effects on macrophage polarization. Funct Integr Genomics 24(6):219
Tang Z, Li Z, Wu G, Li J, Tan J, Zhu L (2025) Long noncoding RNA DHRS4 antisense RNA 1 suppresses osteosarcoma cell proliferation and promotes apoptosis through a competitive endogenous RNA mechanism. Sci Rep 15(1):2891
Usman M, Li A, Wu D, Qinyan Y, Yi LX, He G, Lu H (2024) The functional role of lncRNAs as ceRNAs in both ovarian processes and associated diseases. Noncoding RNA Res 9(1):165–177

Wang Y, Zhang LI, Zheng X, Zhong W, Tian X, Yin B, Tian K, Zhang W (2016) Long non-coding RNA LINC00161 sensitises osteosarcoma cells to cisplatin-induced apoptosis by regulating the miR-645-IFIT2 axis. Cancer Lett 382(2):137–146

Wang H, Yu Y, Fan S, Luo L (2017a) Knockdown of long non-coding RNA NEAT1 inhibits proliferation and invasion and induces apoptosis of osteosarcoma by inhibiting miR-194 expression. Yonsei Med J 58(6):1092

Wang XN, Zhang LH, Cui XD, Wang MX, Zhang GY, Yu PL (2017b) lncRNA HOXA11-AS is involved in fracture healing through regulating mir-124-3p. Eur Rev Med Pharmacol Sci 21(21):4771–4776

Wang Z, Liu Z, Wu S (2017c) Long non-coding RNA CTA sensitizes osteosarcoma cells to doxorubicin through inhibition of autophagy. Oncotarget 8(19):31465

Wang Y, Zhao Z, Zhang S, Li Z, Li D, Yang S, Zhang H, Zeng X, Liu J (2018) LncRNA FAL1 is a negative prognostic biomarker and exhibits pro-oncogenic function in osteosarcoma. J Cell Biochem 119(10):8481–8489

Wang CG, Liao Z, Xiao H, Liu H, Hu YH, Liao QD, Zhong D (2019a) LncRNA KCNQ1OT1 promoted BMP2 expression to regulate osteogenic differentiation by sponging miRNA-214. Exp Mol Pathol 107:77–84

Wang X, Zhao D, Zhu Y, Dong Y, Liu Y (2019b) Long non-coding RNA GAS5 promotes osteogenic differentiation of bone marrow mesenchymal stem cells by regulating the miR-135a-5p/FOXO1 pathway. Mol Cell Endocrinol 496:110534

Wang C, Li L, Li M, Wang W, Liu Y, Wang S (2020a) Silencing long non-coding RNA XIST suppresses drug resistance in acute myeloid leukemia through down-regulation of MYC by elevating microRNA-29a expression. Mol Med 26:1–11

Wang L, Jiang J, Sun G, Zhang P, Li Y (2020b) Effects of lncRNA TUSC7 on the malignant biological behavior of osteosarcoma cells via regulation of miR-375. Oncol Lett 20(5):133

Wang B, Li J, Tian F (2021a) Downregulation of lncRNA SNHG14 attenuates osteoarthritis by inhibiting FSTL-1 mediated NLRP3 and TLR4/NF-κB pathway through miR-124-3p. Life Sci 270:119143

Wang X, Li Z, Cui Y, Cui X, Chen C, Wang Z (2021b) Exosomes isolated from bone marrow mesenchymal stem cells exert a protective effect on osteoarthritis via lncRNA LYRM4-AS1-GRPR-miR-6515-5p. Front Cell Dev Biol 9:644380

Wang H, Xu W, Chen X, Mei X, Guo Z, Zhang J (2023) LncRNA LINC00205 stimulates osteoporosis and contributes to spinal fracture through the regulation of the miR-26b-5p/KMT2C axis. BMC Musculoskelet Disord 24(1):262

Wang D, Cao L, Zhang H, Wang X, You W (2024a) LncRNA NKILA attenuates the progression of osteoarthritis through the targeted inhibition of the NF-κB pathway. Int Immunopharmacol 143:113417

Wang Q, Yang J, Pan R, Zha Z (2024b) LncRNA SNHG1 overexpression alleviates osteoarthritis via activating PI3K/Akt signal pathway and suppressing autophagy. Immunobiology 229(3):152799

Wang Y, Hu K, Liao C, Han T, Jiang F, Gao Z, Yan J (2024c) Exosomes-Shuttled lncRNA SNHG7 by bone marrow mesenchymal stem cells alleviates osteoarthritis through targeting miR-485-5p/FSP1 axis-mediated chondrocytes ferroptosis and inflammation. Tissue Eng Regener Med 21(8):1203–1216

Wen JF, Jiang YQ, Li C, Dai XK, Wu T, Yin WZ (2020) LncRNA-SARCC sensitizes osteosarcoma to cisplatin through the miR-143-mediated glycolysis inhibition by targeting Hexokinase 2. Cancer Biomark 28(2):231–246

Wu Y, Fang L, Gao Y, Zhao Z, Zhou L, Zhang G (2022) lncRNA FGD5-AS1 regulates bone marrow stem cell proliferation and apoptosis by affecting miR-296-5p/STAT3 Axis in steroid-induced osteonecrosis of the femoral head. J Healthcare Eng 2022(1):9364467

Wu X, Shen T, Ji W, Huang M, Sima J, Li J, Song H, Xiong W, Cen M (2024) lncRNA CASC11 regulates the progress of delayed fracture healing via sponging miR-150-3p. J Orthop Surg Res 19(1):757

Xiao T, Yan A, Tan L, Zhu H, Gao W (2024) LncRNA HOXA-AS2 is a prognostic and clinicopathological predictor in patients with cancer: a meta-analysis. Oncol Lett 27(5):226

Xie X, Liu W, Duan Z, Li X, Zhang L, Yang G (2020) LncRNA NORAD targets miR-410-3p to regulate drug resistance sensitivity of osteosarcoma. Cell Mol Biol 66(3):143–148

Xing X, Rodeo SA (2024) Emerging roles of non-coding RNAs in fibroblast to myofibroblast transition and fibrotic diseases. Front Pharmacol 15:1423045

Xu JY, Lv YF, Cao Y, Ma HM, Hao XL, Huang L, Tang XF, Guo QN (2023) Long noncoding RNA XLOC_006786 inhibits the proliferation, invasion and metastasis of osteosarcoma cells through NOTCH3 signaling pathway by targeting miR-491-5p. Hum Cell 36(6):2140–2151

Xu P, Yuan J, Li K, Wang Y, Wu Z, Zhao J, Li T, Wu T, Miao X, He D, Cheng X (2024) Development and validation of a novel endoplasmic reticulum stress-related lncRNAs signature in osteosarcoma. Sci Rep 14(1):25590

Yang N, Chen J, Zhang H, Wang X, Yao H, Peng Y, Zhang W (2017) LncRNA OIP5-AS1 loss-induced microRNA-410 accumulation regulates cell proliferation and apoptosis by targeting KLF10 via activating PTEN/PI3K/AKT pathway in multiple myeloma. Cell Death Dis 8(8):e2975–e2975

Yang Y, Fei M, Zhou X, Li Y, Jin D (2019) The potential value of lncRNA-BC050642 in osteosarcoma origination and outcomes. Artif Cells Nanomed Biotechnol 47(1):1859–1866

Yang Z, Liu X, Zhao F, Yao M, Lin Z, Yang Z, Liu C, Liu Y, Chen X, Du C (2022) Bioactive glass nanoparticles inhibit osteoclast differentiation and osteoporotic bone loss by activating lncRNA NRON expression in the extracellular vesicles derived from bone marrow mesenchymal stem cells. Biomaterials 283:121438

Yao X, Li X, Luo Y, Xu X, Liu J, Bu J (2020) LncRNA GAS5 regulates osteosarcoma cell proliferation, migration, and invasion by regulating RHOB via sponging miR-663a. Cancer Manag Res Volume 12:8253–8261

Yao W, Hou J, Liu G, Wu F, Yan Q, Guo L, Wang C (2022) LncRNA CBR3-AS1 promotes osteosarcoma progression through the network of miR-140-5p/DDX54-NUCKS1-mTOR signaling pathway. Mol Ther Oncolytics 25:189–200

Yin J, Zheng Z, Zeng X, Zhao Y, Ai Z, Yu M, Wu YO, Jiang J, Li J, Li S (2022) lncRNA MALAT1 mediates osteogenic differentiation of bone mesenchymal stem cells by sponging miR-129-5p. PeerJ 10:e13355

Yoon JH, Abdelmohsen K, Gorospe M (2013) Posttranscriptional gene regulation by long noncoding RNA. J Mol Biol 425(19):3723–3730

Yu Y, Zhang X, Li Z, Kong L, Huang Y (2018) LncRNA HOTAIR suppresses TNF-α induced apoptosis of nucleus pulposus cells by regulating miR-34a/Bcl-2 axis. Biomed Pharmacother 107:729–737

Yu H, Zhang B, Qi L, Han J, Guan M, Li J, Meng Q (2023) AP003352. 1/miR-141-3p axis enhances the proliferation of osteosarcoma by LPAR3. PeerJ 11:e15937

Zhang L, Zhao G, Ji S, Yuan Q, Zhou H (2020a) Downregulated long non-coding RNA MSC-AS1 inhibits osteosarcoma progression and increases sensitivity to cisplatin by binding to MicroRNA-142. Med Sci Monit 26:e921594–e921591

Zhang Y, Chen XF, Li J, He F, Li X, Guo Y (2020b) lncRNA Neat1 stimulates osteoclastogenesis via sponging miR-7. J Bone Miner Res 35(9):1772–1781

Zhang GF, Zhou BS, An XC, An FM, Li SH (2021a) LINC01278 is highly expressed in osteosarcoma and participates in the development of tumors by mediating the miR-134-5p/KRAS Axis. Onco Targets Ther Volume 14:683–695

Zhang H, Liu S, Tang L, Ge J, Lu X (2021b) Long non-coding RNA (LncRNA) MRPL23-AS1 promotes tumor progression and carcinogenesis in osteosarcoma by activating Wnt/β-catenin signaling via inhibiting microRNA miR-30b and upregulating myosin heavy chain 9 (MYH9). Bioengineered 12(1):162–171

Zhang H, Wu Y, Li W, Chen H (2021c) Clinical significance and mechanism of LncRNA GAS-5 in osteoarthritis. Am J Transl Res 13(7):8465

Zhang DW, Wang HG, Zhang KB, Guo YQ, Yang LJ, Lv H (2022a) LncRNA XIST facilitates S1P-mediated osteoclast differentiation via interacting with FUS. J Bone Miner Metab 40(2):240–250

Zhang F, Peng W, Wang T, Zhang J, Dong W, Wang C, Xie Z, Luo H, Liu G (2022b) Lnc Tmem235 promotes repair of early steroid-induced osteonecrosis of the femoral head by inhibiting hypoxia-induced apoptosis of BMSCs. Exp Mol Med 54(11):1991–2006

Zhang H, Zhou L, Hu S, Gu W, Li Z, Sun J, Wei X, Wang Y (2022c) The crosstalk between LINC01089 and hippo pathway inhibits osteosarcoma progression. J Bone Miner Metab 40(6):890–899

Zhang Z, Yang P, Wang C, Tian R (2022d) LncRNA CRNDE hinders the progression of osteoarthritis by epigenetic regulation of DACT1. Cell Mol Life Sci 79(8):405

Zhang J, Luo X, Guo C, Dai Z, Tang X, Zhang F, Jiao Q, Lin S, Zou L, Zhang Z, Lv XB (2024) LncRNA GClnc1 promotes osteosarcoma progression by stabilizing NONO and blocking FBXW7-mediated ubiquitination. BMC Cancer 24(1):1375

Zhang L, Song J, Xu X, Sun D, Huang H, Chen Y, Zhang T (2025a) Silencing long non-coding RNA linc00689 suppresses the growth and invasion of osteosarcoma cells by targeting miR-129-5p/NUSAP1. Int J Exp Pathol 106(2):e12524

Zhang Y, Lun H, Zhu N, Yang N, Ding K, Chen B, Chang C, Gu H, Liu Y (2025b) Deciphering the oncogenic network: how C1QTNF1-AS1 modulates osteosarcoma through miR-34a-5p and glycolytic pathways. Front Oncol 14:1485605

Zhao H, Wang Y, Hou W, Ding X, Wang W (2022) Long non-coding RNA MALAT1 promotes cell proliferation, migration and invasion by targeting miR-590-3p in osteosarcoma. Exp Ther Med 24(5):672

Zhao F, Wei C, Pu Y, Zang C (2024a) LAMTOR5-AS1 regulates chemotherapy-induced oxidative stress via the miR-34a-3p/SIRT1/HNF4A axis in osteosarcoma cells. Curr Med Chem 32(30):6633–6645

Zhao P, Chang J, Chen Y, Sun X, Ma X, Zhou C, Zhou L, Wang Y, Yang Y (2024b) Cellular senescence-related long non-coding RNA signatures predict prognosis in juvenile osteosarcoma. Phenomics 4(5):430–452

Zhao Y, Ning J, Teng H, Deng Y, Sheldon M, Shi L, Martinez C, Zhang J, Tian A, Sun Y, Nakagawa S (2024c) Long noncoding RNA Malat1 protects against osteoporosis and bone metastasis. Nat Commun 15(1):2384

Zheng D, Yang K, Chen T, Lv S, Wang L, Gui J, Xu C (2023) Inhibition of LncRNA SNHG14 protects chondrocyte from injury in osteoarthritis via sponging miR-137. Autoimmunity 56(1):2270185

Zhou FC, Zhang YH, Liu HT, Song J, Shao J (2020) LncRNA LINC00588 suppresses the progression of osteosarcoma by acting as a ceRNA for miRNA-1972. Front Pharmacol 11:255

Zhou Y, Xu Z, Wang Y, Song Q, Yin R (2023) LncRNA MALAT1 mediates osteogenic differentiation in osteoporosis by regulating the miR-485-5p/WNT7B axis. Front Endocrinol 13:922560

Zhou J, Yang M, Zhao W, Zhang H, Cao L, Li Q, Wang G (2025a) Lnc-PHF3-3 aggravates the chemoresistance of osteosarcoma cells to doxorubicin via the miR-142-3p/HMGB1 axis. Transl Oncol 53:102328

Zhou X, Cao H, Liao T, Hua W, Zhao R, Wang D, Deng H, Yang Y, Liu S, Ni G (2025b) Mechanosensitive lncRNA H19 promotes chondrocyte autophagy, but not pyroptosis, by targeting miR-148a in post-traumatic osteoarthritis. Noncoding RNA Res 10:163–176

Zhu KP, Ma XL, Zhang CL (2017) LncRNA ODRUL contributes to osteosarcoma progression through the miR-3182/MMP2 Axis. Mol Ther 25(10):2383–2393

Zhu C, Huang L, Xu F, Li P, Li P, Hu F (2020a) LncRNA PCAT6 promotes tumor progression in osteosarcoma via activation of TGF-β pathway by sponging miR-185-5p. Biochem Biophys Res Commun 521(2):463–470

Zhu S, Liu Y, Wang X, Wang J, Xi G (2020b) lncRNA SNHG10 promotes the proliferation and invasion of osteosarcoma via wnt/β-catenin signaling. Mol Ther Nucleic Acids 22:957–970

Chapter 11
Long Non-coding RNAs in Infectious Diseases

Rajesh Das, Md Mujibur Rahman Sheikh, Vigneshwar Suriya Prakash Sinnarasan, Dahrii Paul, and Amouda Venkatesan

Abstract Long non-coding RNAs (lncRNAs) have emerged as critical regulators of gene expression and cellular processes, playing important roles in host-pathogen interactions and immune modulation. This chapter explores their diverse functions in infectious diseases, highlighting how lncRNAs influence pathogen replication, immune signaling, and disease progression across viral, bacterial, parasitic, and fungal infections. It also examines their diagnostic and therapeutic potential, emphasizing the advantages of their tissue-specific expression as non-invasive biomarkers and outlining emerging strategies such as antisense oligonucleotides, CRISPR-based editing, and RNA interference. Additionally, the chapter discusses the relevance of lncRNAs in zoonotic and emerging infectious diseases, their contributions to pandemic preparedness, and the integration of computational tools and databases for functional prediction. Together, these insights underscore the expanding role of lncRNAs in infection biology and their promise for future translational applications.

Keywords lncRNA · Host-pathogen interactions · Biomarkers · Immune modulation · RNA therapeutics · Infectious diseases

Abbreviations

ASO	Antisense Oligonucleotide
CRISPR	Clustered Regulatory Interspaced Short Palindromic Repeats
GC	Gastric Cancer
HBV	Hepatitis B Virus
HIV	Human Immunodeficiency Virus
JAK-STAT	Janus Kinase-Signal Transducer and Activator of Transcription

R. Das · Md M. R. Sheikh · V.S.P. Sinnarasan · D. Paul · A. Venkatesan (✉)
Department of Bioinformatics, Pondicherry University, Kalapet, Puducherry, India
e-mail: amouda@bicpu.edu.in

L. M. Vaschetto (ed.), *Long Non-Coding RNAs, Human Health, and Diseases*,
Epigenetics and Human Health 16, https://doi.org/10.1007/978-3-032-19668-2_11

lncRNA	long non-coding RNA
miRNA	microRNA
NF-κB	Nuclear Factor kappa B
RNAi	RNA interference
SARS-CoV-2	Severe Acute Respiratory Syndrome- Coronavirus- 2
siRNA	silencing RNA
TNF	Tumor Necrosis Factor

11.1 Introduction

Long non-coding RNAs (lncRNAs) represent a fascinating class of RNA molecules that have emerged as key regulators of gene expression and cellular processes. Despite lacking protein-coding capacity, lncRNAs play crucial roles in numerous biological functions, including development, cell differentiation, and disease pathogenesis (Statello et al. 2021; Mattick et al. 2023). Recent years have witnessed an explosion of research into the functions and mechanisms of lncRNAs, revealing their complex involvement in various aspects of cellular biology. These non-coding transcripts can regulate gene expression through multiple mechanisms, including chromatin remodeling, transcriptional regulation, and post-transcriptional processing (Marchese et al. 2017; Fernandes et al. 2019). LncRNAs are typically longer than 200 nucleotides and often display tissue- and cell-type-specific expression patterns that reflect their specialized biological functions (Zheng 2019; Kern et al. 2018; Seifuddin et al. 2020; Chen et al. 2022). Although traditionally considered non-coding, some lncRNAs encode functional micropeptides that participate in innate immunity, metabolism, and tumorigenesis (Xiao et al. 2024). Interestingly, lncRNAs have also been implicated in the host-pathogen interaction landscape, suggesting their potential role in modulating immune responses and disease outcomes in infectious diseases (Arunima et al. 2023). The emerging area of research has opened up new avenues for understanding disease mechanisms, identifying novel biomarkers, and developing innovative therapeutic strategies against infectious pathogens (Beermann et al. 2016; Alvarez-Dominguez and Lodish 2017; Xiao et al. 2024). This chapter aims to provide an overview of the role of lncRNAs in infectious diseases, focusing on the mechanisms of action, potential as biomarkers, and therapeutic applications. The current knowledge on lncRNA functions in immune regulation and the involvement in various infectious diseases has been explored. The challenges and opportunities in harnessing lncRNAs for diagnostic and therapeutic purposes have been discussed in the context of infectious disease research.

11.2 lncRNAs in Host-Pathogen Interactions

11.2.1 General Roles in Infection

LncRNAs are primarily transcribed by RNA Polymerase II, undergoing typical RNA processing events such as 5′ capping, splicing, and polyadenylation. These modifications ensure stability and functionality (Zhang et al. 2019b). Figure 11.1 depicts lncRNA-mediated regulation of gene transcription in the nucleus, while Fig. 11.2 illustrates the diverse functional roles of lncRNAs in gene regulation. As signaling molecules, lncRNAs influence gene transcription by acting alone or interacting with transcription factors. They can also function as decoys, sequestering regulatory proteins and preventing the binding to target DNA or mRNA. Additionally, lncRNAs act as sponges by binding and neutralizing miRNAs, thereby reducing the post-transcriptional gene-silencing effects. Another key function of lncRNAs is scaffolding, providing a structural platform for assembling protein complexes involved in shared biological pathways. Collectively, these multifunctional roles underscore the complexity and importance of lncRNA-mediated gene regulation. Beyond general cellular regulation, lncRNAs have emerged as pivotal players in host-pathogen interactions, where they fine-tune immune signaling and influence infection outcomes.

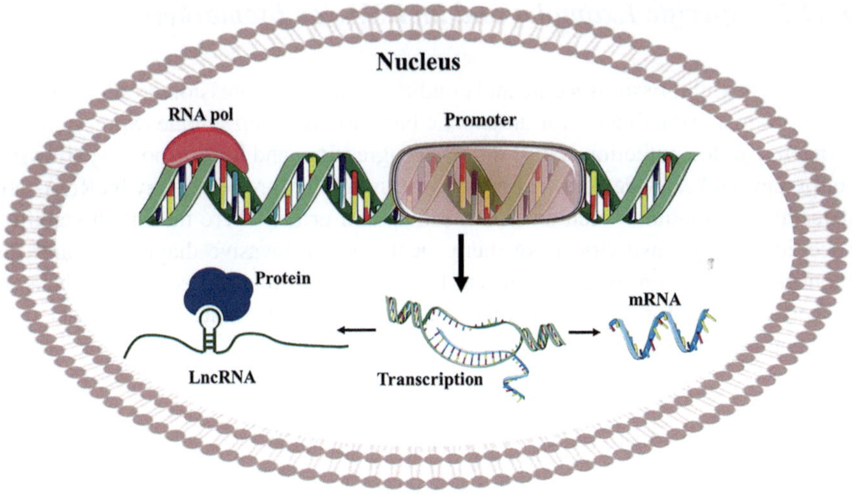

Fig. 11.1 Schematic representation of lncRNA-mediated regulation of gene transcription in the nucleus. LncRNAs influence transcriptional activity within the nucleus. RNA polymerase binds to the DNA template at the promoter region to initiate transcription. LncRNAs interact with specific nuclear proteins, forming regulatory complexes that can modulate promoter accessibility or influence RNA polymerase activity. These regulatory interactions ultimately affect the transcription of target genes, leading to altered mRNA production

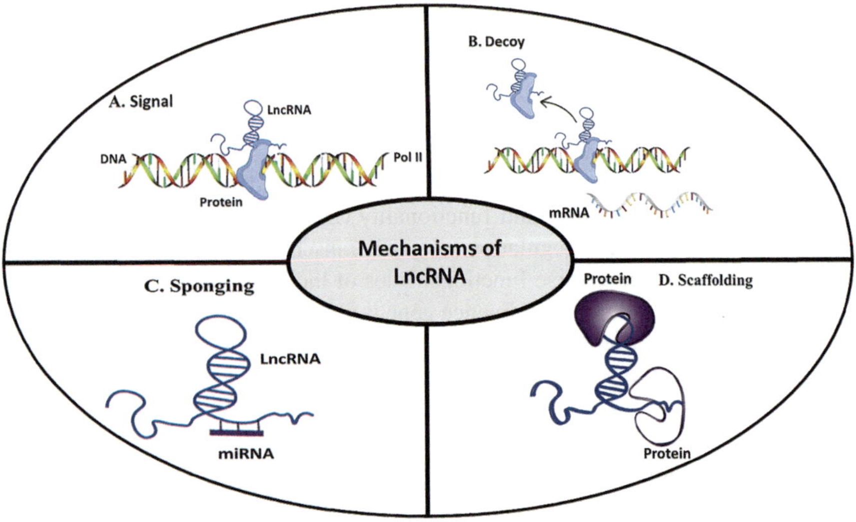

Fig. 11.2 Functional mechanisms of lncRNAs. (**a**) lncRNAs act as signaling molecules, regulating gene transcription independently or in coordination with transcription factors. (**b**) lncRNAs function as decoys by sequestering specific proteins, preventing the interaction with DNA or mRNA. (**c**) lncRNAs serve as molecular sponges by binding complementary miRNAs, reducing their regulatory effects. (**d**) lncRNAs also function as scaffolds, facilitating interactions between proteins in the same biological pathway

11.2.2 Specific Examples and lncRNAs as Biomarkers

LncRNAs exhibit tissue-specific and condition-specific expression profiles, making them promising candidates for diagnostic biomarkers. Their unique expression patterns may reflect infection status, disease progression, and host responses, allowing for precise and early detection of infectious diseases. Moreover, many lncRNAs are detectable in biofluids such as blood, saliva, and urine, where their high stability, specificity, and sensitivity make them ideal for non-invasive diagnostics and for monitoring infection dynamics in real time. These attributes position lncRNAs as powerful tools for precision medicine in infectious disease diagnosis and management.

11.2.2.1 Viral Infections

In viral infections, several lncRNAs have been identified as potential diagnostic indicators. For instance, HOX transcript antisense intergenic RNA (HOTAIR) is overexpressed in hepatitis B virus (HBV)-induced hepatocellular carcinoma, indicating its potential as a biomarker for HBV-related liver damage (Wu et al. 2017; Ren et al. 2020). Nuclear paraspeckle assembly transcript 1 (NEAT1) plays a key role in the replication of HIV and influenza viruses; its altered levels serve as a diagnostic marker (Zhang et al. 2013; Milcamps and Michiels 2024). Non-coding repressor of NFAT (NRON) regulates HIV latency and reactivation, making it a potential therapeutic and diagnostic target (Li et al. 2016; Amir and Taube 2024). HIV utilizes the lncRNA NEAT1 to facilitate viral replication by modulating host chromatin architecture (Zhang et al. 2013). In SARS-CoV-2 infection, lncRNAs such as H19, TUG1, and CRNDE were significantly overexpressed compared to healthy controls (Heydari et al. 2024). These lncRNAs correlate with COVID-19 severity, positioning TUG1 as a potential diagnostic biomarker and predictor of outcomes like survival. Therapeutically, targeting them (e.g., via antisense oligonucleotides) could mitigate inflammation, fibrosis, and disease progression. Overall, they highlight host factors driving pathogenesis, aiding biomarker and treatment development (Tayel et al. 2022; Yu et al. 2025). In influenza, lncRNA NRAV suppresses interferon-stimulated genes, aiding viral replication (Ouyang et al. 2014). The lncRNAs IPAN and PAAN interact with the viral proteins PB2 and PA, respectively, to facilitate viral replication (Wang et al. 2018). LncRNA AVAN strengthens the interaction between TRIM25 and RIG-I, thereby enhancing type I interferon (IFN) response (Bamunuarachchi et al. 2021). In contrast, NRAV appears to suppress early transcription of key interferon-stimulated genes (ISGs), including MxA and IFITM3, by modulating the establishment of histone modifications such as H3K4me3 and H3K27me3 at these loci. This suppression ultimately supports influenza A virus (IAV) replication (Sarma et al. 2025).

11.2.2.2 Bacterial Infections

In bacterial infections, lncRNAs are also implicated as indicators of disease activity and immune regulation. P50-associated cyclooxygenase-2 extragenic RNA (PACER) regulates inflammatory responses in bacterial infections by modulating COX-2 expression, a marker for sepsis (Krawczyk and Emerson 2014; Cheng et al. 2024). TNFα and HNRNPL-related immunoregulatory LincRNA (THRIL) correlate with TNFα expression in *Mycobacterium tuberculosis* infections as a marker for inflammation and disease progression (Ngongang et al. 2022; Wang et al. 2023). In *Helicobacter pylori (H. pylori)* infection, lncRNA-AF147447 is significantly downregulated, leading to the loss of its tumor-suppressive function. This reduction promotes gastric carcinoma (GC) cell proliferation and invasion, underscoring its diagnostic potential in *H. pylori*-induced GC (Zhou et al. 2016). In tuberculosis, the lncRNA MEG3 enhances macrophage-mediated immunity by regulating autophagy-related genes (Pawar et al. 2016). On the other hand, *H. pylori* infection induces lncRNA H19, which may contribute to gastric inflammation and carcinogenesis (Zhang et al. 2019c; He et al. 2024).

11.2.2.3 Parasitic Infections

Similarly, differential lncRNA expression provides insight into host immune modulation in parasitic infections. TARE-3 and TARE-6 are two lncRNAs produced by *Plasmodium falciparum* infections linked to host immune modulation (Lodde et al. 2022). LncRNA HIF1A, HIF1A-AS1, and HIF1A-AS3 are altered in *Leishmania infantum* infected macrophages, where they help regulate the immune response. Their altered expression levels may also serve as indicators of infection and disease progression (Fernandes et al. 2022). LncRNA NONSHAT022487 is upregulated in *Toxoplasma gondii* infections, potentially influencing immune escape and chronic infection (Liu et al. 2018). Specific host lncRNAs are differentially expressed in malaria during *Plasmodium* infection, modulating erythrocyte function (Chen et al. 2022). Similarly, lncRNAs regulate macrophage polarization in leishmaniasis, affecting the host's ability to control parasitic burden (Fernandes et al. 2023).

11.2.2.4 Fungal Infections

In fungal infections, lncRNAs are emerging as regulators of host-pathogen interactions with biomarker potential. HAX1 is identified in *Trichoderma reesei* and plays a regulatory role in filamentous fungal-induced immune responses (Till et al. 2018). Moreover, Hou et al. found that lnc-CCL3L3–1:1 might be involved in the NF-kappa B pathway in *Candida albicans* infection (Hou et al. 2021). LncRNA RPL10A is differentially expressed in *Trichophyton mentagrophytes* infected keratinocytes and has a role in fungal pathogenesis as a potential biomarker (Wang Z et al. 2022b). In *Candida albicans* infections, lncRNAs regulate immune responses by modulating pro-inflammatory cytokine expression (Wang Y et al. 2022a). The lncRNA RZE1 regulates the yeast-to-hypha transition in the human fungal pathogen *Cryptococcus neoformans* by controlling the zinc finger transcription factor ZNF2 (Chacko et al. 2015).

Examples as outlined above underscore the multifaceted and context-dependent roles of lncRNAs in infection biology. Across viral, bacterial, parasitic, and fungal systems, pathogens exploit or modulate host lncRNAs to influence immune signaling, inflammation, and cellular homeostasis (Fig. 11.3). To understand conserved mechanisms underlying these responses, it is essential to examine how lncRNAs intersect with major immune regulatory pathways.

11.3 Therapeutic Potential of lncRNAs

Emerging therapeutic approaches aim to modulate lncRNA activity for treating infectious diseases. Antisense oligonucleotides (ASOs) can bind to specific lncRNAs and thus silence their expression or alter their functions (Dhorne-Pollet et al. 2022; Çakan et al. 2024). The CRISPR/Cas9 system allows for precise editing

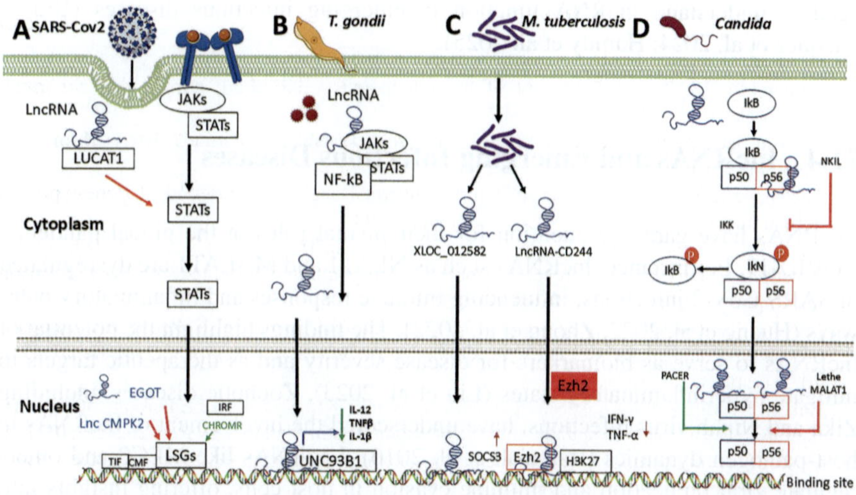

Fig. 11.3 Regulatory roles of lncRNAs in immune responses during infection. (**a**) In COVID-19, lncRNAs modulate the JAK-STAT pathway, with LncCMPK2 downregulation and CHROR upregulation influencing immune responses. (**b**) In *Toxoplasma gondii* infection, lncRNAs suppress UNC93B1 expression in macrophages and activate STAT3 in dendritic cells by preventing dephosphorylation. (**c**) In B and T cells, XLOC_012582 promotes SOCS3 expression, while lncRNA-CD244 recruits Ezh2 to trimethylate H3K27 at the IFN-γ/TNF-α locus, reducing inflammatory cytokine production. (**d**) In cytoplasmic and nuclear NF-κB regulation, NKILA and Mirt2 inhibit IκB degradation, suppressing NF-κB, whereas HOTAIR promotes it. Nuclear lncRNAs, including PACER, Lethe, MALAT1, LincRNA-Cox2, lincRNA-Tnfaip3, and THRIL, regulate NF-κB target genes

of lncRNA loci, while the RNA interference (RNAi) pathway employs small RNA molecules to degrade target lncRNAs, reducing the pathogenic effects. Although these methods show substantial therapeutic promise, several biological, technical, and delivery-related challenges still hinder the successful translation of lncRNA-targeting strategies into clinical practice (Goyal et al. 2017). Developing efficient delivery systems to ensure lncRNA modulators reach target tissues remains a significant obstacle. Off-target effects pose risks of unintended gene regulation, while the stability and immunogenicity of RNA-based therapies require optimization for safe and sustained use.

Although lncRNA-targeted therapeutics for infectious diseases are still in early stages, numerous biological advances in oncology have demonstrated the feasibility of such approaches (Wang 2019). ASOs designed to silence oncogenic lncRNAs have shown efficacy in preclinical and early clinical studies, providing a model for future anti-infective interventions (Montero et al. 2024). Similar approaches can be adapted to suppress lncRNAs that aid viral replication or bacterial survival, paving the way for novel anti-infective therapies. Therapeutic advances in diseases like cancer suggest that lncRNA modulation can address immune dysregulation and pathogen-host interactions (Bandesh et al. 2021; Marima et al. 2024). Building on these therapeutic insights, recent global health crises have underscored the urgent need to understand lncRNA function in emerging infectious diseases (Bravo-Vázquez et al. 2024; Hamdy et al. 2025).

11.4 lncRNAs and Emerging Infectious Diseases

LncRNAs have garnered attention for their pivotal roles in the global pandemic COVID-19. For instance, lncRNAs such as NEAT1 and MALAT1 are dysregulated in SARS-CoV-2 infections, influencing immune responses and inflammatory pathways (Huang et al. 2022; Zhong et al. 2022). The findings highlight the potential of lncRNAs to serve as biomarkers for disease severity and as therapeutic targets to mitigate hyperinflammatory states (Liu et al. 2023). Zoonotic diseases, including Zika and Nipah virus infections, have underscored the involvement of lncRNAs in host-pathogen dynamics (Ramaiah et al. 2016). LncRNAs like MEG3 and others regulate viral replication and immune evasion in host cells, offering insights into how pathogens adapt to human hosts. Understanding the mechanisms can inform strategies to control future zoonotic outbreaks (Xu et al. 2024). Studying lncRNAs can bolster pandemic preparedness by identifying novel biomarkers and therapeutic targets for emerging pathogens (Venkatesan et al. 2022). LncRNA-based diagnostic tools and therapies could provide rapid, non-invasive, and effective responses to newly identified infectious agents, enhancing global health security and outbreak management (Sharma et al. 2024).

11.5 Databases and Computational Approaches for lncRNA Study

Table 11.1 depicts numerous databases that provide comprehensive resources for lncRNA research. LNCipedia (Volders et al. 2019) and NONCODE (Zhao et al. 2020) catalogue annotated lncRNAs with details on the sequences, structures, and potential functions. The LncRNA Disease database catalogs known interactions between specific lncRNAs and diverse human diseases, including infectious diseases, helping researchers understand their pathological roles and identify potential biomarkers and therapeutic targets. Broad lncRNA databases like LNCipedia, NONCODE, and LncRNADisease remain essential in infectious disease research, as they provide foundational sequences, structures, and annotations that researchers adapt for pathogen contexts via RNA-seq integration and custom analyses. These broad databases help address limitations in specialized resources such as VirhostlncR. This database focuses on virus-lncRNA interactions and currently contains only 2802 entries from 33 viruses (Thatai et al. 2023). Researchers therefore use broad lncRNA repositories to supplement virus-specific databases,

Table 11.1 Databases for lncRNA research. A collection of publicly available databases providing information on lncRNA annotations, expression profiles, disease associations, interactions, and functional insights across various species, including virus-host interactions

Database	Description	Link	Reference
VirhostlncR	A database to explore lncRNAs and their targets in viral infections	http://ciods.in/VirhostlncR/	Thatai et al. (2023)
LNCipedia	A comprehensive collection of human lncRNAs with annotations and expression profiles	https://lncipedia.org/	Volders et al. (2019)
lncRNADisease	Links lncRNAs to various human diseases, including infectious diseases	http://www.cuilab.cn/lncrnadisease	Lin et al. (2023)
DIANA-LncBase	Provides experimentally validated miRNA-lncRNA interactions	https://diana.e-ce.uth.gr/lncbasev3	Karagkouni et al. (2020)
RNAcentral	Integrates ncRNA data from multiple sources, including lncRNAs	https://rnacentral.org/	RNAcentral Consortium (2021)
NONCODE	A curated database of non-coding RNAs, including lncRNAs from various species	http://www.noncode.org/	Zhao et al. (2020)
lncATLAS	Provides tissue-specific expression patterns of lncRNAs	https://lncatlas.crg.eu/	Mas-Ponte et al. (2017)
LncExpDB	A large-scale expression database for human lncRNAs	https://ngdc.cncb.ac.cn/lncexpdb/	Li et al. (2021)
ViRBase	A database of virus-host ncRNA interactions, including lncRNAs	http://www.rna-society.org/virbase/	Li et al. (2015)

allowing them to investigate additional pathogens and host responses that are not fully captured in specialized resources.

Advanced computational tools are indispensable for deciphering lncRNA functions. Algorithms for sequence analysis, secondary structure prediction, and functional annotation help to predict lncRNA roles in biological processes. Machine learning models help predict how lncRNAs interact with proteins, DNA, and RNAs by learning patterns from experimental data in online resources (Yang et al. 2015). Network-based methods create interaction maps that show how lncRNAs may regulate genes, for example, by binding transcription factors or miRNAs (Peng et al. 2020). Together, these computational tools help researchers identify promising lncRNAs to study and understand their possible roles in processes like immune responses (Zhang et al. 2019b). Moreover, RNA-seq and single-cell RNA-seq allow high-resolution mapping of lncRNA expression across tissues and disease states. Coupling these data with proteomics and metabolomics enhances our understanding of lncRNA-regulated pathways, providing a holistic view of their roles in health and disease. Various bioinformatics tools to analyze lncRNA are listed in Table 11.2.

Table 11.2 Bioinformatics tools for lncRNA analysis. A list of bioinformatics tools used for lncRNA classification, coding potential prediction, and interaction analysis with miRNAs and mRNAs, aiding in functional characterization and regulatory network studies

Tool	Description	Link	Reference
FEELnc	Classifies lncRNAs based on sequence features	https://github.com/tderrien/FEELnc	Wucher et al. (2017)
CPC2 (Coding Potential Calculator 2)	Predicts coding potential of lncRNAs	https://cpc2.gao-lab.org/	Kang et al. (2017)
IntaRNA	Predicts lncRNA-mRNA and lncRNA-miRNA interactions	https://rna.informatik.uni-freiburg.de/IntaRNA/	Mann et al. (2017)
miRcode	Predicts miRNA-lncRNA interactions	http://www.mircode.org/	Jeggari et al. (2012)

11.6 Challenges and Future Perspectives

The study of lncRNAs in infectious diseases faces several challenges, including the complexity of lncRNA mechanisms, limited functional annotations, and the lack of standardized methods for validation. Additionally, the distinction between infection-induced lncRNAs and those with inherent roles in host biology remains a critical hurdle. Addressing these challenges necessitates interdisciplinary collaborations among bioinformatics, molecular biology, immunology, and clinical research. Integrating computational predictions with experimental validation can enhance the functional understanding of lncRNAs. Partnerships between academia and industry can expedite the development of lncRNA-based diagnostics and therapeutics. As our understanding deepens, lncRNAs are poised to drive personalized approaches to infectious disease management, offering high specificity and minimal off-target effects through RNA-based and antisense oligonucleotide strategies.

11.7 Conclusion

lncRNAs are increasingly recognized as important regulators in the molecular landscape of infectious diseases. Through the diverse roles in gene regulation, immune response, and pathogen-host interactions, lncRNAs have substantially advanced our understanding of disease mechanisms. Progresses in high-throughput sequencing, bioinformatics, and systems biology have enabled the identification of numerous infection-associated lncRNAs, supporting their relevance across infectious diseases. The interplay between lncRNAs and infectious agents illustrates their important role in shaping host-pathogen dynamics. It offers new avenues for combating viral infections and bacterial and parasitic diseases. The growing body of evidence highlights the significant potential of lncRNAs in infectious disease research, supporting the development of new therapeutic strategies and precision medicine approaches. Future work will require systematic integration of computational predictions with experimental validation to determine causative roles rather than correlative associations. Another important direction is evaluating how lncRNAs can be effectively targeted or leveraged therapeutically. Although RNA-based interventions and antisense oligonucleotide platforms show promise, challenges such as delivery efficiency, off-target effects, and context-specific functions remain significant barriers. Addressing these limitations will help determine whether lncRNA modulation can complement existing antimicrobial strategies or contribute to personalized treatment approaches. As the field evolves, integrating lncRNA studies into clinical research holds considerable promise for addressing global health challenges and advancing personalized treatment.

Acknowledgments The authors of the article would like to thank the Department of Bioinformatics, Pondicherry University, for providing high-performance computation to perform the analysis. RD would like to thank NSFDC, Ministry of Social Justice and Empowerment, Govt. of India, and UGC for providing a Senior Research Fellowship.

Compliance with Ethical Standards

Funding No funding was used for the study.

Conflict of Interest Authors declare that they have no conflict of interest.

Ethical Approval This chapter contains no studies with human/animal participants performed by the author.

References

Alvarez-Dominguez JR, Lodish HF (2017) Emerging mechanisms of long noncoding RNA function during normal and malignant hematopoiesis. Blood 130(18):1965–1975. Available at: https://doi.org/10.1182/blood-2017-06-788695

Amir N, Taube R (2024) Role of long noncoding RNA in regulating HIV infection-a comprehensive review. MBio 15(2):e0192523. Available at: https://doi.org/10.1128/mbio.01925-23

Arunima A, van Schaik EJ, Samuel JE (2023) The emerging roles of long non-coding RNA in host immune response and intracellular bacterial infections. Front Cell Infect Microbiol 13. Available at: https://doi.org/10.3389/fcimb.2023.1160198

Bamunuarachchi G, Pushparaj S, Liu L (2021) Interplay between host non-coding RNAs and influenza viruses. RNA Biol 18(5):767–784. Available at: https://doi.org/10.1080/15476286.2021.1872170

Bandesh K et al (2021) Harnessing the potential of long non-coding RNAs to manage metabolic diseases. Curr Pharm Des 27(35):3668–3685. Available at: https://doi.org/10.2174/1381612827666210315145254

Beermann J et al (2016) Non-coding RNAs in development and disease: background, mechanisms, and therapeutic approaches. Physiol Rev 96(4):1297–1325. Available at: https://doi.org/10.1152/physrev.00041.2015

Bravo-Vázquez LA et al (2024) Exploring the therapeutic significance of microRNAs and lncRNAs in kidney diseases. Genes:123. Available at: https://doi.org/10.3390/genes15010123

Çakan E et al (2024) Therapeutic antisense oligonucleotides in oncology: from bench to bedside. Cancers. Available at: https://doi.org/10.3390/cancers16172940
Chacko N et al (2015) The lncRNA RZE1 controls Cryptococcal morphological transition. PLoS Genet 11(11):e1005692. Available at: https://doi.org/10.1371/journal.pgen.1005692
Chen G et al (2022) Plasmodium manipulates the expression of host long non-coding RNA during red blood cell intracellular infection. Parasit Vectors 15(1):182. Available at: https://doi.org/10.1186/s13071-022-05298-4
Cheng Y et al (2024) Host long noncoding RNAs in bacterial infections. Front Immunol 15. Available at: https://doi.org/10.3389/fimmu.2024.1419782
Dhorne-Pollet S et al (2022) Antisense oligonucleotides targeting ORF1b block replication of severe acute respiratory syndrome coronavirus 2 (SARS-CoV-2). Front Microbiol 13. Available at: https://doi.org/10.3389/fmicb.2022.915202
Fernandes JCR et al (2019) Long non-coding RNAs in the regulation of gene expression: physiology and disease. Noncoding RNA. Available at: https://doi.org/10.3390/ncrna5010017
Fernandes JCR et al (2022) Comparative transcriptomic analysis of long noncoding RNAs in Leishmania-infected human macrophages. Front Genet 13:1051568. Available at: https://doi.org/10.3389/fgene.2022.1051568
Fernandes JCR et al (2023) Comparative transcriptomic analysis of long noncoding RNAs in Leishmania-infected human macrophages. Front Genet 13. Available at: https://doi.org/10.3389/fgene.2022.1051568
Goyal A et al (2017) Challenges of CRISPR/Cas9 applications for long non-coding RNA genes. Nucleic Acids Res 45(3):e12. Available at: https://doi.org/10.1093/nar/gkw883
Hamdy NM et al (2025) Comprehensive insights and in silico analysis into the emerging role of LincRNAs in lung diseases pathogenesis; a step toward ncRNA precision. Funct Integr Genomics 25(1):34. Available at: https://doi.org/10.1007/s10142-025-01540-1
He X et al (2024) A prominent role of LncRNA H19 in H. pylori CagA induced DNA damage response and cell malignancy. Sci Rep 14(1):14185. Available at: https://doi.org/10.1038/s41598-024-65221-y
Heydari R et al (2024) Long non-coding RNAs in biomarking COVID-19: a machine learning-based approach. Virol J 21(1):134. Available at: https://doi.org/10.1186/s12985-024-02408-9
Hou J et al (2021) Gene expression profile and long noncoding RNA analysis in Candida albicans insoluble β-glucan-stimulated CD14+ monocytes and THP-1 cells. Microb Pathog 157:104963. Available at: https://doi.org/10.1016/j.micpath.2021.104963
Huang K et al (2022) Long non-coding RNAs (lncRNAs) NEAT1 and MALAT1 are differentially expressed in severe COVID-19 patients: an integrated single-cell analysis. PLoS One 17(1):e0261242. Available at: https://doi.org/10.1371/journal.pone.0261242
Jeggari A, Marks DS, Larsson E (2012) miRcode: a map of putative microRNA target sites in the long non-coding transcriptome. Bioinformatics 28(15):2062–2063. Available at: https://doi.org/10.1093/bioinformatics/bts344
Kang Y-J et al (2017) CPC2: a fast and accurate coding potential calculator based on sequence intrinsic features. Nucleic Acids Res 45(W1):W12–W16. Available at: https://doi.org/10.1093/nar/gkx428
Karagkouni D et al (2020) DIANA-LncBase v3: indexing experimentally supported miRNA targets on non-coding transcripts. Nucleic Acids Res 48(D1):D101–D110. Available at: https://doi.org/10.1093/nar/gkz1036
Kern et al (2018) Genome-wide identification of tissue-specific long non-coding RNA in three farm animal species BMC Genomics 19(1). Available at: https://doi.org/10.1186/s12864-018-5037-7
Krawczyk M, Emerson BM (2014) p50-associated COX-2 extragenic RNA (PACER) activates COX-2 gene expression by occluding repressive NF-κB complexes. eLife 3:e01776. Available at: https://doi.org/10.7554/eLife.01776
Li Y et al (2015) ViRBase: a resource for virus–host ncRNA-associated interactions. Nucleic Acids Res 43(D1):D578–D582. Available at: https://doi.org/10.1093/nar/gku903

Li J et al (2016) Long noncoding RNA NRON contributes to HIV-1 latency by specifically inducing tat protein degradation. Nat Commun 7(1):11730. Available at: https://doi.org/10.1038/ncomms11730

Li Z et al (2021) LncExpDB: an expression database of human long non-coding RNAs. Nucleic Acids Res 49(D1):D962–D968. Available at: https://doi.org/10.1093/nar/gkaa850

Lin X et al (2023) LncRNADisease v3.0: an updated database of long non-coding RNA-associated diseases. Nucleic Acids Res 52(D1):D1365–D1369. Available at: https://doi.org/10.1093/nar/gkad828

Liu W et al (2018) Microarray analysis of long non-coding RNA expression profiles uncovers a Toxoplasma-induced negative regulation of host immune signaling. Parasit Vectors 11(1):174. Available at: https://doi.org/10.1186/s13071-018-2697-8

Liu X et al (2023) Non-coding RNAs expression in SARS-CoV-2 infection: pathogenesis, clinical significance and therapeutic targets. Signal Transduct Target Ther 8(1):441. Available at: https://doi.org/10.1038/s41392-023-01669-0

Lodde V et al (2022) Non-coding RNAs in malaria infection. Wiley Interdiscip Rev RNA 13(3):e1697. Available at: https://doi.org/10.1002/wrna.1697

Mann M, Wright PR, Backofen R (2017) IntaRNA 2.0: enhanced and customizable prediction of RNA–RNA interactions. Nucleic Acids Res 45(W1):W435–W439. Available at: https://doi.org/10.1093/nar/gkx279

Marchese FP, Raimondi I, Huarte M (2017) The multidimensional mechanisms of long non-coding RNA function. Genome Biol 18(1):206. Available at: https://doi.org/10.1186/s13059-017-1348-2

Marima R et al (2024) Exosomal long non-coding RNAs in cancer: interplay, modulation, and therapeutic avenues. Noncoding RNA Res 9(3):887–900. Available at: https://doi.org/10.1016/j.ncrna.2024.03.014

Mas-Ponte D et al (2017) LncATLAS database for subcellular localization of long noncoding RNAs. RNA (New York, N.Y.) 23(7):1080–1087. Available at: https://doi.org/10.1261/rna.060814.117

Mattick JS et al (2023) Long non-coding RNAs: definitions, functions, challenges and recommendations. Nat Rev Mol Cell Biol 24(6):430–447. Available at: https://doi.org/10.1038/s41580-022-00566-8

Milcamps R, Michiels T (2024) Involvement of paraspeckle components in viral infections. Nucleus (Austin, Tex.) 15(1):2350178. Available at: https://doi.org/10.1080/19491034.2024.2350178

Montero JJ et al (2024) Genome-scale pan-cancer interrogation of lncRNA dependencies using CasRx. Nat Methods 21(4):584–596. Available at: https://doi.org/10.1038/s41592-024-02190-0

Ngongang NN et al (2022) TNF and HNRNPL related Immunoregulatory Long non-coding RNA (THRIL) and long intergenic noncoding RNA-p21 (lincRNA-p21) as potential useful biomarkers for the diagnosis of tuberculosis. Front Trop Dis 3. Available at: https://doi.org/10.3389/fitd.2022.969307

Ouyang J et al (2014) NRAV, a long noncoding RNA, modulates antiviral responses through suppression of interferon-stimulated gene transcription. Cell Host Microbe 16(5):616–626. Available at: https://doi.org/10.1016/j.chom.2014.10.001

Pawar K et al (2016) Down regulated lncRNA MEG3 eliminates mycobacteria in macrophages via autophagy. Sci Rep 6(1):19416. Available at: https://doi.org/10.1038/srep19416

Peng L et al (2020) Probing lncRNA–protein interactions: data repositories, models, and algorithms. Front Genet 10-2019. Available at: https://doi.org/10.3389/fgene.2019.01346

Ramaiah A et al (2016) Dysregulation of long non-coding RNA (lncRNA) genes and predicted lncRNA-protein interactions during Zika virus infection. bioRxiv. [Preprint]. Available at: https://doi.org/10.1101/061788

Ren F et al (2020) LncRNA HOTAIR modulates hepatitis B virus transcription and replication by enhancing SP1 transcription factor. Clin Sci 134(22):3007–3022. Available at: https://doi.org/10.1042/CS20200970

RNAcentral Consortium (2021) RNAcentral 2021: secondary structure integration, improved sequence search and new member databases. Nucleic Acids Res 49(D1):D212–D220. Available at: https://doi.org/10.1093/nar/gkaa921

Sarma A et al (2025) An emphasis on the role of long non-coding RNAs in viral gene expression, pathogenesis, and innate immunity in viral chicken diseases. Noncoding RNA 11(3) Available at: https://doi.org/10.3390/ncrna11030042

Seifuddin et al (2020) lncRNAKB a knowledgebase of tissue-specific functional annotation and trait association of long noncoding RNA Abstract Scientific Data 7(1). Available at: https://doi.org/10.1038/s41597-020-00659-z

Sharma K, Singh M, Sharma SC (2024) Revolutionizing antiviral therapeutics: *in silico* approaches for emerging and neglected RNA viruses. Curr Pharm Des:3276–3290. Available at: https://doi.org/10.2174/0113816128322226240815063730

Statello L et al (2021) Gene regulation by long non-coding RNAs and its biological functions. Nat Rev Mol Cell Biol 22(2):96–118. Available at: https://doi.org/10.1038/s41580-020-00315-9

Tayel SI et al (2022) Interplay of LncRNAs NEAT1 and TUG1 in incidence of cytokine storm in appraisal of COVID-19 infection. Int J Biol Sci 18(13):4901–4913. Available at: https://doi.org/10.7150/ijbs.72318

Thatai AKS et al (2023) VirhostlncR: a comprehensive database to explore lncRNAs and their targets in viral infections. Comput Biol Med 164:107279. Available at: https://doi.org/10.1016/j.compbiomed.2023.107279

Till P et al (2018) A long noncoding RNA promotes cellulase expression in Trichoderma reesei. Biotechnol Biofuels 11(1):78. Available at: https://doi.org/10.1186/s13068-018-1081-4

Venkatesan A et al (2022) Identification of novel lncRNA by reanalysis of RNA-seq data in Zika virus infected hiNPCs. Virus 33(2):185–193. Available at: https://doi.org/10.1007/s13337-022-00771-1

Volders P-J et al (2019) LNCipedia 5: towards a reference set of human long non-coding RNAs. Nucleic Acids Res 47(D1):D135–D139. Available at: https://doi.org/10.1093/nar/gky1031

Wang P (2019) The opening of Pandora's box: an emerging role of long noncoding RNA in viral infections. Front Immunol 9. Available at: https://doi.org/10.3389/fimmu.2018.03138

Wang J et al (2018) Host long noncoding RNA lncRNA-PAAN regulates the replication of influenza a virus. Viruses 10(6) Available at: https://doi.org/10.3390/v10060330

Wang Y et al (2022a) LncRNA: a potential target for host-directed therapy of Candida infection. Pharmaceutics. Available at: https://doi.org/10.3390/pharmaceutics14030621

Wang Z et al (2022b) Comprehensive analysis of long non-coding RNA expression profiles in Trichophyton mentagrophytes-infected keratinocytes. Microb Pathog 167:105565. Available at: https://doi.org/10.1016/j.micpath.2022.105565

Wang L-J et al (2023) Association of lncRNA THRIL, HOTAIR genes variations and expression levels with pulmonary tuberculosis. BMC Med Genet 16(1):326. Available at: https://doi.org/10.1186/s12920-023-01770-x

Wu L, Zhang L, Zheng S (2017) Role of the long non-coding RNA HOTAIR in hepatocellular carcinoma (review). Oncol Lett 14(2):1233–1239. Available at: https://doi.org/10.3892/ol.2017.6312

Wucher V et al (2017) FEELnc: a tool for long non-coding RNA annotation and its application to the dog transcriptome. Nucleic Acids Res 45(8):e57–e57. Available at: https://doi.org/10.1093/nar/gkw1306

Xiao Y et al (2024) Long non-coding RNA-encoded micropeptides: functions, mechanisms and implications. Cell Death Discov 10(1):450. Available at: https://doi.org/10.1038/s41420-024-02175-0

Xu B et al (2024) Unveiling the hidden regulators: the impact of lncRNAs on Zoonoses. Int J Mol Sci 25(6) Available at: https://doi.org/10.3390/ijms25063539

Yang Y, Wen L, Zhu H (2015) Unveiling the hidden function of long non-coding RNA by identifying its major partner-protein. Cell Biosci 5:59. Available at: https://doi.org/10.1186/s13578-015-0050-x

Yu T et al (2025) From correlation to causation: unraveling the role of long non-coding RNAs in COVID-19 pathogenesis. Virol J England:293. Available at: https://doi.org/10.1186/s12985-025-02660-7

Zhang Q et al (2013) NEAT1 long noncoding RNA and paraspeckle bodies modulate HIV-1 post-transcriptional expression. mBio 4(1):e00596–e00512. Available at: https://doi.org/10.1128/mBio.00596-12

Zhang X et al (2019b) Mechanisms and functions of long non-coding RNAs at multiple regulatory levels. Int J Mol Sci 20(22) Available at: https://doi.org/10.3390/ijms20225573

Zhang Y et al (2019c) LncRNA H19 induced by helicobacter pylori infection promotes gastric cancer cell growth via enhancing NF-κB-induced inflammation. J Inflamm (London, England) 16:23. Available at: https://doi.org/10.1186/s12950-019-0226-y

Zhao L et al (2020) NONCODEV6: an updated database dedicated to long non-coding RNA annotation in both animals and plants. Nucleic Acids Res 49(D1):D165–D171. Available at: https://doi.org/10.1093/nar/gkaa1046

Zheng et al (2019) Expression Profiles of Long Noncoding RNA and mRNA in Epicardial Adipose Tissue in Patients with Heart Failure BioMed Research International 20191-10. Available at: https://doi.org/10.1155/2019/394547

Zhong Y et al (2022) Assessing the suitability of long non-coding RNAs as therapeutic targets and biomarkers in SARS-CoV-2 infection. Front Mol Biosci 9. Available at: https://doi.org/10.3389/fmolb.2022.975322

Zhou X et al (2016) Helicobacter pylori infection related long noncoding RNA (lncRNA) AF147447 inhibits gastric cancer proliferation and invasion by targeting MUC2 and up-regulating miR-34c. Oncotarget 7(50):82770–82782. Available at: https://doi.org/10.18632/oncotarget.13165

Chapter 12
Long Noncoding RNAs in Rare Diseases

Alessio Coppola, María Isidoro-García, Pablo Prieto-Matos, and Jesus Lacal

Abstract This chapter examines the growing body of evidence establishing long noncoding RNAs (lncRNAs) as key molecular players in the pathogenesis of rare diseases. Rare diseases, although individually uncommon, affect millions of people worldwide, often leaving patients with limited treatment options. In the quest to understand these complex conditions, scientists are investigating the role of lncRNAs. Once considered "genomic noise," these molecules are now emerging as critical regulators of health and disease, offering new avenues for diagnosis and therapy. This section explores how lncRNAs influence rare diseases by fine-tuning gene expression, shaping cellular behavior, and interacting with key biological pathways. Across diverse disease categories, including neurodevelopmental disor-

A. Coppola · J. Lacal (✉)
Laboratory of Functional Genetics of Rare Diseases, Department of Microbiology and Genetics, University of Salamanca (USAL), Salamanca, Spain

GIR of Biomedicine of Rare Diseases, University of Salamanca, Salamanca, Spain

Institute of Biomedical Research of Salamanca (IBSAL), Salamanca, Spain
e-mail: acoppola@usal.es; jlacal@usal.es

M. Isidoro-García
Institute of Biomedical Research of Salamanca (IBSAL), Salamanca, Spain

Department of Clinical Biochemistry, University Hospital of Salamanca, Salamanca, Spain

Reference Unit for Rare Diseases (DiERCyL), Castilla León, Spain

Department of Medicine, University of Salamanca, Salamanca, Spain
e-mail: misidoro@saludcastillayleon.es

P. Prieto-Matos
GIR of Biomedicine of Rare Diseases, University of Salamanca, Salamanca, Spain

Institute of Biomedical Research of Salamanca (IBSAL), Salamanca, Spain

Reference Unit for Rare Diseases (DiERCyL), Castilla León, Spain

Pediatrics Area, Department of Biomedical and Diagnostic Sciences, University of Salamanca (USAL), Salamanca, Spain
e-mail: pabloprieto@usal.es

L. M. Vaschetto (ed.), *Long Non-Coding RNAs, Human Health, and Diseases*, Epigenetics and Human Health 16, https://doi.org/10.1007/978-3-032-19668-2_12

ders, neuromuscular conditions, rare malignancies, and metabolic diseases, lncRNAs demonstrate critical regulatory functions through their tissue-specific expression patterns and disease-relevant molecular interactions. Mechanistic studies have revealed how specific lncRNAs participate in pathogenic cascades, often showing conserved functions across experimental models. What makes lncRNAs especially exciting is their dual potential as both biomarkers and therapeutic targets. Their presence, or absence, can signal disease onset or progression, while their ability to interact with cellular pathways opens doors for precision medicine. Although challenges remain in translating these discoveries into treatments, the study of lncRNAs represents a thrilling frontier in rare disease research.

Keywords Long noncoding RNA · Rare diseases · Gene regulation · Biomarkers · Therapeutic targets

Abbreviations

AM	Acromegaly
ANRIL	Antisense Noncoding RNA in the INK4 Locus
AOSD	Adult-Onset Still's Disease
APAP	Autoimmune Pulmonary Alveolar Proteinosis
BMD	Becker Muscular Dystrophy
CCNA2	Cyclin A2 (Cell-cycle regulator)
ceRNA	Competitive Endogenous RNA
DMD	Duchenne Muscular Dystrophy
EMPD	Extramammary Paget's Disease
ENS	Enteric Nervous System
FSHD	Facioscapulohumeral Muscular Dystrophy
GEP-NENs	Gastro-Entero-Pancreatic Neuroendocrine Neoplasms
GR	Glucocorticoid Receptor
HSCR	Hirschsprung Disease
IGF1R	Insulin-Like Growth Factor 1 Receptor
KS	Klinefelter Syndrome
lncRNA	Long Noncoding RNA
MIAT	myocardial infarction-associated transcript
MMD	Moyamoya Disease
NEAT1	nuclear-enriched abundant transcript 1
NF1	Neurofibromatosis Type 1
NTT	noncoding transcript in T-cells
PACERR	p50-associated COX-2 extragenic RNA
PAH	Pulmonary Artery Hypertension
PASMCs	Pulmonary Artery Smooth Muscle Cells
PDGF-BB	Platelet-Derived Growth Factor BB
PI3K	Phosphoinositide 3-Kinase

RMRP	RNA component of mitochondrial RNA-processing endoribonuclease
RTT	Rett Syndrome
sALS	Sporadic Amyotrophic Lateral Sclerosis
TAK	Takayasu Arteritis
TGF-β	Transforming Growth Factor Beta
THRIL	TNF-α/hnRNPL-Related Immunoregulatory lincRNA.

12.1 Introduction

12.1.1 Long Noncoding RNAs in the Pathogenesis of Rare Diseases

Rare diseases are defined as life-threatening or chronically debilitating conditions that affect at most one person in 2000. Although each rare disease is uncommon on its own, collectively they are widespread, with over 10,000 identified conditions impacting approximately 350 million people globally (Fu et al. 2023). Most rare diseases (around 80%) have a genetic origin, often caused by mutations in protein-coding genes. However, recent research highlights the crucial role of long noncoding RNAs (lncRNAs) in rare genetic disorders, even though they were once considered "genomic junk" (Finotti et al. 2019; Arunkumar 2024). These RNA molecules, typically longer than 200 nucleotides and lacking protein-coding potential, are now recognized for their roles in diverse cellular processes such as chromatin remodeling, transcriptional and post-transcriptional regulation, and nuclear organization, acting both in cis and in trans to regulate gene expression across development, differentiation, and disease states (Schmitz et al. 2016; Arunkumar 2024) (Fig. 12.1). Structurally similar to mRNAs, lncRNAs are often spliced, capped, and

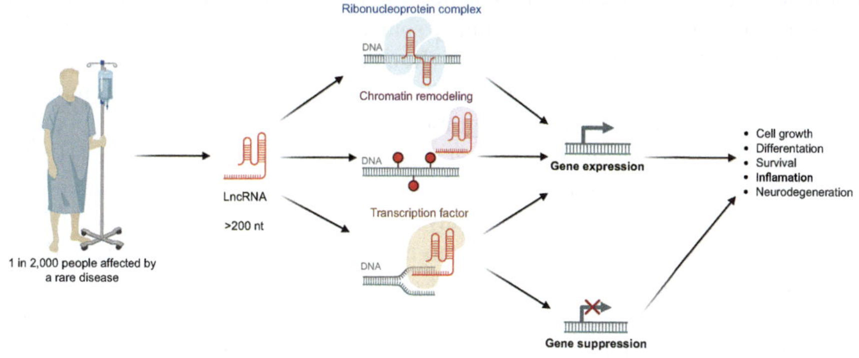

Fig. 12.1 Role of lncRNAs in rare diseases. The main functions of lncRNAs in the pathogenesis of rare disorders are reported

polyadenylated, although expressed at lower levels and exhibiting higher tissue specificity (Constanty and Shkumatava 2021). Their ability to fold into complex secondary and tertiary structures enables precise interactions with proteins, DNA, and other RNAs (Constanty and Shkumatava 2021). They can act as master regulators, controlling genes involved in cell growth, differentiation, and survival (Chen et al. 2016) (Fig. 12.1). They can serve as signaling hubs, modulating pathways linked to inflammation, fibrosis, and neurodegeneration (Walther and Schulte 2021; Ni et al. 2022), and they can function as biomarkers, helping in early diagnosis and tracking disease progression (Taghizadeh et al. 2021; Najafi et al. 2022) (Fig. 12.1). Ultimately, lncRNAs represent a largely untapped layer of regulatory complexity in cellular biology. Their capacity to act as scaffolds, guides, decoys, and enhancers in gene expression underscores their significance in both normal physiology and disease, positioning them as promising candidates for biomarker development and targeted therapeutics. The interplay between lncRNAs and the human genome is detailed in Chap. 1 by Palma.

This chapter examines the growing body of evidence establishing lncRNAs as key molecular players in the pathogenesis of rare diseases. Several well-characterized examples illustrate how lncRNA dysregulation contributes to disease phenotypes through various molecular mechanisms, including epigenetic regulation, modulation of signaling pathways, and interactions with protein complexes. The systematic investigation of lncRNAs in rare diseases provides a framework for connecting fundamental RNA biology with translational medicine, offering new avenues for developing targeted interventions. This synthesis of current knowledge highlights lncRNAs as integral components of rare disease biology while identifying critical gaps that warrant future investigation.

12.1.2 Acromegaly

Acromegaly (AM) is a rare hormonal disorder characterized by excessive growth hormone levels, often due to tumors in the pituitary gland (Rolla et al. 2023). As one of the main symptoms of AM is pituitary somatotroph adenoma, it was thought to use lncRNA H19 as a diagnostic marker (Katznelson et al. 2014; Rolla et al. 2023) (Fig. 12.2; Table 12.1). H19 is encoded by the *H19* gene located on chromosome 11p15.5, and it is involved in cell migration, invasion, angiogenesis, and apoptosis inhibition (Yoshimura et al. 2018; Wu et al. 2021; Rolla et al. 2023). While H19 has been linked to various forms of cancer and disease progression, its relevance to noncancerous conditions like acromegaly had yet to be clearly understood (Rolla et al. 2023). However, while AM leads to multiple systemic problems, H19 likely plays a limited role in their development or detection (Rolla et al. 2023). In fact, while H19 expression does not appear to play a major role in the general pathology of AM or tumor behavior, it may be linked to age-related changes and specific complications like cholelithiasis.

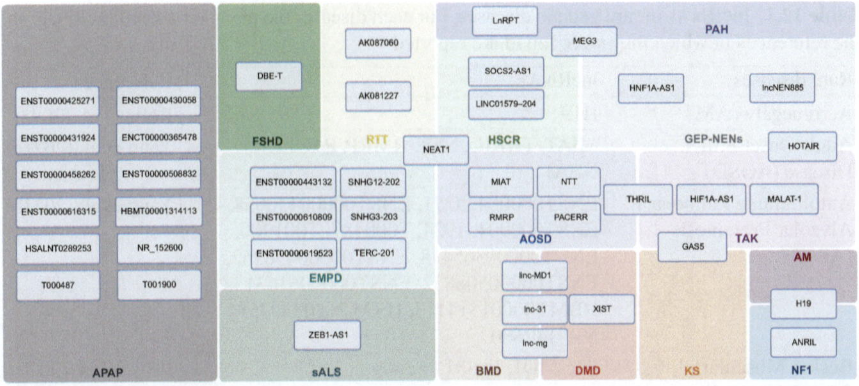

Fig. 12.2 Functional landscape of lncRNAs in rare diseases. Clustering of disease-associated lncRNAs. Some lncRNAs are common to multiple diseases. *AOSD* Adult-Onset Still's Disease, *AM Acromegaly*, *APAP* Autoimmune Pulmonary Alveolar Proteinosis, *BMD* Becker Muscular Dystrophy, *DMD* Duchenne Muscular Dystrophy, *EMPD* Extramammary Paget's Disease, *FSHD* Facioscapulohumeral Muscular Dystrophy, *GEP-NENs* Gastro-Entero-Pancreatic Neuroendocrine Neoplasms, *HSCR* Hirschsprung Disease, *KS* Klinefelter Syndrome, *NF1* Neurofibromatosis Type 1, *PAH* Pulmonary Artery Hypertension, *RTT* Rett Syndrome, *sALS* Sporadic Amyotrophic Lateral Sclerosis, *TAK* Takayasu Arteritis

12.1.3 Adult-Onset Still's Disease

Adult-onset Still's disease (AOSD) represents a complex autoinflammatory disorder characterized by systemic inflammation and cytokine dysregulation (Efthimiou et al. 2021). Recent investigations have identified several lncRNAs as key modulators of immune pathways involved in AOSD pathogenesis including myocardial infarction-associated transcript (MIAT), TNF-α/hnRNPL-related immunoregulatory lincRNA (THRIL), noncoding transcript in T-cells (NTT), RNA component of mitochondrial RNA-processing endoribonuclease (RMRP), p50-associated COX-2 extragenic RNA (PACERR), and nuclear-enriched abundant transcript 1 (NEAT1) (Yang et al. 2021a) (Fig. 12.2; Table 12.1). MIAT exhibits elevated expression in AOSD patients and demonstrates dual regulatory capacity—while it activates Phosphoinositide 3-Kinase (PI3K)/Akt signaling in some contexts such as atherosclerosis and melanoma (Yang et al. 2021a) (Fig. 12.3). It also suppresses TNF-α and IL-1β production in macrophage models (Yang et al. 2021a) (Fig. 12.3). THRIL forms a transcriptional complex with heterogeneous nuclear ribonucleoprotein L (hnRNPL) to enhance TNF-α promoter activity yet shows decreased expression in AOSD (Yang et al. 2021a) (Fig. 12.3). This lncRNA additionally activates PI3K/Akt signaling during inflammatory responses (Yang et al. 2021a) (Fig. 12.3). NTT, identified in activated CD4+ T cells, is known to regulate inflammation and contribute to the progression of rheumatoid arthritis (Yang et al. 2021a). RMRP contributes to Th17 cell differentiation by facilitating RORγt-DDX5 complex assembly at effector gene loci and modulates PI3K/Akt/mTOR pathway (Yang et al. 2021a) (Fig. 12.3). PACERR positively regulates COX-2 expression in macrophage-like cells

Table 12.1 lncRNAs in rare genetic diseases. For each disease, the main associated lncRNAs and the references in which they were found are reported

Rare diseases	lncRNAs	References
Acromegaly (AM)	H19	Rolla et al. (2023)
Adult-Onset Still's Disease (AOSD)	MIAT, THRIL, NTT, RMRP, PACERR, NEAT1	Yang et al. (2021a)
Autoimmune Pulmonary Alveolar Proteinosis (APAP)	ENST00000425271, ENST00000430058, ENST00000431924, T000487, T001900, ENCT00000365478, ENST00000458262, ENST00000508832, ENST00000616315, HBMT00001314113, HSALNT0289253, NR_152600	Yang et al. (2021b)
Becker Muscular Dystrophy (BMD)	linc-MD1, lnc-31, lnc-mg	Brusa et al. (2020)
Duchenne Muscular Dystrophy (DMD)	linc-MD1, lnc-31, lnc-mg, XIST	Brusa et al. (2020) and Xu et al. (2020)
Extramammary Paget's Disease (EMPD)	SNHG3-203, TERC-201, SNHG12202, ENST00000610809, ENST00000443132, ENST00000619523, NEAT1	Zheng et al. (2021)
Facioscapulohumeral Muscular Dystrophy (FSHD)	DBE-T	Vizoso and Esteller (2012)
Gastro-entero-pancreatic Neuroendocrine Neoplasms (GEP-NENs)	HOTAIR, MEG3, lncNEN885, HNF1A-AS1	Modali et al. (2015), Wei et al. (2018), Xue et al. (2021), and Di Mauro et al. (2021)
Hirschsprung Disease (HSCR)	SOCS2-AS1, MEG3, NEAT1, LINC01579–204	Torroglosa et al. (2020) and Qin et al. (2023)
Klinefelter Syndrome (KS)	GAS5	Salemi et al. (2019)
Moyamoya Disease (MMD)	308 differentially expressed lncRNAs (not individually named)	Mamiya et al. (2022)
Neurofibromatosis Type 1 (NF1)	ANRIL, H19	Tritto et al. (2019) and Khosravi and Oladnabi (2023)
Pulmonary Artery Hypertension (PAH)	LnRPT, MEG3	Sun et al. (2017) and Chen et al. (2018)
Rett Syndrome (RTT)	NEAT1, AK081227, AK087060	Petazzi et al. (2013) and Siqueira et al. (2025)
Sporadic Amyotrophic Lateral Sclerosis (sALS)	ZEB1-AS1	Rey et al. (2023)
Takayasu Arteritis (TAK)	THRIL, HIF1A-AS1, GAS5, MALAT-1, HOTAIR	Espinosa-Bautista et al. (2024)

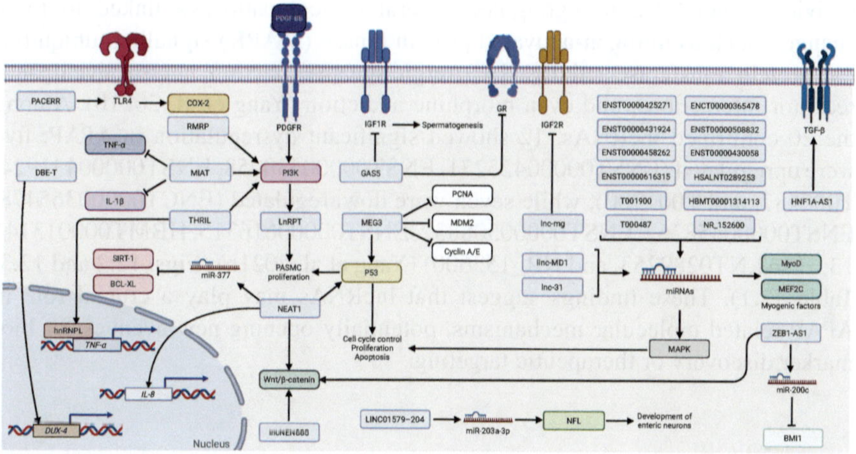

Fig. 12.3 Overview of lncRNAs involved in key regulatory pathways. The lncRNAs whose effects are known at the cellular level are reported

following TLR4 activation, and NEAT1 is essential for paraspeckle formation and IL-8 transcriptional activation (Yang et al. 2021a) (Fig. 12.3). The coordinated changes in NEAT1, MIAT, and THRIL levels following treatment highlight their potential as dynamic biomarkers of therapeutic response (Yang et al. 2021a). The observed inverse relationship in AOSD patients—where MIAT expression is increased while THRIL is decreased—hints at a complex and possibly compensatory interaction within the lncRNA network that controls inflammation. These lncRNAs appear to work together in an interconnected regulatory network, closely tied to key inflammatory pathways in AOSD. By influencing critical processes like PI3K/Akt signaling, the production of inflammatory cytokines (TNF-α, IL-1β, IL-8), and Th17 cell differentiation, they act as crucial regulators in the development and progression of the disease. This suggests that these lncRNAs are not just bystanders but active players in shaping AOSD's inflammatory response, making them potential targets for further research and therapeutic intervention.

12.1.4 Autoimmune Pulmonary Alveolar Proteinosis

Autoimmune pulmonary alveolar proteinosis (APAP) is a rare lung disorder characterized by the overaccumulation of surfactant proteins in the alveoli, primarily due to autoantibodies targeting granulocyte-macrophage colony-stimulating factor (McCarthy et al. 2022). Patients with APAP exhibit significant changes in lncRNA expression in peripheral blood, as demonstrated in a study using microarray analysis (Yang et al. 2021b). The researchers identified 12,459 differentially expressed lncRNAs and later validated 20 key candidates through qRT-PCR (Yang et al. 2021b). Although the study did not detail the exact expression patterns of individual lncRNAs, it highlighted several critical pathways linked to these changes, such as mitogen-activated protein kinase (MAPK) signaling, ubiquitin-like protein transferase activity, GABAergic synapse function, neuroactive ligand-receptor interactions, and even morphine addiction (Yang et al. 2021b). Among the 20 confirmed lncRNAs, 12 showed significant dysregulation in APAP: five were upregulated (ENST00000425271, ENST00000430058, ENST00000431924, T000487, and T001900), while seven were downregulated (ENCT00000365478, ENST00000458262, ENST00000508832, ENST00000616315, HBMT00001314113, HSALNT0289253, and NR_152600) (Yang et al. 2021b) (Figs. 12.2 and 12.3; Table 12.1). These findings suggest that lncRNAs may play a crucial role in APAP-related molecular mechanisms, potentially opening new avenues for biomarker discovery or therapeutic targeting.

12.1.5 Duchenne and Becker Muscular Dystrophy

Duchenne (DMD) and Becker (BMD) muscular dystrophies are X-linked disorders caused by mutations in the dystrophin gene, with disease severity determined by mutation type (Flanigan 2014). Out-of-frame mutations in DMD lead to complete dystrophin loss and rapid progression, while in-frame mutations in BMD produce partially functional dystrophin and a milder course (Finotti et al. 2019; Brusa et al. 2020). Emerging evidence implicates lncRNAs as critical regulators of disease progression through their effects on muscle regeneration pathways. Linc-MD1, typically downregulated in DMD, normally promotes muscle differentiation by sequestering miR-133 and miR-135 that inhibit myogenic factors (MyoD and MEF2C), whereas lnc-31, which is upregulated in dystrophic muscle, actively suppresses differentiation, exacerbating the regenerative deficit phenotype (Finotti et al. 2019; Brusa et al. 2020) (Figs. 12.2 and 12.3; Table 12.1). Additional lncRNAs like lnc-mg further modulate muscle growth through pathways such as IGF2 signaling (Brusa et al. 2020) (Figs. 12.2 and 12.3; Table 12.1). Additionally, X-inactive specific transcript (XIST), a well-known lncRNA involved in X-chromosome inactivation, was found to be significantly downregulated in DMD patients (Xu et al. 2020) (Fig. 12.2; Table 12.1). This finding is interesting as XIST's function in regulating X-chromosome inactivation could have broader implications for gender differences observed in DMD prevalence and severity (Xu et al. 2020). Together, these results reveal a lncRNA network that either compensates for or exacerbates dystrophic pathology, offering potential therapeutic targets to restore myogenic balance in muscular dystrophies.

12.1.6 Extramammary Paget's Disease

Extramammary Paget's disease (EMPD) is a rare cutaneous malignancy characterized by the proliferation of Paget cells in apocrine gland-rich skin regions (Asel and LeBoeuf 2019; Zheng et al. 2021). While the molecular underpinnings of EMPD remain incompletely understood, emerging evidence implicates lncRNAs as key regulators of disease pathogenesis through their influence on oncogenic signaling pathways, cell proliferation, and tumor microenvironment modulation (Zheng et al. 2021). In six patients (12 samples total, with tumor and normal skin), high-throughput sequencing identified 785 differentially expressed lncRNAs and 997 differentially expressed mRNAs (Zheng et al. 2021). lncRNAs SNHG3–203, TERC-201, and SNHG12–202 are upregulated, while ENST00000610809, ENST00000443132, and ENST00000619523 are downregulated in the EMPD (Zheng et al. 2021) (Fig. 12.2; Table 12.1). Notably, the lncRNA NEAT1 was shown to positively regulate the mRNAs PGAP1, FKBP5, and CDON, which are linked to epidermal development and cell adhesion (Zheng et al. 2021). Increased NEAT1 expression activates the Wnt/β-catenin signaling pathway, driving the progression of colorectal cancer (Zheng et al. 2021) (Fig. 12.3). On the other hand, when NEAT1 is knocked down, the upregulation of miR-377 leads to a decrease in SIRT1 and BCL-XL expression levels (Zheng et al. 2021) (Fig. 12.3). Furthermore, NEAT1 is a p53-regulated large intergenic noncoding RNA (lincRNA) that plays a crucial role in preventing cellular transformation and the onset of cancer (Zheng et al. 2021) (Fig. 12.3). These findings support that lncRNAs—particularly NEAT1—are involved in the molecular landscape of EMPD.

12.1.7 Facioscapulohumeral Muscular Dystrophy

Facioscapulohumeral muscular dystrophy (FSHD) is a rare genetic disorder affecting skeletal muscles and causing progressive weaknesses, particularly in the facial and shoulder muscles (Mul 2022). FSHD is mainly caused by deletions in the D4Z4 region on chromosome 4, which normally contain multiple repeats (Mul 2022). In FSHD patients, the number of these repeats is reduced, leading to chromatin relaxation and the abnormal activation of genes that promote muscle degeneration (Vizoso and Esteller 2012). One key player in this process is DBE-T, a lncRNA produced from the D4Z4 region (Vizoso and Esteller 2012) (Fig. 12.2; Table 12.1). When the D4Z4 repeats are fewer than normal, DBE-T becomes active and alters the chromatin structure, switching it from a tightly packed, inactive state to a more open and transcriptionally active one (Vizoso and Esteller 2012). This epigenetic change is critical because it allows the harmful expression of *DUX4*, a gene that triggers muscle cell damage and death when activated (Vizoso and Esteller 2012). Thus, the reduction in D4Z4 repeats, combined with the effects of DBE-T, creates a cascade of events that ultimately leads to muscle weakness and degeneration seen in

FSHD (Vizoso and Esteller 2012). DBE-T recruits the ASH1L protein and facilitates the deposition of activator histone marks, such as H3K36me2 and H3K4me3, which open the chromatin and promote the transcription of genes like *DUX4* (Vizoso and Esteller 2012) (Fig. 12.3). The identification of DBE-T as a key regulator in this process opens potential therapeutic avenues for FSHD, targeting the molecular mechanisms that lead to inappropriate gene activation.

12.1.8 Gastro-Entero-Pancreatic Neuroendocrine Neoplasms

Gastro-entero-pancreatic neuroendocrine neoplasms (GEP-NENs) represent a heterogeneous group of tumors arising from neuroendocrine cells throughout the digestive tract (Di Mauro et al. 2021). Growing evidence implicates lncRNAs as critical regulators of GEP-NEN pathogenesis through their involvement in key oncogenic processes (Di Mauro et al. 2021). Three independent studies report that lncRNAs participate in gastro-entero-pancreatic neuroendocrine neoplasms (Fig. 12.2; Table 12.1). Modali et al. (2015) show that Maternally Expressed Gene 3 (MEG3), activated by menin-induced epigenetic modifications in pancreatic neuroendocrine tumors, blocks cell proliferation and delays cell cycle progression while reducing migration and invasion. Wei et al. (2018) document that lncNEN885 is markedly decreased in human gastric neuroendocrine samples and that its modulation of key markers—including Snail, vimentin, N-cadherin, and E-cadherin—impacts cell migration, invasion, and aspects of epithelial-mesenchymal transition via the Wnt/β-catenin pathway (Fig. 12.3). Xue et al. (2021) report that HNF1A-AS1 suppresses proliferation, migration, and invasion in both cell lines and animal models by engaging Transforming Growth Factor Beta (TGF-β) signaling and altering the balance of epithelial and mesenchymal markers (Fig. 12.3). Each lncRNA is associated with tumor-suppressive activities through distinct pathways—epigenetic regulation, Wnt/β-catenin signaling, and TGF-β signaling—and collectively they regulate processes critical to tumor progression (Curt et al. 2020; Di Mauro et al. 2021). It was found that also the lncRNA HOX Transcript Antisense Intergenic RNA (HOTAIR) expression increases progressively with GEP-NEN's tumor grade and that functions as a modulator of chromatin states by interacting with PRC2 at its 5′ end and with LSD1 at its 3′ end, facilitating the epigenetic regulation of gene expression, either through activation or repression; this suggests a pivotal role for HOTAIR in promoting aggressive tumor characteristics through epigenetic regulation (Di Mauro et al. 2021). Moreover, the inverse relationship between HOTAIR and HOX paralogous genes underscores a potential mechanism where HOTAIR downregulates HOX gene expression, facilitating tumor progression (Di Mauro et al. 2021). Thus, these studies support the involvement of lncRNAs in GEP-NENs.

12.1.9 Hirschsprung Disease

Hirschsprung disease (HSCR) is a rare congenital disorder characterized by defective migration, proliferation, and differentiation of enteric neural crest cells, resulting in aganglionosis of the distal intestine and severe digestive dysfunction in newborns (Langer 2013). Emerging evidence highlights the critical involvement of several lncRNAs in HSCR pathogenesis through their regulatory roles in enteric nervous system (ENS) development (Torroglosa et al. 2020; Qin et al. 2023). Key lncRNAs implicated in HSCR include: MEG3, NEAT1, SOCS2-AS1, and LINC01579-204 (Torroglosa et al. 2020; Qin et al. 2023) (Fig. 12.2; Table 12.1). MEG3 and NEAT1 are upregulated, whereas SOCS2-AS1 and LINC01579-204 are downregulated in HSCR patients (Torroglosa et al. 2020; Qin et al. 2023). In particular, LINC01579-204 appears to function as a competitive endogenous RNA (ceRNA) by sponging miR-203a-3p—a microRNA elevated in HSCR that targets both LINC01579-204 and NEFL (Qin et al. 2023) (Fig. 12.3). The dysregulation of these lncRNAs—along with their interactions with key genes and miRNAs—provides a framework for understanding the disrupted neurodevelopmental processes in HSCR and opens avenues for future therapeutic exploration.

12.1.10 Klinefelter Syndrome

Klinefelter syndrome (KS) is a chromosomal disorder characterized by the presence of an additional X chromosome (47, XXY) (Blackburn et al. 2025). The multifunctional lncRNA GAS5 plays a critical role in KS pathogenesis through its regulation of key biological processes (Salemi et al. 2019) (Figs. 12.2 and 12.3; Table 12.1). By competitively inhibiting glucocorticoid receptor (GR) signaling, GAS5 represses glucocorticoid-mediated transcription, potentially exacerbating the immune dysregulation and inflammatory conditions frequently observed in KS patients (Kino et al. 2010; Salemi et al. 2019) (Fig. 12.3). Moreover, in reproductive pathology, GAS5-mediated suppression of Insulin-Like Growth Factor 1 Receptor (IGF1R)—a gene crucial for Sertoli cell function and spermatogenesis—provides a molecular basis for the germ cell loss and infertility characteristic of KS (Dong et al. 2015; Salemi et al. 2019) (Fig. 12.3). GAS5's known roles in promoting endothelial apoptosis and its upregulation in atherosclerotic plaques suggest a direct contribution to the elevated cardiovascular risk in KS patients, particularly through mechanisms involving endothelial dysfunction and atherogenesis signaling (Calogero et al. 2017; Chen et al. 2017; Salemi et al. 2019). These findings position GAS5 as a central regulator connecting the immune, reproductive, and cardiovascular manifestations of KS.

12.1.11 Moyamoya Disease

Moyamoya disease (MMD) is a rare cerebrovascular condition characterized by progressive stenosis of the internal carotid artery and the formation of abnormal collateral vessels (Fujimura et al. 2016). Despite advancements in surgical interventions, the molecular mechanisms underlying MMD remain poorly understood (Fujimura et al. 2016). Microarray analyses of blood samples and intracranial arteries from MMD patients showed significant differences in lncRNA expression profiles compared to healthy controls (Gao et al. 2016; Wang et al. 2017; Gu et al. 2020; Mamiya et al. 2022). These studies have identified hundreds of differentially expressed lncRNAs in MMD, with the vast majority being upregulated. Bioinformatics analyses suggest that these lncRNAs regulate multiple signaling pathways involved in inflammation, immune response, and vascular development (Wang et al. 2017). Among these pathways, the MAPK signaling pathway has been highlighted as a core regulatory axis in the pathogenesis of MMD (Gao et al. 2016; Wang et al. 2017). In addition, the construction of a comprehensive ceRNA network has revealed key regulatory miRNAs and critical mRNAs associated with immune and inflammatory responses, further underscoring their central role in MMD progression (Gu et al. 2020). Specifically, 308 differentially expressed lncRNAs have been identified, with 306 upregulated and only 2 downregulated in MMD samples, reflecting strong associations with biological processes critical to disease pathology (Mamiya et al. 2022) (Table 12.1). The pathways enriched in these analyses, particularly those related to blood vessel branching and endothelial adhesion, closely mirror the vascular abnormalities observed in MMD (Mamiya et al. 2022). Collectively, these findings significantly advance our understanding of MMD pathophysiology and may offer valuable insights for the development of future therapeutic strategies.

12.1.12 Neurofibromatosis Type 1

Neurofibromatosis type 1 (NF1) is a rare genetic disorder caused by mutations in the *NF1* gene, which encodes the neurofibromin protein, involved in cell growth and tumor suppression (Báez-Flores et al. 2023). lncRNA Antisense Noncoding RNA in the INK4 Locus (ANRIL) has been linked to NF1 progression, particularly in the context of plexiform neurofibromas (Khosravi and Oladnabi 2023) (Fig. 12.2; Table 12.1). ANRIL is known for its involvement in regulating key tumor suppressor genes such as *CDKN2A*, *CDKN2B*, and *ARF* through a polycomb-mediated epigenetic mechanism (Tritto et al. 2019). Previous studies have linked alterations in the 9p21.3 region, where ANRIL resides, to various cancers, including gliomas and neurofibromas (Tritto et al. 2019). A specific genetic variant of ANRIL, rs2151280, is associated with the presence of optic gliomas and the severity of NF1 phenotypes (Tritto et al. 2019). Rs2151280 T allele in ANRIL is linked to reduced ANRIL expression in peripheral blood mononuclear cells and is correlated with a milder NF1 phenotype (Tritto et al. 2019). lncRNA H19 has been implicated also in non-small cell lung cancer and may regulate NF1 in the context of cancer progression (Khosravi and Oladnabi 2023) (Fig. 12.2; Table 12.1). These insights highlight the significant role of lncRNAs like ANRIL and H19 in the progression of NF1 and cancer, pointing to their potential as novel therapeutic targets for improving treatment outcomes in these conditions.

12.1.13 Pulmonary Artery Hypertension

Pulmonary artery hypertension (PAH) is a rare dyspnea-fatigue syndrome defined by an increase of pulmonary artery pressure and an increase of pulmonary vascular resistance; the proliferation of pulmonary artery smooth muscle cells (PASMCs) is a key feature of this disease (Sun et al. 2017; Chen et al. 2018; D'Alto and Naeije 2022). Emerging evidence reveals that the lncRNAs LnRPT (lncRNA regulated by platelet-derived growth factor [PDGF] and transforming growth factor β) and MEG3 function as critical counterbalancing regulators of PASMC proliferation in PAH, through distinct molecular mechanisms (Sun et al. 2017; Chen et al. 2018) (Fig. 12.2; Table 12.1). LnRPT, whose expression is suppressed by PDGF-BB via PI3K signaling (Fig. 12.3), acts as a gatekeeper of vascular proliferation by tonically inhibiting the Notch3/Jag1/Cyclin A2 (CCNA2) axis; its downregulation unleashes this pro-proliferative pathway, while its overexpression can override mitogenic stimuli (Chen et al. 2018). Conversely, MEG3—a p53-stabilizing lncRNA markedly downregulated in PAH patient vasculature—normally constrains cell cycle progression by maintaining nuclear p53 activity and suppressing PCNA, Cyclin A/E, and MDM2 (Sun et al. 2017) (Fig. 12.3). When depleted, MEG3 loss triggers a proliferative cascade: p53 degradation and cytoplasmic mislocalization permit G1/S transition, while upregulated cyclins drive S/G2/M progression (Sun et al. 2017). The convergent phenotypic impact of LnRPT downregulation and MEG3 deficiency creates a permissive environment for pathological vascular remodeling, suggesting these lncRNAs operate as complementary brakes on PASMC hyperproliferation. Their coordinated dysregulation in PAH—LnRPT by growth factor signaling and MEG3 potentially by epigenetic silencing—highlights lncRNAs as nodal points where multiple pathogenic signals integrate to drive disease progression. Therapeutic strategies targeting either lncRNA (e.g., LnRPT augmentation or MEG3 restoration) could exploit these mechanisms to reverse vascular remodeling, positioning them as promising targets for precision intervention in PAH.

12.1.14 Rett Syndrome

Rett syndrome (RTT) is a rare neurodevelopmental disorder primarily caused by loss-of-function mutations in the *MECP2* gene, a key epigenetic regulator crucial for proper neuronal development and function (Petazzi et al. 2013; Siqueira et al. 2025). lncRNA NEAT1 is consistently downregulated in *MECP2* knockout cells across various stages of neural differentiation (Siqueira et al. 2025) (Fig. 12.2; Table 12.1). This downregulation correlates with abnormal activation of the mTOR signaling pathway, a key regulator of cellular metabolism and growth, leading to defective protein metabolism, impaired autophagic flux, and accumulation of protein aggregates—all hallmark features of RTT (Siqueira et al. 2025). Also, NEAT1 directly binds to specific autophagy-related mRNAs (ATG mRNAs) and controls their subcellular localization (Siqueira et al. 2025). In MECP2-KO cells, the lack of NEAT1 leads to the nuclear retention of these mRNAs, disrupting the formation of autophagic complexes, including autophagosomes and autolysosomes, essential for cellular cleanup processes (Siqueira et al. 2025). Moreover, NEAT1 is downregulated during neuronal differentiation in normal cells, but this regulation is disrupted in RTT models (Siqueira et al. 2025). Interestingly, while NEAT1 is reduced in progenitor cells and early differentiation stages, it is upregulated in mature postmortem brain tissues from RTT patients, indicating context-dependent dysregulation influenced by developmental stage and cell type (Siqueira et al. 2025). Other two lncRNAs, AK081227 and AK087060, emerged as significantly upregulated by *MECP2* loss (Petazzi et al. 2013) (Fig. 12.2; Table 12.1). These studies open potential avenues for therapeutic intervention by targeting NEAT1, AK081227, and AK087060 or its regulatory pathways to restore cellular homeostasis in RTT.

12.1.15 Sporadic Amyotrophic Lateral Sclerosis

Sporadic Amyotrophic Lateral Sclerosis (sALS) is a progressive rare neurological disorder that affects motor neurons, causing muscle weakness and atrophy, and occurs without a family history of the disease (Hulisz 2018). lncRNA ZEB1-AS1 was found to be significantly reduced in the blood cells of patients with sALS (Rey et al. 2023) (Fig. 12.2; Table 12.1). Previously known in cancer research for regulating gene expression via interactions with microRNAs, ZEB1-AS1 had not been studied in the context of ALS until recently (Rey et al. 2023). In cancer, ZEB1-AS1 is known to bind and “sponge” miR-200c, preventing this microRNA from degrading its target RNAs, but in ALS models, when it is silenced, leads to increased levels of miR-200c, which suppressed BMI1, a gene important for cell survival and development (Rey et al. 2023) (Fig. 12.3). This silencing also affected the β-Catenin, a key player in neuronal development and protein aggregation in neurodegenerative diseases, which showed an altered cellular distribution, suggesting a link between this lncRNA and sALS pathology (Rey et al. 2023) (Fig. 12.3). As a result, neuronal differentiation was disrupted, with a notable reduction in neurite length—an abnormality common in neurons from sALS patients (Rey et al. 2023). Interestingly, the effects of ZEB1-AS1 loss varied depending on the stage of neuronal maturity, highlighting its context-dependent role in maintaining neuronal health (Rey et al. 2023). These findings suggest that ZEB1-AS1 plays a critical, stage-dependent role in neuronal health and may offer new insights into potential therapeutic targets for sALS.

12.1.16 Takayasu Arteritis

Takayasu arteritis (TAK) is a rare systemic vasculitis that mainly affects the aorta and its major branches (Keser et al. 2018). Recent research has identified significant dysregulation in several lncRNAs in TAK patients, including THRIL, HIF1A-AS1, GAS5, MALAT-1, and HOTAIR, all of which were found at much lower levels compared to healthy individuals (Espinosa-Bautista et al. 2024) (Fig. 12.2; Table 12.1). Among these, HOTAIR, HIF1A-AS1, and THRIL stood out for their high diagnostic potential, with specificity rates around 95%, making them promising biomarkers for the disease (Espinosa-Bautista et al. 2024). THRIL is especially interesting due to its connection with TNF, a key player in the abnormal immune responses seen in TAK (Espinosa-Bautista et al. 2024). However, while these lncRNAs could help in diagnosing the disease, no clear link was found between their levels and disease activity (Espinosa-Bautista et al. 2024). Moving forward, studies should focus on understanding how these lncRNAs contribute to TAK development and whether targeting them could open new treatment options for patients.

12.2 Conclusion

lncRNAs studies across diverse rare diseases have revealed their roles in pathogenesis, offering insights into disease mechanisms and therapeutic opportunities. In pulmonary arterial hypertension (PAH), the counterbalancing actions of LnRPT (Notch3/Jag1/CCNA2 axis suppression) and MEG3 (p53-mediated cell cycle control) demonstrate how lncRNAs integrate proliferative signals to drive vascular remodeling. Similarly, in Klinefelter syndrome, GAS5 emerges as a pleiotropic regulator linking immune dysregulation (via GR antagonism), reproductive deficits (through IGF1R suppression), and cardiovascular risk (via endothelial apoptosis). For muscular dystrophies, the antagonistic roles of linc-MD1 (pro-myogenic) and lnc-31 (antidifferentiation) illustrate how lncRNA networks modulate muscle regeneration, while in Hirschsprung disease, LINC01579–204 and SOCS2-AS1 orchestrate enteric nervous system development via miRNA sponging (miR-203a-3p/NEFL) and immune crosstalk. Autoimmune and inflammatory disorders like APAP and AOSD further highlight lncRNAs (NEAT1, THRIL, and MIAT) as nodal points in cytokine signaling (TNF-α and IL-17) and macrophage dysfunction, with NEAT1 serving as a dynamic biomarker of treatment response. In cancer-associated rare diseases (EMPD and GEP-NENs), lncRNAs such as HOTAIR (epigenetic reprogramming via PRC2/LSD1) and MEG3 (cell cycle arrest) converge on oncogenic pathways (Wnt/β-catenin and PI3K/AKT), whereas in neurodevelopmental disorders (Rett syndrome, sALS), NEAT1 and ZEB1-AS1 regulate autophagy (ATG mRNA nuclear retention) and neuronal integrity (miR-200c/β-Catenin). Notably, facioscapulohumeral muscular dystrophy (FSHD) and neurofibromatosis type 1 (NF1) underscore lncRNA-mediated epigenetic dysregulation, with DBE-T activating DUX4 through chromatin remodeling and ANRIL modulating CDKN2A/B in tumor suppression.

Despite significant advances in understanding the functions of lncRNAs in rare diseases, critical knowledge gaps remain that must be addressed to fully harness their diagnostic and therapeutic potential. A primary challenge lies in elucidating the precise regulatory mechanisms of disease-associated lncRNAs, including their cis/trans interactions and spatiotemporal expression dynamics. Mapping lncRNA expression patterns across developmental stages and cellular compartments is crucial to uncovering their context-dependent functions and regulatory networks. Moving beyond association studies, robust functional validation is essential to establish causal relationships between lncRNAs and disease processes. This demands the development of sophisticated in vivo models to dissect the complex and often pleiotropic effects of these molecules. Complementary approaches, such as crosslinking immunoprecipitation sequencing and proteomic analyses, will be key to revealing the intricate interaction networks of lncRNAs and their binding partners, providing structural and functional insights into their molecular mechanisms. Translational applications require parallel progress in biomarker optimization and therapeutic delivery strategies. Standardization of detection methods and validation across diverse patient cohorts are necessary steps for establishing circulating lncRNAs as reliable biomarkers. Therapeutic approaches must also overcome delivery challenges, through the engineering of targeted vehicles for antisense oligonucleotides, RNA therapeutics, or CRISPR-based modulators, ensuring selective manipulation of lncRNA levels within affected tissues. Comparative studies across diseases could help identify conserved lncRNA pathways with broad therapeutic relevance, opening new avenues for intervention.

Fundamental questions about lncRNA biology persist, particularly regarding their contribution to disease heterogeneity and phenotypic variability. Understanding how lncRNA diversity influences clinical manifestations and exploring the potential of correcting disease-modifying polymorphisms could pave the way for future precision medicine strategies. The study of lncRNAs in rare diseases has evolved from descriptive observations to mechanistic and translational research. As we continue to uncover their multifaceted roles as epigenetic regulators, molecular sponges, and scaffolding molecules, we gain insight into both shared principles of pathogenesis and disease-specific regulatory nuances. The coming decade must focus on bridging molecular discoveries with clinical innovation, transforming our understanding of lncRNA biology into tangible benefits for patients. Achieving this goal will require interdisciplinary collaboration across molecular biology, bioengineering, and clinical medicine to overcome the unique challenges posed by targeting these noncoding architects of cellular function.

Acknowledgments The authors sincerely thank the University of Salamanca for its constant commitment to excellence in research and for providing the intellectual environment that made this work possible. We extend our appreciation to the Institute of Biomedical Research of Salamanca (IBSAL) for its collaborative spirit, which has been instrumental in advancing our understanding of rare diseases. We are deeply grateful to the Junta de Castilla y León for its support of the Biomedicine of Rare Diseases Research Group (GIR). Their funding and dedication to promoting biomedical innovation have been vital to the progress of our research. Finally, we thank our colleagues, students, and collaborators for their contribution, which has enriched this work and continues to drive our mission to improve the lives of patients affected by rare diseases.

Compliance with Ethical Standards

Conflict of Interest The authors declare no conflict of interest.

Ethical Approval This chapter does not contain any studies with human participants or animals performed by any of the authors.

Disclosure All figures presented herein are original to this work. Some complex English sentences were refined with the assistance of artificial intelligence (AI) tools (OpenAI, ChatGPT, 4o, 2025). Every section was carefully reviewed, revised, and approved by authors to ensure that the text is appropriate concerning content and scientific correctness and maintains the integrity of the original content.

References

Arunkumar G (2024) LncRNAs: the good, the bad, and the unknown. Biochem Cell Biol 102:9–27. https://doi.org/10.1139/BCB-2023-0155

Asel M, LeBoeuf NR (2019) Extramammary Paget's disease. Hematol Oncol Clin North Am 33:73–85. https://doi.org/10.1016/J.HOC.2018.09.003

Báez-Flores J, Rodríguez-Martín M, Lacal J (2023) The therapeutic potential of neurofibromin signaling pathways and binding partners. Commun Biol 6:436. https://doi.org/10.1038/S42003-023-04815-0

Blackburn J, Ramakrishnan A, Graham C et al (2025) Klinefelter syndrome: a review. Clin Endocrinol 102:565–573. https://doi.org/10.1111/CEN.15200

Brusa R, Magri F, Bresolin N et al (2020) Noncoding RNAs in Duchenne and Becker muscular dystrophies: role in pathogenesis and future prognostic and therapeutic perspectives. Cell Mol Life Sci 77:4299–4313. https://doi.org/10.1007/S00018-020-03537-4

Calogero AE, Giagulli VA, Mongioì LM et al (2017) Klinefelter syndrome: cardiovascular abnormalities and metabolic disorders. J Endocrinol Investig 40:705–712. https://doi.org/10.1007/S40618-017-0619-9

Chen LX, Wang W, Cao L et al (2016) Long non-coding RNA CCAT1 acts as a competing endogenous RNA to regulate cell growth and differentiation in acute myeloid leukemia. Mol Cells 39:330–336. https://doi.org/10.14348/MOLCELLS.2016.2308

Chen L, Yang W, Guo Y et al (2017) Exosomal lncRNA GAS5 regulates the apoptosis of macrophages and vascular endothelial cells in atherosclerosis. PLoS One 12:e0185406. https://doi.org/10.1371/JOURNAL.PONE.0185406

Chen J, Guo J, Cui X et al (2018) The long noncoding RNA LnRPT is regulated by PDGF-BB and modulates the proliferation of pulmonary artery smooth muscle cells. Am J Respir Cell Mol Biol 58:181–193. https://doi.org/10.1165/RCMB.2017-0111OC

Constanty F, Shkumatava A (2021) lncRNAs in development and differentiation: from sequence motifs to functional characterization. Development 148:dev182741. https://doi.org/10.1242/DEV.182741

Curt AM, Ilie IRP, Căinap C et al (2020) MicroRNAs and treatment with somatostatin analogs in gastro- Entero-pancreatic neuroendocrine neoplasms: challenges in personalized medicine. J Gastrointestin Liver Dis 29:647–658. https://doi.org/10.15403/JGLD-2866

D'Alto M, Naeije R (2022) Pulmonary artery pressure-directed therapies in pulmonary arterial hypertension? Vasc Pharmacol 147:107124. https://doi.org/10.1016/J.VPH.2022.107124

Di Mauro A, Scognamiglio G, Aquino G et al (2021) Aberrant expression of Long non coding RNA HOTAIR and de-regulation of the paralogous 13 HOX genes are strongly associated with aggressive behavior of gastro-entero-pancreatic neuroendocrine tumors. Int J Mol Sci 22:7049. https://doi.org/10.3390/IJMS22137049

Dong S, Qu X, Li W et al (2015) The long non-coding RNA, GAS5, enhances gefitinib-induced cell death in innate EGFR tyrosine kinase inhibitor-resistant lung adenocarcinoma cells with wide-type EGFR via downregulation of the IGF-1R expression. J Hematol Oncol 8:43. https://doi.org/10.1186/S13045-015-0140-6

Efthimiou P, Kontzias A, Hur P et al (2021) Adult-onset still's disease in focus: clinical manifestations, diagnosis, treatment, and unmet needs in the era of targeted therapies. Semin Arthritis Rheum 51:858–874. https://doi.org/10.1016/j.semarthrit.2021.06.004

Espinosa-Bautista F, Salazar-Sánchez MI, Brianza-Padilla M et al (2024) Dysregulation of long non-coding RNAs in Takayasu arteritis: a proof-of-concept study. Clin Rheumatol 43:1253–1259. https://doi.org/10.1007/S10067-024-06880-0

Finotti A, Fabbri E, Lampronti I et al (2019) MicroRNAs and long non-coding RNAs in genetic diseases. Mol Diagn Ther 23:155–171. https://doi.org/10.1007/S40291-018-0380-6

Flanigan KM (2014) Duchenne and becker muscular dystrophies. Neurol Clin 32:671–688. https://doi.org/10.1016/j.ncl.2014.05.002

Fu MP, Merrill SM, Sharma M et al (2023) Rare diseases of epigenetic origin: challenges and opportunities. Front Genet 14:1113086. https://doi.org/10.3389/FGENE.2023.1113086

Fujimura M, Bang OY, Kim JS (2016) Moyamoya disease. Front Neurol Neurosci 40:204–220. https://doi.org/10.1159/000448314

Gao F, Yu L, Zhang D et al (2016) Long noncoding RNAs and their regulatory network: potential therapeutic targets for adult Moyamoya disease. World Neurosurg 93:111–119. https://doi.org/10.1016/J.WNEU.2016.05.081

Gu X, Gu X, Jiang D et al (2020) Construction and comprehensive analysis of dysregulated long noncoding RNA-associated competing endogenous RNA network in Moyamoya disease. Comput Math Methods Med 2020:2018214. https://doi.org/10.1155/2020/2018214

Hulisz D (2018) Amyotrophic lateral sclerosis: disease state overview. Am J Manag Care 24:S320

Katznelson L, Laws ER, Melmed S et al (2014) Acromegaly: an endocrine society clinical practice guideline. J Clin Endocrinol Metab 99:3933–3951. https://doi.org/10.1210/JC.2014-2700

Keser G, Aksu K, Direskeneli H (2018) Takayasu arteritis: an update. Turk J Med Sci 48:681–697. https://doi.org/10.3906/SAG-1804-136

Khosravi T, Oladnabi M (2023) The role of miRNAs and lncRNAs in neurofibromatosis type 1. J Cell Biochem 124:17–30. https://doi.org/10.1002/JCB.30349

Kino T, Hurt DE, Ichijo T et al (2010) Noncoding RNA gas5 is a growth arrest- and starvation-associated repressor of the glucocorticoid receptor. Sci Signal 3:ra8. https://doi.org/10.1126/SCISIGNAL.2000568

Langer JC (2013) Hirschsprung disease. Curr Opin Pediatr 25:368–374. https://doi.org/10.1097/MOP.0B013E328360C2A0

Mamiya T, Kanamori F, Yokoyama K et al (2022) Long noncoding RNA profile of the intracranial artery in patients with moyamoya disease. J Neurosurg 138:709–716. https://doi.org/10.3171/2022.5.JNS22579

McCarthy C, Carey BC, Trapnell BC (2022) Autoimmune pulmonary alveolar proteinosis. Am J Respir Crit Care Med 205:1016–1035. https://doi.org/10.1164/RCCM.202112-2742SO

Modali SD, Parekh VI, Kebebew E, Agarwal SK (2015) Epigenetic regulation of the lncRNA MEG3 and its target c-MET in pancreatic neuroendocrine tumors. Mol Endocrinol 29:224–237. https://doi.org/10.1210/ME.2014-1304

Mul K (2022) Facioscapulohumeral muscular dystrophy. Continuum (Minneap Minn) 28:1735–1751. https://doi.org/10.1212/CON.0000000000001155

Najafi S, Khatami SH, Khorsand M et al (2022) Long non-coding RNAs (lncRNAs); roles in tumorigenesis and potentials as biomarkers in cancer diagnosis. Exp Cell Res 418:113294. https://doi.org/10.1016/J.YEXCR.2022.113294

Ni YQ, Xu H, Liu YS (2022) Roles of long non-coding RNAs in the development of aging-related neurodegenerative diseases. Front Mol Neurosci 15:844193. https://doi.org/10.3389/FNMOL.2022.844193

Petazzi P, Sandoval J, Szczesna K et al (2013) Dysregulation of the long non-coding RNA transcriptome in a Rett syndrome mouse model. RNA Biol 10:1197–1203. https://doi.org/10.4161/RNA.24286

Qin D, Yang W, Zhu X et al (2023) LINC01579-204 involved in the development of Hirschsprung's disease maybe by regulating the expression of miR-203a-3p and NEFL. Clin Res Hepatol Gastroenterol 47:102240. https://doi.org/10.1016/J.CLINRE.2023.102240

Rey F, Maghraby E, Messa L et al (2023) Identification of a novel pathway in sporadic amyotrophic lateral sclerosis mediated by the long non-coding RNA ZEB1-AS1. Neurobiol Dis 178:106030. https://doi.org/10.1016/J.NBD.2023.106030

Rolla M, Jawiarczyk-Przybyłowska A, Kolačkov K et al (2023) Is H19 RNA a useful marker of acromegaly and its complications? A preliminary study. Biomedicines 11:1211. https://doi.org/10.3390/BIOMEDICINES11041211

Salemi M, Cannarella R, Condorelli RA et al (2019) Evidence for long noncoding RNA GAS5 up-regulationin patients with Klinefelter syndrome. BMC Med Genet 20:4. https://doi.org/10.1186/S12881-018-0744-0

Schmitz SU, Grote P, Herrmann BG (2016) Mechanisms of long noncoding RNA function in development and disease. Cell Mol Life Sci 73:2491–2509. https://doi.org/10.1007/S00018-016-2174-5

Siqueira E, Velasco CD, Tarrasón A et al (2025) NEAT1-mediated regulation of proteostasis and mRNA localization impacts autophagy dysregulation in Rett syndrome. Nucleic Acids Res 53:gkaf074. https://doi.org/10.1093/NAR/GKAF074

Sun Z, Nie X, Sun S et al (2017) Long non-coding RNA MEG3 downregulation triggers human pulmonary artery smooth muscle cell proliferation and migration via the p53 signaling pathway. Cell Physiol Biochem 42:2569–2581. https://doi.org/10.1159/000480218

Taghizadeh E, Gheibihayat SM, Taheri F et al (2021) LncRNAs as putative biomarkers and therapeutic targets for Parkinson's disease. Neurol Sci 42:4007–4015. https://doi.org/10.1007/S10072-021-05408-7

Torroglosa A, Villalba-Benito L, Fernández RM et al (2020) Identification of new potential LncRNA biomarkers in Hirschsprung disease. Int J Mol Sci 21:1–13. https://doi.org/10.3390/IJMS21155534

Tritto V, Ferrari L, Esposito S et al (2019) Non-coding RNA and tumor development in neurofibromatosis type 1: ANRIL Rs2151280 is associated with optic glioma development and a mild phenotype in neurofibromatosis type 1 patients. Genes (Basel) 10:892. https://doi.org/10.3390/GENES10110892

Vizoso M, Esteller M (2012) The activatory long non-coding RNA DBE-T reveals the epigenetic etiology of facioscapulohumeral muscular dystrophy. Cell Res 22:1413–1415. https://doi.org/10.1038/CR.2012.93

Walther K, Schulte LN (2021) The role of lncRNAs in innate immunity and inflammation. RNA Biol 18:587–603. https://doi.org/10.1080/15476286.2020.1845505

Wang W, Gao F, Zhao Z et al (2017) Integrated analysis of LncRNA-mRNA co-expression profiles in patients with Moyamoya disease. Sci Rep 7:42421. https://doi.org/10.1038/SREP42421

Wei YL, Hua J, Liu XY et al (2018) LncNEN885 inhibits epithelial-mesenchymal transition by partially regulation of Wnt/β-catenin signalling in gastroenteropancreatic neuroendocrine neoplasms. Cancer Sci 109:3139–3148. https://doi.org/10.1111/CAS.13747

Wu B, Zhang Y, Yu Y et al (2021) Long noncoding RNA H19: a novel therapeutic target emerging in oncology via regulating oncogenic signaling pathways. Front Cell Dev Biol 9:796740. https://doi.org/10.3389/FCELL.2021.796740

Xu X, Hao Y, Xiong S, He Z (2020) Comprehensive analysis of Long non-coding RNA-associated competing endogenous RNA network in Duchenne muscular dystrophy. Interdiscip Sci 12:447–460. https://doi.org/10.1007/S12539-020-00388-2

Xue J, Bai J, Long Q et al (2021) TCF-3-mediated transcription of lncRNA HNF1A-AS1 targeting oncostatin M expression inhibits epithelial-mesenchymal transition via TGFβ signaling in gastroenteropancreatic neuroendocrine neoplasms. Aging 13:14065–14077. https://doi.org/10.18632/AGING.203024

Yang CA, Chen PK, Lan JL et al (2021a) Expression signature of inflammation-associated long non-coding RNAs in adult-onset still's disease. Clin Exp Rheumatol 39(Suppl 132):S67–S74. https://doi.org/10.55563/CLINEXPRHEUMATOL/4JX6ZY

Yang Y, Xu W, Xiang RL et al (2021b) Expression profiles and potential functions of long non-coding RNAs and mRNAs in autoimmune pulmonary alveolar proteinosis patients. Aging 13:10535–10554. https://doi.org/10.18632/AGING.202818

Yoshimura H, Matsuda Y, Yamamoto M et al (2018) Expression and role of long non-coding RNA H19 in carcinogenesis. Front Biosci (Landmark) 23:614–625. https://doi.org/10.2741/4608/LANDMARK4608

Zheng D c, Shen Y t, Wei Z w et al (2021) Transcriptome sequencing reveals a lncRNA-mRNA interaction network in extramammary Paget's disease. BMC Med Genet 14:291. https://doi.org/10.1186/S12920-021-01135-2

GPSR Compliance

The European Union's (EU) General Product Safety Regulation (GPSR) is a set of rules that requires consumer products to be safe and our obligations to ensure this.

If you have any concerns about our products, you can contact us on ProductSafety@springernature.com

In case Publisher is established outside the EU, the EU authorized representative is:

Springer Nature Customer Service Center GmbH
Europaplatz 3
69115 Heidelberg, Germany

Batch number: 10755216

Printed by Printforce, the Netherlands